Lecture Notes in Control and Information Sciences

Edited by M. Thoma and A. Wyner

Vol. 62: Analysis and Optimization
of Systems
Proceedings of the Sixth International
Conference on Analysis and Optimization
of Systems
Nice, June 19–22, 1984
Edited by A. Bensoussan, J. L. Lions
XIX, 591 pages. 1984.

Vol. 63: Analysis and Optimization
of Systems
Proceedings of the Sixth International
Conference on Analysis and Optimization
of Systems
Nice, June 19–22, 1984
Edited by A. Bensoussan, J. L. Lions
XIX, 700 pages. 1984.

Vol. 64: Arunabha Bagchi
Stackelberg Differential Games
in Economic Models
VIII, 203 pages, 1984

Vol. 65: Yaakov Yavin
Numerical Studies
in Nonlinear Filtering
VIII, 273 pages, 1985.

Vol. 66: Systems and Optimization
Proceedings of the Twente Workshop
Enschede, The Netherlands, April 16–18, 1984
Edited by A. Bagchi, H. Th. Jongen
X, 206 pages, 1985.

Vol. 67: Real Time Control of Large Scale Systems
Proceedings of the First European Workshop
University of Patras, Greece, Juli 9–12, 1984
Edited by G. Schmidt, M. Singh, A. Titli,
S. Tzafestas
XI, 650 pages, 1985.

Vol. 68: T. Kaczorek
Two-Dimensional Linear Systems
IX, 397 pages, 1985.

Vol. 69: Stochastic Differential Systems –
Filtering and Control
Proceedings of the IFIP-WG 7/1 Working Conference
Marseille-Luminy, France, March 12-17, 1984
Edited by M. Metivier, E. Pardoux
X, 310 pages, 1985.

Vol. 70: Uncertainty and Control
Proceedings of a DFVLR International Colloquium
Bonn, Germany, March, 1985
Edited by J. Ackermann
IV, 236 pages, 1985.

Vol. 71: N. Baba
New Topics in Learning Automata
Theory and Applications
VII, 231 pages, 1985.

Vol. 72: A. Isidori
Nonlinear Control Systems:
An Introduction
VI, 297 pages, 1985.

Vol. 73: J. Zarzycki
Nonlinear Prediction
Ladder-Filters for Higher-Order
Stochastic Sequences
V, 132 pages, 1985.

Vol. 74: K. Ichikawa
Control System Design based on
Exact Model Matching Techniques
VII, 129 pages, 1985.

Vol. 75: Distributed Parameter
Systems
Proceedings of the 2nd International
Conference, Vorau, Austria 1984
Edited by F. Kappel, K. Kunisch,
W. Schappacher
VIII, 460 pages, 1985.

Vol. 76: Stochastic Programming
Edited by F. Archetti, G. Di Pillo,
M. Lucertini
V, 285 pages, 1986.

Vol. 77: Detection of
Abrupt Changes in Signals
and Dynamical Systems
Edited by M. Basseville,
A. Benveniste
X, 373 pages, 1986.

Vol. 78: Stochastic
Differential Systems
Proceedings of the 3rd Bad Honnef
Conference, June 3–7, 1985
Edited by N. Christopeit, K. Helmes,
M. Kohlmann
V, 372 pages, 1986.

Vol. 79: Signal
Processing for Control
Edited by K. Godfrey, P. Jones
XVIII, 413 pages, 1986.

Vol. 80: Artificial Intelligence
and Man-Machine Systems
Edited by H. Winter
IV, 211 pages, 1986.

For information about Vols. 1–61 please contact your bookseller or Springer-Verlag.

Lecture Notes in Control and Information Sciences

Edited by M. Thoma and A. Wyner

136

J. Zabczyk (Editor)

Stochastic Systems and Optimization

Proceedings of the 6th IFIP WG 7.1 Working Conference
Warsaw, Poland, September 12–16, 1988

Springer-Verlag
Berlin Heidelberg GmbH

Editor
Jerzy Zabczyk
Institute of Mathematics
Polish Academy of Sciences
Sniadeckich 8
00-950 Warsaw
Poland

ISBN 978-3-540-51619-4

Library of Congress Cataloging-in-Publication Data

IFIP WG 7.1 Working Conference (6th : 1988 : Warsaw, Poland)
Stochastic systems and optimization :
proceedings of the 6th IFIP WG 7.1 Working Conference, Warsaw, Poland, September 12–16, 1988 / J. W. Zabczyk, editor.
(Lecture notes in control and information sciences ; 136)
ISBN 978-3-540-51619-4 ISBN 978-3-540-46719-9 (eBook)
DOI 10.1007/978-3-540-46719-9
1. Control theory--Congresses. 2. Stochastic processes--Congresses.
3. Mathematical optimization--Congresses.
I. Zabczyk, Jerzy. II. Title. III. Series.
QA402.3.I4539 1988
629.8'312--dc20 89-21791

Originally published by Springer-Verlag Berlin Heidelberg New York in 1989

2161/3020-543210 Printed on acid-free paper.

P R E F A C E

This volume presents the scripts of most of the lectures given at the Sixth IFIP Working Conference on Stochastic Systems and Optimization which took place in Warsaw, Poland, September 12-16, 1988. The conference was held under the auspices of IFIP WG 7.1 and was organized by the Institute of Mathematics of the Polish Academy of Sciences in cooperation with the Institute of Mathematics of Warsaw University.

The programme of the conference was prepared with the help of the International Programme Committee: A.Bensoussan, A.V.Balakrishnan (Committee Chairman), M.H.A. Davis, H.Engelbert, W.Fleming, B.Grigelionis, A.Halanay, K.Helmes, L.Telksnys and J.Zabczyk. The Organizing Committee consisted of T.Bielecki, T.Bojdecki, B.Gołdys, K.Malanowski, W.Smoleński, Ł.Stettner (Secretary), J.Zabczyk (Conference Chairman) and P.Zaremba.

The meeting was a continuation of the foregoing conferences in Kyoto (1976), Vilnius (1978), Visegrad (1980), Marseille-Luminy (1984) and Eisenach (1986) and focused on topics of current interest in the field of stochastic systems and optimization. Particular emphasis was placed on stochastic differential equations both in finite and infinite dimensional spaces, stochastic control and estimation, asymptotic methods and periodic systems. During a special evening session Professor T.Hida and Professor G.Kallianpur shared with the participants their views on the white noise theory.

On behalf of the organizers I want to thank all the participants for making the conference an interesting and a memorable event.

J. Zabczyk

CONTENTS

2. STOCHASTIC INFINITE-DIMENSIONAL SYSTEMS

3. STOCHASTIC CONTROL AND ESTIMATION

1. STOCHASTIC FINITE DIMENSIONAL SYSTEMS

SOME RESULTS ABOUT TWO-MODE STOCHASTIC COMPARTMENTAL MODELS

C. Bruni [1], G. Koch [2]

[1]) Dipartimento di Informatica e Sistemistica, Universitá di Roma "La Sapienza"
via Eudossiana 18, 00184 Roma

[2]) Dipartimento di Matematica, Universita di Roma "La Sapienza"
piazzale A. Moro 5, 00186 Roma

1. Models for two-mode compartmental systems.

In this paper we consider a class of population models, which will be named "two-mode" and is suitable to represent phenomena characterized by the interaction of two compartment groups. The two groups, both containing the same number of compartments, exhibit some difference in their behaviour ("mode"), and in the following we conventionally denote them as "active" and "inactive" compartments.

Let 2n be the total number of compartments, and let X^a_{kt}, X^i_{kt}, k=1,2,...n, denote the number of individuals at time t, respectively in the active and inactive k-th compartment. By X^a_t and X^i_t we denote the n-dimensional vectors respectively with components X^a_{kt}, X^i_{kt}, and by X_t we denote the 2n-dimensional aggregate vector :

$$X_t = \begin{pmatrix} X^a_t \\ X^i_t \end{pmatrix} \tag{1.1}$$

We assume that within both active and inactive subsystems the exchanges of individuals may occur in all possible ways, and we denote by ν^a_{kj}, ν^i_{kj} the exchange rates from k-th to j-th compartment, k,j=1,2,...n, respectively in the active and inactive subsystem. We also assume that between the active and inactive subsystems exchanges only occur in a pairwise fashion, that is each active compartment communicates in both direction with just one inactive compartment (by a suitable ordering, any two communicating compartments will be given the same index). Moreover a special feature of our model is that these last exchanges occur much faster than those ones within each subsystem. For this reason, the exchange rate from j-th active (inactive) compartment to j-th inactive (active) one will be denoted by $c\cdot\nu^{ai}_j$ ($c\cdot\nu^{ia}_j$), where c is a gain factor, c>>1.

In order to account for possible birth/reproduction and death phenomena, each exchange between communicating compartments is attached a reproduction factor α, which is a non negative integer expressing the number of individuals appearing into the arrival compartment whenever one individual disappears from the departure one. This factor will be indexed in the same way as the corresponding exchange rate. By assigning to a reproduction factor of the type α^a_{kk} (α^i_{kk}) a value greater than 1, one can model reproduction phenomena occurring within the k-th active (inactive) compartment itself. No reproduction is allowed in the exchanges between an active (inactive) compartment and the inactive (active) corresponding one; thus $\alpha^{ai}_k = \alpha^{ia}_k = 1$. This is motivated by the fact that such exchanges have indeed to be considered as changes of "status" (of mode) for the individual, rather than real compartmental transitions; besides, they will be supposed to occur at a very high rate, which cannot be matched by necessarily slower reproduction processes. Deaths can be accounted for by simply giving one compartment (let us say the n-th one) in each subsystem the role of death compartment, defined as the arrival compartment for all (and only those) exchanges with a zero reproduction factor. Assuming for the initial condition X^a_{n0}, X^i_{n0}, the zero value., the death compartments stay permanently empty and death corresponds indeed to a disappearing of individuals.

Possible nonlinear interactions among compartments will be included in the model by letting the exchange rates to depend on the sum $X^a_t + X^i_t$. As a special case we mention crowding effects, which are likely to induce a dependence of those rates on the total number of individuals $\sum_{k=1}^{n} (X^a_{kt}+X^i_{kt})$. On the contrary it does not seem interesting to introduce any dependence of this kind for the reproduction factors (this, if desired, could anyway be done without much difficulty). In some cases, it could be requested that the crowding effect also depend on the dead individuals; this can be obtained by suitably redefining the role of the death compartments, which now may well be nonempty.

The initial value X0 is a random vector with nonnegative integer finite variance components (zero for X^a_{n0}, X^i_{n0}).

As far as the dynamics of the model is concerned, we assume X to be a 2n dimensional jump Markov process characterized, for each possible exchange, by a jump size equal to -1 for the departure component and to the reproduction factor for the arrival one, and by a jump rate given by the corresponding exchange rate times the departure component. This is equivalent to assume that, conditioned on the knowledge of the initial number of individuals in each compartment and of all past exchanges, each individual behaves independently from the others and all times of its possible transition to the communicating compartments are independent exponential random variables, each one with parameter equal to the corresponding exchange rate.

It appears from above that the qualifying features of the class of models here considered are the following:

i) the existence of two different dynamics, the first one describing the evolution of "active" individuals and the second the "inactive" ones;

ii) the symmetry of the system, that is to each compartment in the active subsystem it corresponds one compartment in the inactive one: the two subsystems only communicate through the pairs of corresponding compartments;

iii) the velocity of the exchanges between pairs of corresponding compartments, which is much higher than the one of the exchanges within each subsystem.

The range of applications of this model is quite large and includes applicative areas of different disciplines, such as population dynamics, demography, traffic modeling, queueing theory, cell biology, chemistry, osmotic phenomena. Indeed the model finds a use whenever individuals (men, cells, interacting molecules, vehicules,...) may have two different dynamical behaviours according to their "mode" (active/inactive, fast/slow, univalent/multivalent,...) and the transition from one mode to the other occurs almost instantaneously (by the effect of some catalyst, enzyme, membrane crossing, stimulus,...). A limit case is that one in which the mode switches from cycling to quiescent and viceversa, with the quiescent individuals frozen in their compartments, unless they go back to the cycling mode or die ($\nu^{i}_{kj} = 0,\ j \neq n$). An instance of that, which indeed first motivated this research, is the model proposed in [1] for the cell replicative cycle.

The aim of the present paper is to investigate the behaviour of the model here considered when the gain factor c diverges to infinity and to check to which extent it can be substituted by a reduced model of lower dimension. This problem may clearly be framed into the wider area of singular perturbation theory [2,3,4]. The intuitive feeling is that, when $c \to \infty$, an instantaneous relationship link the number of individuals X^{a}_{kt} and X^{i}_{kt} in each pair of corresponding active and inactive compartments. Indeed, if we define as "unbalance variables" the difference between the mean values of the number of individuals exchanged in the two directions per unit time conditioned on the past, divided by the gain factor c, that is $\nu^{ai}_{k} X^{a}_{kt} - \nu^{ia}_{k} X^{i}_{kt}$, it will be proved that the mean value of these unbalance variables goes to zero as $c \to \infty$.

A stronger result holds for the "pooled variables" $X^{a}_{kt} + X^{i}_{kt}$. Obviously, for any finite c, one can easily deduce a model for them from the original 2n-dimensional one; however such a model does not allow any effective dimensionality reduction since in it the variables X^{a}_{kt}, X^{i}_{kt} keep entering separately. But we shall prove that, as $c \to \infty$, $X^{a}_{kt} + X^{i}_{kt}$ converges weakly (in the sense of distributions in the path space) to the solution of a reduced n-dimensional model which is obtained from the previous one by simply putting:

$$(\nu_k^{ai} + \nu_k^{ia})\, X_{kt}^{i} = \nu_k^{ai}\,(X_{kt}^{a} + X_{kt}^{i}) \tag{1.2}$$

$$(\nu_k^{ai} + \nu_k^{ia})\, X_{kt}^{a} = \nu_k^{ia}\,(X_{kt}^{a} + X_{kt}^{i}) \tag{1.3}$$

as would follow from the instantaneous relationships which correspond to the vanishing of the unbalances.
Equations (1.2),(1.3), which hold true in the mean value sense as c diverges, can be exploited within the reduced model to evaluate the variables X_{kt}^{a} , X_{kt}^{i} from the pooled ones.

The mathematical model for a two mode compartmental system incorporates the features already described. Let Q denote the lattice in $\mathbb{R}^{2n}$ of the points with non negative integer components. We consider a time interval [0,T], $T<\infty$, and we introduce the probability space $(\Omega,\mathcal{F}_T,P)$, where Ω is the space $D_Q[0,T]$ of all Q-valued right-continuous functions on [0,T] with left limits, $\mathcal{F}_T$ is the appropriate σ-algebra induced on Ω by the Skorokhod topology [5] and P is a probability measure on $(\Omega,\mathcal{F}_T)$.

We now consider the 2n-dimensional process X defined on $(\Omega,\mathcal{F},P)$ by the evaluation functional: $X_t(\omega) = \omega_t$, $\forall(\omega,t)\in\Omega\times[0,T]$. Obviously, from the definition of X, by equation (1.1) we get the following definition of the processes X^a and X^i, which describe the behaviour of active and inactive compartments:

$$X_k^a = X_k \qquad X_k^i = X_{k+n} , \qquad k=1,2,\ldots n \tag{1.4}$$

We also introduce the family of increasing σ-algebras $\{\mathcal{F}_t\}$, $t\in[0,T]$, where $\mathcal{F}_t$ is defined as the trace of $\mathcal{F}_T$ on $D_Q[0,t]$.

For k,j = 1,2,...,n, let $\alpha_{kj}^{a},\alpha_{kj}^{i}$ denote integer numbers, strictly positive unless j=n, in which case they are zero. And let $\nu_{kj}^{a},\nu_{kj}^{i},\nu_j^{ai},\nu_j^{ia}$ denote non negative bounded measurable functions on the lattice $\bar{Q}$ in $\mathbb{R}^n$ of the points with non negative integer components. Then it is possible to choose P such that the process X is a jump Markov process characterized by the jump sizes and the corresponding jump rates conditioned on $\mathcal{F}_t$ indicated in Tab.1 [6], where e_k denotes the 2n-dimensional vector with the j-th component equal to 0 for $j \neq k$ and equal to 1 for $j = k$, and indexes j and k range from 1 to n.

Transition from compt.	to compt.	Jump size	Jump rate
k-th act.	j-th act.	$-1 \cdot e_k + \alpha^a_{kj} \cdot e_j$	$\nu^a_{kj}(X^a_t + X^i_t) \cdot X_{kt}$
k-th inact.	j-th inact.	$-1 \cdot e_{k+n} + \alpha^i_{kj} \cdot e_{j+n}$	$\nu^i_{kj}(X^a_t + X^i_t) \cdot X_{k+n,t}$
j-th act.	j-th inact.	$-1 \cdot e_j + 1 \cdot e_{j+n}$	$c \cdot \nu^{ai}_j (X^a_t + X^i_t) \cdot X_{jt}$
j-th inact.	j-th act.	$-1 \cdot e_{j+n} + 1 \cdot e_j$	$c \cdot \nu^{ia}_j (X^a_t + X^i_t) \cdot X_{j+n,t}$

Tab. 1.

Observe that the jump sizes have components either positive integer valued or equal to -1, and that, when some of the components of the process vanish, the jump rates which would lead the process out of Q vanish as well. From the above facts it follows the existence of a P with the feature we are looking for.

We further assume that the initial value X_0 is a Q-valued random vector with finite moment up to third order, and with its n-th and 2n-th components equal to 0 (with probability 1).

The generator L of the process X can be immediately deduced from Tab. 1. In particular it follows the representation for the process X, for $t \in [0,T]$:

$$X^a_{kt} = X^a_{k0} + \sum_{j=1}^{n} \int_0^t [\alpha^a_{jk}\, \nu^a_{jk}(X^a_s + X^i_s)\, X^a_{js} - \nu^a_{kj}(X^a_s + X^i_s)\, X^a_{ks}\,]\, ds +$$

$$+\, c \cdot \int_0^t [\nu^{ia}_k (X^a_s + X^i_s)\, X^i_{ks} - \nu^{ai}_k (X^a_s + X^i_s)\, X^a_{ks}\,]\, ds + m^a_{kt}$$

$$X^i_{kt} = X^i_{k0} + \sum_{j=1}^{n} \int_0^t [\alpha^i_{jk}\, \nu^i_{jk}(X^a_s + X^i_s)\, X^i_{js} - \nu^i_{kj}(X^a_s + X^i_s)\, X^i_{ks}\,]\, ds +$$

$$+ c\cdot\int_0^t [\nu_k^{ai}(X_s^a+X_s^i)\, X_{ks}^a - \nu_k^{ia}(X_s^a+X_s^i)\, X_{ks}^i]\, ds + m_{kt}^i \quad , k=1,2,\dots n \tag{1.5}$$

where m_{kt}^a and m_{kt}^i are $\{\mathcal{F}_t\}$-martingales with zero mean values and quadratic covariances easily obtainable from the knowledge of the generator L.

As already pointed out , it is useful to our purpose to consider an alternative model for a two-mode system in terms of pooled and unbalance variables. Let us introduce the two n-dimensional processes Y^p and Y^u, defined as follows:

$$Y_{kt}^p = X_{kt}^a + X_{kt}^i \qquad k=1,2,\dots n \qquad t\in[0,T] \tag{1.6}$$

$$Y_{kt}^u = \nu_k^{ai}(X_t^a+X_t^i)\, X_{kt}^a - \nu_k^{ia}(X_t^a+X_t^i)\, X_{kt}^i \quad k=1,2,\dots n \qquad t\in[0,T] \tag{1.7}$$

and the 2n-dimensional process Y, defined as the aggregate of Y^p and Y^u:

$$Y_t = \begin{pmatrix} Y_t^p \\ Y_t^u \end{pmatrix} \qquad t\in[0,T] \tag{1.8}$$

It is easy to show [7] that if the condition holds:

$$\gamma \equiv \inf_k \inf_{\xi\in\bar{Q}} (\nu_k^{ai}(\xi) + \nu_k^{ia}(\xi)) > 0 \tag{1.9}$$

then Y is a jump Markov process. on $(\Omega,\mathcal{F}_T,P)$, and $\mathcal{F}_t$ is generated by Y_s, $0\le s\le t$. Moreover equations (1.6), (1.7) can be solved in the X variables:

$$X_{kt}^a = \frac{\nu_k^{ia}(Y_t^p)\, Y_{kt}^p + Y_{kt}^u}{\nu_k^{ai}(Y_t^p)+\nu_k^{ia}(Y_t^p)} \qquad k=1,2,\dots n \qquad t\in[0,T] \tag{1.10}$$

$$X^{i}_{kt} = \frac{\nu^{ai}_{k}(Y^{p}_{t})Y^{p}_{kt}-Y^{u}_{kt}}{\nu^{ai}_{k}(Y^{p}_{t})+\nu^{ia}_{k}(Y^{p}_{t})} \qquad k=1,2,\ldots n \qquad t\in[0,T] \tag{1.11}$$

It follows a representation for the process Y, for $t\in[0,T]$, similar to the one deduced in (1.5) for X. For sake of simplicity we shall give here this representation only for Y^p, which indeed will turn out to be useful in the following, when dealing with the reduced model.

$$\begin{aligned} Y^{p}_{kt} = Y^{p}_{k0} &+ \sum_{j=1}^{n}\int_{0}^{t}[\alpha^{a}_{jk}\,\nu^{a}_{jk}(Y^{p}_{s})\,\frac{\nu^{ia}_{j}(Y^{p}_{s})Y^{p}_{js}+Y^{u}_{js}}{\nu^{ai}_{j}(Y^{p}_{s})+\nu^{ia}_{j}(Y^{p}_{s})} - \\ &\qquad - \nu^{a}_{kj}(Y^{p}_{s})\,\frac{\nu^{ia}_{k}(Y^{p}_{s})Y^{p}_{ks}+Y^{u}_{ks}}{\nu^{ai}_{k}(Y^{p}_{s})+\nu^{ia}_{k}(Y^{p}_{s})}]\,ds\, + \\ &+ \sum_{j=1}^{n}\int_{0}^{t}[\alpha^{i}_{jk}\,\nu^{i}_{jk}(Y^{p}_{s})\,\frac{\nu^{ai}_{j}(Y^{p}_{s})Y^{p}_{js}-Y^{u}_{js}}{\nu^{ai}_{j}(Y^{p}_{s})+\nu^{ia}_{j}(Y^{p}_{s})} - \\ &\qquad - \nu^{i}_{kj}(Y^{p}_{s})\,\frac{\nu^{ai}_{k}(Y^{p}_{s})Y^{p}_{ks}-Y^{u}_{ks}}{\nu^{ai}_{k}(Y^{p}_{s})+\nu^{ia}_{k}(Y^{p}_{s})}]\,ds + m^{p}_{kt}\;,\; k=1,2,\ldots n \end{aligned} \tag{1.12}$$

where m^{p}_{kt} is a $\{\mathcal{F}_t\}$-martingale. Note that even if the parameter c does not appear in (1.12), the processes Y^{p}_{k} depend on it via the coupling with the processes Y^{u}_{k}.

Finally, let us consider the reduced model defined in terms of a jump Markov 2n-dimensional process Z, which is the aggregate of Z^v and Z^o

$$Z_t = \begin{pmatrix} Z^v_t \\ Z^o_t \end{pmatrix} \qquad t\in[0,T] \qquad (1.13)$$

The second n variables of the reduced model Z^o are defined as a degenerate process identically equal to zero, while the first n ones Z^v are defined as a $\bar{Q}$-valued jump process which is a Markov process on its own, solution of the equation:

$$\begin{aligned} Z^v_{kt} = Z^v_{k0} &+ \sum_{j=1}^{n} \int_0^t [\alpha^a_{jk}\, \nu^a_{jk}(Z^v_s) \frac{\nu^{ia}_j(Z^v_s) Z^v_{js}}{\nu^{ai}_j(Z^v_s)+\nu^{ia}_j(Z^v_s)} - \\ &- \nu^a_{kj}(Z^v_s) \frac{\nu^{ia}_k(Z^v_s) Z^v_{ks}}{\nu^{ai}_k(Z^v_s)+\nu^{ia}_k(Z^v_s)}]\, ds + \\ &+ \sum_{j=1}^{n} \int_0^t [\alpha^i_{jk}\, \nu^i_{jk}(Z^v_s) \frac{\nu^{ai}_j(Z^v_s) Z^v_{js}}{\nu^{ai}_j(Z^v_s)+\nu^{ia}_j(Z^v_s)} - \\ &- \nu^i_{kj}(Z^v_s) \frac{\nu^{ai}_k(Z^v_s) Z^v_{ks}}{\nu^{ai}_k(Z^v_s)+\nu^{ia}_k(Z^v_s)}]\, ds + m^r_{kt}\,, \quad k=1,2,\dots n \end{aligned}$$

(1.14)

where m^r_{kt} is a $(\mathcal{F}Z^v_t)$-martingale.

As a matter of fact, it is worth remarking that in the reduced model the behaviour of Z^v is the same as the behaviour of the pooled vector $Y^p = X^a+X^i$ when the unbalance vector is identically set equal to 0, while the behaviour of Z^o corresponds to the instantaneous relationships:

$$Y^u_{kt} = \nu^{ai}_k(X^a_t+X^i_t)\, X^a_{kt} - \nu^{ia}_k(X^a_t+X^i_t)\, X^i_{kt} = 0,$$

$$k=1,2,\ldots n,\ t\in[0,T] \qquad (1.15)$$

2. Convergence results.

In the previous section we introduced the model for a two-mode system in terms of the original X^a, X^i processes, and the equivalent model in terms of the transformed Y^p, Y^u processes. To stress dependence on the gain factor c, in the following we will add a superscript c to the notation for these processes. We also introduced the reduced model, independent of c, in terms of the Z^v processes, along with the constrained Z^o processes.

In this section we prove the convergence results of the c-dependent processes X^c, Y^c to the Z process as $c\to\infty$, which have already been anticipated in the introduction. To this aim, we first report a lemma; for its proof we refer to [7].

Lemma 2.1. Let the transition rates ν be non negative bounded measurable functions on the lattice $\bar{Q}$. Let us further assume that the following conditions hold:

$$\gamma \equiv \inf_k \inf_{\xi\in\bar{Q}} (\nu^{ai}_k(\xi) + \nu^{ai}_k(\xi)) > 0 \qquad (2.1)$$

$$\sup_{\xi\in\bar{Q},\Delta\xi} |\nu^{ai}_k(\xi+\Delta\xi) - \nu^{ai}_k(\xi)|\,\xi_k \le M < \infty \qquad k=1,2,\ldots n \qquad (2.2)$$

$$\sup_{\xi\in\bar{Q},\Delta\xi} |\nu^{ia}_k(\xi+\Delta\xi) - \nu^{ia}_k(\xi)|\,\xi_k \le M < \infty \qquad k=1,2,\ldots n \qquad (2.3)$$

where in (2.2), (2.3) $\Delta\xi$ denotes any possible jump size of $Y^{p,c}$ assumed $Y^{p,c} = \xi$. Then, for any f in the space $B(\bar{Q})$ of real measurable bounded functions on $\bar{Q}$, it results:

$$\lim_{c\to\infty} E\left\{\left|\int_0^t f(Y^{p,c}_s)\cdot Y^{u,c}_{ks}\,ds\right|\right\}=0 \quad k=1,2,\dots n \quad t\in[0,T] \qquad (2.4)$$

We are now in the position of proving the convergence results.

Theorem 2.1. Given the two-mode system modeled by the process X^c as in (1.5) with generator L as in Tab.1, under the same assumptions of Lemma 2.1, the process of the pooled variables $Y^{p,c} = X^{a,c} + X^{i,c}$ weakly converges to the process Z^v, solution of the reduced model (1.14).

Proof. We first prove that, if $R_\psi(Y^c_s)$ denotes the difference between the generators of Y^c and Z when applied to a bounded measurable function ψ of $Y^{p,c}$ and computed at Y^c_s, then we have:

$$\lim_{c\to\infty} E\left\{\left|\int_0^t R_\psi(Y^c_s)\,ds\right|\right\}=0 \qquad t\in[0,T] \qquad (2.5)$$

Usually, this difference turns out to be non zero since its second term is like the first one but with the $Y^{u,c}$ components forced to vanish. Thus recalling Tab.1 and (2.3), we see that the left hand side of (2.5) takes the form:

$$\lim_{c\to\infty} E\left\{\left|\int_0^t \sum_{1}^{n}{}_{k,j}[\psi(Y^{p,c}_s-1\cdot\varepsilon_k+\alpha^a_{kj}\cdot\varepsilon_j)-\psi(Y^{p,c}_s)]\frac{\nu^a_{kj}(Y^{p,c}_s)Y^{u,c}_{ks}}{\nu^{ai}_k(Y^{p,c}_s)+\nu^{ia}_k(Y^{p,c}_s)}ds\right|\right\}$$

$$-\int_0^t \sum_{k,j=1}^{n} [\psi(Y^{p,c}_s - 1\cdot\varepsilon_k + \alpha^{i}_{kj}\cdot\varepsilon_j) - \psi(Y^{p,c}_s)] \frac{\nu^{i}_{kj}(Y^{p,c}_s)Y^{u,c}_{ks}}{\nu^{ai}_{k}(Y^{p,c}_s) + \nu^{ia}_{k}(Y^{p,c}_s)} ds \Bigg|\Bigg\} =$$

$$= \lim_{c\to\infty} E\left\{\left| \int_0^t \sum_{k=1}^{n} \Phi_k(Y^{p,c}_s)\, Y^{u,c}_{ks} ds \right|\right\} = \lim_{c\to\infty} E\left\{\left| \sum_{k=1}^{n} \int_0^t \Phi_k(Y^{p,c}_s)\, Y^{u,c}_{ks} ds \right|\right\}$$

$$\le \sum_{k=1}^{n} \lim_{c\to\infty} E\left\{\left| \int_0^t \Phi_k(Y^{p,c}_s)\, Y^{u,c}_{ks} ds \right|\right\} = 0 \quad , t\in[0,T] \tag{2.6}$$

where for $k=1,2,\ldots,n$, ε_k denotes the n-dimensional vector with the j-th component equal to 0 for $j\neq k$ and equal to 1 for $j=k$, and the implicitly defined function Φ_k, depending on ψ, is measurable and bounded due to the assumptions. Then the conclusion in (2.6) follows from Lemma 2.1.

We now apply Thm. 2.1 of [8]. We note that the assumption H_4 in there is a stronger requirement than our (2.5), since the absolute value should be taken inside the integral; but we checked that the validity of the theorem is not affected by the weaker assumption (2.5) (which anyhow in our case could not be strenghtened). All other assumptions requested in the above mentioned theorem are here easily fulfilled due to the properties of uniform boundedness of jump sizes and sublinear growth of jump rates for $Y^{p,c}$ which follow from (2.1)-(2.3) [7], so that the claim of the present theorem is proved.

Theorem 2.2. Given the two-mode system modeled by the process X^c as in (1.5) with generator L as in Tab.1, under the same assumptions of Lemma 2.1, the unbalanced variables converge to 0 as $c\to\infty$ in the following sense:

$$\lim_{c\to\infty} E\left\{ \int_0^t (\nu^{ai}_{k}(X^{a,c}_s + X^{i,c}_s)\cdot X^{a,c}_{ks} - \nu^{ia}_{k}(X^{a,c}_s + X^{i,c}_s)\cdot X^{i,c}_{ks}) ds \right\} = 0$$

$$k=1,2,\ldots n,\ t\in[0,T] \tag{2.7}$$

so that in particular:

$$\lim_{c\to\infty} E\left\{\nu_k^{ai}(X_t^{a,c}+X_t^{i,c})\cdot X_{kt}^{a,c}-\nu_k^{ia}(X_t^{a,c}+X_t^{i,c})\cdot X_{kt}^{i,c}\right\} = 0$$

$$k=1,2,\ldots n, \quad \text{a.a.}\, t\in[0,T] \qquad (2.8)$$

Proof. The limit (2.7) is an immediate consequence of Lemma 3.1 when we take f as the identity and recall the definition (1.7). From (1.6), (1.7) it follows that there exists a suitable constant β_1 such that:

$$| Y_{kt}^{u,c} | \le \beta_1 \cdot Y_{kt}^{p,c} \le \beta_1 \cdot \| Y_t^{p,c} \| , \quad k=1,2,\ldots n , \; t\in[0,T] \qquad (2.9)$$

where $\|\cdot\|$ denotes a norm in $\mathbb{R}^n$. Furthermore the sublinear growth of the jump rates of $Y_{kt}^{p,c}$, k=1,2,...n, implies uniform boundedness, over [0,T], of the moments of the norm of $Y_t^{p,c}$ itself up to third order [6]:

$$E\,(\| Y_t^{p,c} \|^r) \le \delta_r < \infty, \qquad r=1,2,3, \quad t\in[0,T] \qquad (2.10)$$

The limit (2.8) then follows immediately from (2.7), once we take (2.9), (2.10) into account.

Remark 1. As it follows from Thm. 2.1, the reduced model allows for an asymptotic ($c\to\infty$) investigation of pooled variables $Y_k^{p,c}=X_k^{a,c}+X_k^{i,c}$, k=1,2,...,n. It also follows from Thm. 2.2 that:

$$\lim_{c\to\infty} E\left\{[\nu_k^{ai}(Y_t^{p,c})+\nu_k^{ia}(Y_t^{p,c})]\cdot X_{kt}^{a,c}\right\} = \lim_{c\to\infty} E\left\{\nu_k^{ia}(Y_t^{p,c})\cdot Y_{kt}^{p,c}\right\} \qquad (2.11)$$

$$\lim_{c\to\infty} E\left\{[\nu_k^{ai}(Y_t^{p,c})+\nu_k^{ia}(Y_t^{p,c})]\cdot X_{kt}^{i,c}\right\} = \lim_{c\to\infty} E\left\{\nu_k^{ai}(Y_t^{p,c})\cdot Y_{kt}^{p,c}\right\} \qquad (2.12)$$

In the case of constant intersubsystem transition rates, from (2.11), (2.12) one gets the expected values of active and inactive populations in terms of the pooled ones:

$$\lim_{c\to\infty} E\left\{X^{a,c}_{kt}\right\} = \frac{\nu^{ia}_{k}}{\nu^{ai}_{k}+\nu^{ia}_{k}} \lim_{c\to\infty} E\left\{Y^{p,c}_{kt}\right\}, \; k=1,2,\dots n, \quad \text{a.a.} t\in[0,T] \qquad (2.13)$$

$$\lim_{c\to\infty} E\left\{X^{i,c}_{kt}\right\} = \frac{\nu^{ai}_{k}}{\nu^{ai}_{k}+\nu^{ia}_{k}} \lim_{c\to\infty} E\left\{Y^{p,c}_{kt}\right\}, \; k=1,2,\dots n, \quad \text{a.a.} t\in[0,T] \qquad (2.14)$$

Note that a comparison of (1.5) and (1.14) shows that the transition rates of the reduced model turns out to be a weighted average of corresponding transition rates of the original active and inactive subsystems, the weights being the fractions of active and inactive individuals.

Remark 2. It is not true that for a η positive small enough:

$$\lim_{c\to\infty} E\left\{\left| \nu^{ai}_{k}(X^{a,c}_{s}+X^{i,c}_{s})\cdot X^{a,c}_{ks} - \nu^{ia}_{k}(X^{a,c}_{s}+X^{i,c}_{s})\cdot X^{i,c}_{ks}\right|^{\eta}\right\} = 0 \qquad (2.15)$$

This is because as c diverges the unbalance process fluctuates faster and faster but its jump sizes are not rescaled so as to decrease to zero. Thus the limiting process is a discontinuous one, as it should be for a physical model of (fixed size) cell evolution. Therefore sufficient conditions for weak convergence such as in [9] are not applicable.

Remark 3. Eq. (2.4) may be interpreted as an asymptotic uncorrelation property between $f(Y^{p,c}_{t})$ and $Y^{u,c}_{kt}$, for any $f\in B(\mathbb{R}^n)$ and $k=1,2,\dots,n$, which is weaker than independence between $Y^{p,c}_{t}$ and $Y^{u,c}_{t}$, but stronger than their simple uncorrelation.

The above convergence results allow to achieve an effective simplification, in the study of phenomena which fall within the class of two-mode compartmental systems, through a reduction of the model order. At the same time the reduced model keeps full information on all relevant variables of the original model.

References

1. C. Bruni, G. Koch - A stochastic model of the cell replicative cycle. To appear.

2. P.A. Iannou, P.V. Kokotovic - Adaptive systems with reduced models. Lecture Notes in Control and Information Sciences, n. 47, Springer Verlag, 1983.

3. P.V. Kokotovic, H.K. Khalil, J. O'Reilly - Singular perturbation methods in control: analysis and design. Academic Press, 1986.

4. P.V. Kokotovic, A. Bensoussan, G. Blankenship (eds) - Singular perturbation and asymptotic analysis in control systems. Lecture Notes in Control and Information Sciences, n. 90, Springer Verlag, 1987.

5. P. Billingsley - Convergence of probability measures. J. Wiley, 1968.

6. I.I. Gikhman, A.V. Skorokhod - Introduction to the theory of random processes. Saunders, 1968.

7. C. Bruni, G. Koch - Stochastic models for two-mode compartmental systems and their asymptotic reduction. Rapp. 19-88, Dip. Inform. Sistem., Universita di Roma "La Sapienza", Oct. 1988.

8. A. Joffe, M. Métivier - Weak convergence of subsequences of semimartingales with applications to multitype branching processes. Rapp. DMS n. 84-5, Univ. de Montreal, Mars 1984.

9. S.N. Ethier, T.G. Kurtz - Markov processes. J. Wiley, 1988.

ANTICIPATING LINEAR STOCHASTIC DIFFERENTIAL EQUATIONS

Rainer Buckdahn
Sektion Mathematik der Humboldt-Universität
Unter den Linden 6, PSF 1297, 1086 Berlin, GDR

Abstract: We study linear stochastic differential equations with a random initial condition and a drift anticipating the driving Wiener process, and we give fairly general conditions under which they have a unique solution.

1. Introduction

Let $\{W_t\}$ denote a Brownian motion. Recently, progress has been made in developing a useful theory of stochastic integrals $\int u_s(\omega)dW_s$ in which the integrand $\{u_s(\omega)\}$ anticipates $\{W_t\}$. For the integral constructed by Skorohod in [6], an extended stochastic calculus on the basis of the Malliavin calculus has been developed by Gaveau-Trauber [2] and Nualart-Pardoux [4]. This allows to formulate the linear stochastic differential equation (abbr. LSDE)

$$(1) \qquad Y_t = \eta + \int_0^t \sigma_s Y_s dW_s + \int_0^t b_s Y_s ds \quad , \qquad 0 \leq t \leq 1,$$

for a random initial condition η and a drift b which anticipates the driving noise. Such a situation is given if we want to keep the LSDE symmetric w.r.t. time reversal.

Our aim is to show that equation (1) has a uniquely determined solution and to determine this.

The present note presents the results for the special case of a deterministic diffusion coefficient σ, proofs are only sketched here. The complete results and proofs of the more general case of an anticipating diffusion coefficient were published in [1].

2. Skorohod's integral

Let $\Omega = C([0,1])$, $\mathcal{F}$ be the Borel field over Ω, P the standard Wiener measure, $W_t(\omega) = \omega(t)$ the canonical process. Skorohod's integral - see Skorohod [6], Gaveau-Trauber [2] - can be defined as a closed linear unbounded operator δ from $L_2([0,1] \times \Omega)$ into

$L_2(\Omega)$. For that reason, let us first recall the notion of derivation on a Wiener space $(\Omega, \mathcal{F}, P)$.

A random variable of the form $F = f(W_{t_1}, \ldots, W_{t_n})$, with $n \in \mathbb{N}$, $0 \leq t_1, \ldots, t_n \leq 1$, $f \in C^\infty(R^n)$ bounded as well as its derivatives, will be called a smooth Wiener functional. The class of smooth Wiener functionals will be denoted by $\mathcal{S}$. For $F \in \mathcal{S}$ we define its derivative DF as the element of $L_2([0,1] \times \Omega)$ given by:

$$D_t F = \sum_{k=1}^{n} (\frac{\partial}{\partial x_k} f)(W_{t_1}, \ldots, W_{t_n}) I_{[0,t_k)}(t) \ , \qquad 0 \leq t_k \leq 1.$$

Now we define Skorohod's integral δ as the adjoint operator of the derivation D, i.e., the domain $\mathrm{Dom}\,\delta$ is the set of processes $u \in L_2([0,1] \times \Omega)$ for which there exists a random variable $G^u \in L_2(\Omega)$ such that

$$\text{(2)} \qquad E[G^u F] = E[\int_0^1 u_s D_s F ds] \ , \qquad \text{for all } F \in \mathcal{S},$$

and, for $u \in \mathrm{Dom}\,\delta$, $\delta(u)$ is the uniquely determined element $G^u \in L_2(\Omega)$. Finally we define the set L_δ of processes $u \in \mathrm{Dom}\,\delta$ such that $I_{[0,t]} u \in \mathrm{Dom}\,\delta$, for all $0 \leq t \leq 1$. For $u \in L_\delta$ we introduce

$$\int_0^t u_s dW_s = \delta(I_{[0,t]} \cdot u) \ , \qquad 0 \leq t \leq 1.$$

Remark 1: If $u \in L_2([0,1] \times \Omega)$ does not anticipate the driving Wiener process $\{W_t\}$, then $u \in L_\delta$, and $\int_0^t u_s dW_s$ coincides with Itô's integral.

3. The concept of the solution of an LSDE

The concept of Skorohod's integral gives sense to stochastic differential equations with anticipating coefficients.

Let σ and b be processes from $L_\infty([0,1] \times \Omega)$, and η a random variable from $L_2(\Omega)$.

Definition: A process $Y \in L_2([0,1] \times \Omega)$ is called a solution of LSDE (1) if

(i) $\sigma \cdot Y \in L_\delta$, and

(ii) $\int_0^t \sigma_s Y_s dW_s = Y_t - \eta - \int_0^t b_s Y_s ds$, P-a.s., $0 \leq t \leq 1$.

The definition of Skorohod's integral (2) allows to reformulate the above definition of the solution and to give a more practicable criterion.

Lemma 1: A process $Y \in L_2([0,1] \times \Omega)$ is a solution of LSDE (1) if and only if

(3) $$E[Y_t F] = E[\eta F] + E[\int_0^t \sigma_s Y_s D_s F ds] + E[\int_0^t b_s Y_s ds \cdot F]$$
for all $F \in \mathcal{S}$, $0 \leq t \leq 1$.

Remark 2: From Remark 1 we derive that the nonanticipating solution of Itô's LSDE for a nonanticipating drift coefficient b and a deterministic initial condition η is also a solution of the LSDE in the above sense.

4. Girsanov transformation

The method of the Girsanov transformation is the main instrument in the study of LSDE (1). Therefore we recall some basic statements about the Girsanov transformation (cf. [3]).
Let σ be a deterministic process from $L_\infty([0,1])$. We define the family of transformations $U_{s,t}$ over Ω,

$$U_{s,t}\omega = \omega - \int_0^{\cdot} I_{[s,t]}(r)\sigma_r dr , \quad \omega \in \Omega , \quad s \leq t,$$

which maps

$$\Omega \ni \omega \longmapsto \{\omega(u) - \int_0^u I_{[s,t]}(r)\sigma_r dr, \ 0 \leq u \leq 1\} \in \Omega .$$

These transformations $U_{s,t}$ are absolutely continuous, i.e., the induced measures $P \cdot [U_{s,t}]^{-1}$ are absolutely continuous w.r.t. P, and, with the notation

$$X_t = \exp\left\{ \int_0^t \sigma_s dW_s - \frac{1}{2} \int_0^t \sigma_s^2 ds \right\}, \quad 0 \leq t \leq 1,$$

it holds

(4) $$E[f(U_{s,t})X_t \cdot X_s^{-1}] = E[f] , \quad \text{for all } f \in L_2(\Omega), \ 0 \leq s \leq t \leq 1.$$

For all $0 \leq s \leq t \leq 1$, the transformation $U_{s,t}$ is invertible, its inverse transformation has the form

$$T_{s,t}\omega = \omega + \int_0^{\cdot} I_{[s,t]}(r)\sigma_r dr , \quad \omega \in \Omega ,$$

and is also absolutely continuous. For convenience, we introduce the notations $U_t = U_{0,t}$, $T_t = T_{0,t}$.

The transformations introduced above have the following properties:

Lemma 2: For $\sigma \in L_\infty([0,1])$ we have:

(i) $U_{s,t}\omega = T_s U_t \omega, \quad \omega \in \Omega, \quad 0 \le s \le t \le 1.$

(ii) Let $F \in \mathcal{S}$. Then the processes $\{F(T_t\omega)\}$ and $\{F(U_t\omega)\}$ are pathwise absolutely continuous w.r.t. the Lebesgue measure,

$$\frac{d}{dt} F(U_t\omega) = -\sigma_t (D_t F)(U_t\omega), \quad dt\text{-a.e.}$$

$$\frac{d}{dt} F(T_t\omega) = \sigma_t (D_t F)(T_t\omega), \quad dt\text{-a.e.}$$

Statement (i) can be checked easily. The correctness of (ii) follows immediately by using the specific form of F as an element from $\mathcal{S}$ and of the transformations $U_t\omega$ and $T_t\omega$, which allows to calculate the density of $F(U_t\omega)$ and $F(T_t\omega)$ by chain rule.

5. Existence and Uniqueness of the solution

Theorem 1: For $\sigma \in L_\infty([0,1])$ and $b \in L_\infty([0,1]\times\Omega)$, the only solution $Z \in L_2([0,1]\times\Omega)$ of the LSDE

$$(5) \qquad Z_t = \int_0^t \sigma_s Z_s dW_s + \int_0^t b_s Z_s ds, \quad 0 \le t \le 1,$$

is the process $Z_t = 0$, $0 \le t \le 1$.

Proof: Assume that $Z \in L_2([0,1]\times\Omega)$ is a solution of (5). For $F \in \mathcal{S}$, also $F(U_t)$ belongs to $\mathcal{S}$. Hence, Lemma 1 gives:

$$E[Z_t F(U_t)] = E[\int_0^t \sigma_r Z_r D_r F(U_t) dr] + E[\int_0^t b_r Z_r F(U_t) dr], \quad 0 \le t \le 1, \ F \in \mathcal{S}.$$

If $F \in \mathcal{S}$, then also $D_r F$ is a smooth Wiener functional for $0 \le r \le 1$. Due to Lemma 2, $F(U_t)$ and $D_r F(U_t)$ are absolutely continuous and have the densities $[-\sigma_t (D_t F)(U_t)]$ and $[-\sigma_t (D_t D_r F)(U_t)]$, respectively. Hence,

$$(6) \qquad E[Z_t F(U_t)] = E[\int_0^t \sigma_r Z_r D_r F(U_r) dr] + E[\int_0^t b_r Z_r F(U_r) dr]$$

$$- E[\int_0^t \int_r^t \sigma_r Z_r \sigma_s (D_s D_r F)(U_s) ds dr] - E[\int_0^t \int_r^t b_r Z_r \sigma_s (D_s F)(U_s) ds dr], \quad 0 \le t \le 1.$$

We now apply Fubini's theorem and interchange the order of the integrals and the expectation:

$$(7)\quad E[\int_0^t\int_r^t \sigma_r Z_r \sigma_s (D_s D_r F)(U_s)dsdr] + E[\int_0^t\int_r^t b_r Z_r \sigma_s (D_s F)(U_s)dsdr] =$$

$$= \int_0^t \{E[\int_0^s \sigma_r Z_r \sigma_s (D_s D_r F)(U_s)dr] + E[\int_0^s b_r Z_r \sigma_s (D_s F)(U_s)dr]\} ds .$$

Note that

$$\sigma_s (D_s D_r F)(U_s) = D_r[\sigma_s (D_s F)(U_s)] , \quad 0 \leq s,r \leq 1.$$

Thus, we can conclude from (5) that expression (7) coincides with $\int_0^t E[Z_s \sigma_s (D_s F)(U_s)]ds$.

For equation (6) this provides

$$(8)\qquad E[Z_t F(U_t)] = E[\int_0^t b_r Z_r F(U_r)dr] , \quad 0 \leq t \leq 1.$$

Relation (8) is satisfied for all $F \in \mathcal{S}$ and, consequently, for all $F \in L_2(\Omega)$. For $F = \text{sign } Z_t(T_t)$ relation (8) yields

$$E[|Z_t|] = E[\int_0^t b_r Z_r \text{ sign } Z_t(T_{r,t})dr]$$

$$\leq E[\int_0^t |b_r| \, |Z_r| dr] , \quad 0 \leq t \leq 1.$$

For $b \in L_\infty([0,1] \times \Omega)$, the only square integrable process that satisfies such an inequality is $Z_t = 0$, $0 \leq t \leq 1$. □

Now we show that there exists a solution and determine its form.

<u>Theorem 2:</u> Let $\sigma \in L_\infty([0,1])$, $b \in L_\infty([0,1] \times \Omega)$ and $\eta \in L_p(\Omega)$, for some $p > 2$. Then LSDE (1) has a uniquely determined solution. This solution hat the form

$$Y_t = \eta(U_t) \exp\{\int_0^t b_s(U_{s,t})ds\} X_t , \quad P\text{-a.e.}, \quad 0 \leq t \leq 1.$$

<u>Proof:</u> Obviously, the process Y defined above is square integrable. Thus, we have to prove that relation (3) is satisfied. Let $F \in \mathcal{S}$. Then we can conclude from Lemma 2,

$$E[\eta \exp\{\int_0^t b_s(T_s)ds\} F(T_t)] = E[\eta F] + E[\int_0^t \eta \exp\{\int_0^s b_r(T_r)dr\} \sigma_s(T_s)(D_s F)(T_s)ds]$$

$$+ E[\int_0^t \eta \exp\{\int_0^s b_r(T_r)dr\} b_s(T_s) F(T_s)ds].$$

From Fubini's theorem we get

$$E[\eta \exp\{\int_0^t b_s(T_s)ds\}F(T_t)] = E[\eta F] + \int_0^t E[\eta \exp\{\int_0^s b_r(T_r)dr\}\sigma_s(T_s)(D_sF)(T_s)]ds$$
$$+ \int_0^t E[\eta \exp\{\int_0^s b_r(T_r)d_r\}b_s(T_s)F(T_s)]ds.$$

Hence, relation (4) can be applied to both sides and the equation takes the form

$$E[\eta(U_t)\exp\{\int_0^t b_s(U_{s,t})ds\}X_tF] = E[\eta F] +$$
$$+ \int_0^t E[\sigma_s \eta(U_s)\exp\{\int_0^s b_r(U_{r,s})dr\} X_sD_sF]ds$$
$$+ \int_0^t E[b_s \eta(U_s) \exp\{\int_0^s b_r(U_{r,s})dr\} X_sF]ds .$$

A renewed application of Fubini's theorem provides that (3) holds for the process Y defined above.
The uniqueness of the solution follows from Theorem 1. □

Remark 3: We want to refer to the fact that the solution of LSDE (1) need not be continuous. Put, for instance,

$\sigma_t = 1$, $b_t = 0$, $0 \le t \le 1$, and $\eta = \mathrm{sign}(W_1)$.

Then the solution of LSDE (1) has the form

$$Y_t = \mathrm{sign}(W_1 - t) \exp\{W_t - \frac{1}{2} t\} , \quad 0 \le t \le 1,$$

and is discontinuous for all $\omega \in \Omega$ such that $0 \le \omega(1) \le 1$.
On the other hand, this example also shows that, in general, the process $\int_0^t u_s dW_s$, $u \in L_\delta$, is not continuous: If we put

$$u_s = \mathrm{sign}(W_1 - s)\exp\{W_s - \frac{1}{2}s\}, \quad 0 \le s \le 1,$$

then

$$\int_0^t u_s dW_s = \mathrm{sign}(W_1 - t)\exp\{W_t - \frac{1}{2}t\} - \mathrm{sign}(W_1), \quad 0 \le t \le 1.$$

References

[1] R. Buckdahn: Girsanov Transformation and linear Stochastic Differential Equations without Nonanticipating Requirement, preprint.
[2] B. Gaveau, P. Trauber: L'integrale stochastique comme opérateur de divergence dans l'espace fonctionnel, J.Funct.Anal.46,230-238, 1982.
[3] G. Kallianpur: Stochastic Filtering Theory, Springer-Verlag New York Inc., 1980.
[4] D. Nualart, E. Pardoux: Stochastic Calculus associated with Skorohod's Integral, LN in Control and Information Sciences 96, Proceedings of the IFIP-WG 7/1, Working Conference Eisenach, GDR, 1986.
[5] Y. Shiota: A linear stochastic integral equation containing the extended Itô integral, Rep. Toyama Univ. 9, 43-65, 1986.
[6] A. Skorohod: On a generalization of a stochastic integral, Theory of Prob. and Appl. XX, 219-233, 1975.

Nonlinear Filtering for Signal Correlated with the Noise[*)]

Ognian Enchev

Dept. of Mathematics

"D.Blagoev" Institute of National Economy, Varna-9002, Bulgaria

1. Introduction

The informal description of the filtering model we adopt is this. An *arbitrary* Abstract Wiener Space (AWS) (i,H,E) is assumed to be given. The Hilbert space $H \subset E$ is conceived as the space of the signal and the Banach space E is conceived as the space of the noise. We then have an observation y, which, being the sum of the signal component $h \in H$ and the noise component $\omega \in E$, becomes a member of E, i.e.

(1.1) $$y = \omega + h \ .$$

The most important component of the theory is the distribution p of $\omega \in E$, which is assumed to be the standard Gaussian measure on E with variance parameter 1 (white noise over E). The following particular case of AWS is usually referred to as "classical": $E = C_o[0,1]$ – the space of continuous functions $f:[0,1] \to \mathbb{R}$ with $f(0)=0$; $H = C'$ – the subspace of functions $\varphi(t) = \int_0^t f(x)dx$, $f \in L^2[0,1]$. In that case $p = \mu$ is the Wiener measure on $C_o[0,1]$. Therefore in the classical case one writes the model (1.1) as

(1.1′) $$y(.) = \omega(.) + \int^{\bullet} f(x)dx \ , \quad f \in L^2[0,1] \ , \quad y(.), \omega(.) \in C_o[0,1] \ .$$

In the model adopted by Kallianpur and Karandikar [4] all the three components h, ω and y are assumed to belong to H and the distribution of ω is described by a *finitely additive* cylinder Gaussian measure on H.

Customarily the signal h is assumed to be a function of some parameter u, which is interpreted as "communication" and runs through a *general* probability space $(U, \mathcal{U}, \Pi)$, $u \in U$. We allow the signal h, except on the parameter $u \in U$, to depend also on the noise component $\omega \in E$. Thus, the actual model considered below

---

*) Research supported by the Bulgarian Ministery of Culture Science and Education, under grants #1006 and #1035

is this

(1.2) $y[u,\omega] = \omega + h[u,\omega]$, $u\in U$, $\omega\in E$,

where $h[u,\omega]\in H$ is the signal, $\omega\in E$ is the noise, and $y[u,\omega]\in E$ is the observation. We perform this model on the product probability space

$$(S,\mathcal{S},\mathbb{P}) = (U\times E, \mathfrak{U}\otimes\mathcal{E}, \Pi\otimes p) ,$$

where $\mathcal{E}=\mathcal{B}(E)$ is the Borel σ-field on E .

The problem of nonlinear filtering is this. For a given (generally nonlinear) function $f:H\to\mathbb{R}$ one has to construct another function $[\mathfrak{F}f]:E\to\mathbb{R}$, called "filter", such that (s.t.)

$$[\mathfrak{F}f](\omega') = \mathbb{E}_{\Pi\otimes p}\{f(h[u,\omega]) \mid y[u,\omega]=\omega'\} \quad \text{for } p\text{-a.e. } \omega'\in E .$$

One should note that although we consider standard countably additive probability measures, and, therefore the conventional probability technique can be employed in its full, as G. Kallianpur pointed out to the author, by no means our model substitutes the finitely additive one. This is for various reasons - for example the canonical Gaussian measure p on E is a countably additive extension only of "a part" of the cylinder Gaussian measure on H. In the model considered in [4], as well as in the Kallianpur-Striebel formula [5], the signal and the noise components are assumed to be independent, that is to write in (1.2) above $h[u,\omega]=h[u]$.

The present exposition is a part of a larger text, which will appear elsewhere. The proofs were cut because of the lack of space, but not the proof of the main statement #5.1, in which an explicit formula for the nonlinear filter is derived. Our main concept is explained in #3.12.

2. General Notions and Notations

The present section provides the necessary background on White Noise Calculus. In this end we follow closely [1,3,6 and 8].

#2.1® We fix once and for all an AWS (i,H,E). The Hilbert space H is assumed separable and its norm and scalar product we denote respectively by $|.|$ and $(.|.)$. We shall write $|.|_H$ when the norm $|.|$ might be confused with the absolute value of scalars. The norm of the Banach space E we denote by $\|.\|$, i.e. $\|.\|$ is some measurable norm on H (see [6]) and E is the completion of H with respect to (w.r.t.) $\|.\|$. Thus H is identified with a dense linear subspace of E and it is this identification what we denote above by i. In the classical AWS $(i,C',C_0[0,1])$ (see Section.1) the scalar product in C' is

$$\left(\int^{\cdot} f(x)dx \mid \int^{\cdot} g(x)dx \right) = \int_0^1 f(x)g(x)dx , \quad f,g\in L^2[0,1],$$

and the corresponding norm $\|.\|$ in $C_0[0,1]$ is the standard sup-norm. We have the following continuous dense embeddings

$$E^* \xhookrightarrow{i^*} H \xhookrightarrow{i} E ,$$

where E^* is the topological dual of E and we identify H with its dual H^*. The canonical bilinear form that links E^* and E we shall denote by $\langle z,\omega\rangle$, $z\in E^*$, $\omega\in E$. We shall not distinguish between $h\in H$ and $ih\in E$ and also between $z\in E^*$ and $i^*z\in H$. Thus for $z\in E^*\subset H$ and for $h\in H$ we write

(2.1) $$\langle z,h\rangle \equiv \langle z,ih\rangle = (i^*z|h) \equiv (z|h) .$$

Usually the vectors in E we shall denote by ω, the vectors in H we shall denote by h, and those in $E^*\subset H$ by z. ∎

2.2 We also assume that it is given a continuous orthogonal curve (Z) in H. By this we mean a collection of vectors $(Z) = \{ Z_t\in E^*\subset H: 0\le t\le 1 \}$, which satisfies the following conditions:

A1. $\{ Z_t\in E^* : 0\le t\le 1 \}$ is dense subset of H and $Z_0 = 0$;

(A) A2. (orthogonality) $(Z_t - Z_{t'} | Z_s - Z_{s'}) = 0$, $0\le s\le s'\le t\le t'\le 1$;

A3. (continuity) $t\longmapsto |Z_t|_H$ is continuous function on $[0,1]$.

For any simple function on the interval $[0,1]$ (we always write $1_A(.)$ for the indicator function of the set A)

$$f(x) = \Sigma\, \alpha_i 1_{(s_i,t_i]}(x) , \quad \alpha_i\in\mathbb{R}, \quad x\in[0,1],$$

define

$$Z[f] = \Sigma\, \alpha_i (Z_{t_i} - Z_{s_i}) \in H .$$

Because of the orthogonal property of (Z) for every two simple functions f and g we have that

(2.2) $$(Z[f]|Z[g]) = \int_0^1 f(x)g(x)\nu(dx) ,$$

where ν is the only positive Borel measure on $[0,1]$ for which

$$\nu[s,t] = |Z_t - Z_s|^2 , \quad 0 \le s \le t \le 1 .$$

Note that $\nu(.)$ is correctly defined nonatomic measure, because $t \to |Z_t|^2 = \nu[0,t]$ is continuous (condition A3) and nondecreasing (condition A2) function on $[0,1]$. Formally one can write $\nu(dt) = |dZ_t|^2$. The simple functions form a dense subset of $\mathcal{H}_\nu = L^2([0,1],d\nu)$. Then (2.2) shows that $Z[.]$ is densely defined isometry from $\mathcal{H}_\nu$ into H , the range of which is also dense (condition A1.). Then $Z[.]$ can be naturally extended to an unitary operator between $\mathcal{H}_\nu$ and H and we denote this extension also by $Z[.]$. Thus (2.2) actually holds for arbitrary $f,g \in \mathcal{H}_\nu$. With a slight abuse of notation for any Borel set $A \subset [0,1]$ we write $Z(A)$ for $Z[1_A]$. Then $Z(.)$ is easily seen to be a vector-valued measure on $[0,1]$ with values in H, and the integral w.r.t. this measure is nothing but the operator $Z[.]$. Therefore we can write

$$Z[f] \equiv \int f dZ \equiv \int f(x) Z(dx) , \quad f \in \mathcal{H}_\nu .$$

For the sake of simplicity through the whole paper, as in the last expression, we shall write simply $\int$ when the domain of integration is the entire interval $[0,1]$, and we shall write $\int^t$ when the domain is $[0,t]$; other domains will be written explicitly. The integral

$$\int_s^t f dZ \equiv \int_s^t f(x) Z(dx) , \quad f \in \mathcal{H}_\nu , \quad 0 \le s \le t \le 1 ,$$

will be obviously understood as $Z[1_{(s,t]} f]$. Since $Z[.]$ is unitary operator, for every $h \in H$ there exists unique function (in fact a class of ν-equivalent functions) $\delta h \equiv \delta_x h$, $x \in [0,1]$, s.t. $\delta h \in \mathcal{H}_\nu$ and $Z[\delta h] = h$. Also for $h,h' \in H$

(2.3) $$(h|h') = \int (\delta_x h)(\delta_x h') \nu(dx) .$$

Note that $\delta_x (Z_t - Z_s) = 1_{(s,t]}(x)$.

With the curve (Z) we associate a *nonincreasing* family of Hilbert subspaces $(H_t \subseteq H)$, $t \in [0,1]$, where H_t is spanned by $\{ Z_s - Z_{s'} : t \le s,s' \le 1 \}$. Condition A3 implies that the family (H_t) is continuous from the right and from the left. Note also that $H_0 = H$ and $H_1 = \{0\}$. Then $H_t^\perp$ is the subspace spanned by $\{ Z_s - Z_{s'} : 0 \le s,s' \le t \}$ (remind that (Z) spans the whole H -

condition A1). For $t\in[0,1]$ we write P_t for the orthogonal projector in H with Range$(P_t) = H_t^{\perp}$. It is easy to check that

$$P_t Z[f] \equiv P_t(\int f(x)Z(dx)) = \int^t f(x)Z(dx) \quad . \quad \#$$

#2.3® The following "classical" example clarifies the role of the integrator (Z) introduced in #2.2®. The classical AWS $(i, C', C_0[0,1])$ is considered. Then E^* as a subspace of $H \equiv C'$ coincides with the manifold of all functions of the form $\int m[x,1]dx$, where $m\in M[0,1]$ – the set of all bounded signed measures on $[0,1]$. For $t\in[0,1]$ let $\delta_t \in M[0,1]$ be the Dirac measure at the point $\{t\}$, i.e. $\delta_t(A) = 1_A(t)$, $A\subseteq[0,1]$. Define

$$(2.4) \qquad Z_t = \int^{\cdot} \delta_t[x,1]dx \in C' \ , \quad t\in[0,1] \ .$$

As a function from C', each Z_t can be written as

$$Z_t(x) = \begin{cases} x & x\le t \\ t & x>t \end{cases} \quad .$$

Then (Z) is easily seen to obey all conditions (A). In this case $\nu(dt) = |dZ_t|^2 = dt$ is simply the Lebeasgue measure and $H_\nu = L^2[0,1]$. One can prove that for $f\in L^2[0,1]$

$$(2.5) \qquad Z[f] = \int^{\cdot} f(x)dx \quad .$$

It will be very instructive always to have in mind this last example when general integrals $Z[f]$, $f\in H_\nu$, for general H occur. It should be noted that when (i,H,E) is a general AWS, via the unitary operator $Z[.]$, H inherits some "order" from $H_\nu = L^2([0,1],d\nu)$. This order in H is however an *arbitrary* one, for (Z) itself is arbitrary. Thus, if one considers another curve (Z) in C', different than those given by (2.4), then the integral $Z[.]$ might lack the interpretation (2.5). #

#2.4® In this subsection we shall describe the σ-fields and probability distributions that are of interest.

By $B[s,t]$ we denote the Borel σ-field on the interval $[s,t]$ and we write simply B for $B[0,1]$. The Borel σ-field in H, w.r.t. the Hilbertian topology, we denote by $B(H)$. The Borel σ-field in E w.r.t. the norm $\|.\|$ we denote by ξ. We have that ξ is generated by all functionals $E\ni\omega \to \langle z,\omega\rangle$, $z\in E^*$, (see

[1],[6]). The canonical Gaussian probability measure with variance parameter 1, defined on the measurable space $(E,\mathcal{E})$, (i.e. the white noise over E) we denote by p . Then for every $z\in E^*$, the mapping $E\ni\omega \rightarrow \langle z,\omega\rangle$ has Gaussian distribution w.r.t. p with $\int_E \langle z,\omega\rangle p(d\omega) = 0$ and

$$\int_E \langle z,\omega\rangle\langle z',\omega\rangle p(d\omega) = (z|z') , \quad z,z'\in E^*\subset H .$$

On $(E,\mathcal{E})$ define the filtration $(\mathcal{E}_t)$, $t\in[0,1]$, given by $\mathcal{E}_t = \sigma\{ \langle Z_s,.\rangle : 0\le s\le t \}$. In fact this is the natural filtration of the process

(2.6) $$N_t(\omega) = \langle Z_t,\omega\rangle , \quad 0\le t\le 1 ,$$

which will play an important role in our analysis. Note that σ-fields $\mathcal{E}_t$ are not completed with p-null sets and $N_t(\omega)$ in (2.6) is regarded as an *everywhere defined* function on E. Because $Z_0 = 0$, we have that $\mathcal{E}_0 = \{\emptyset ,E\}$.

Also given is another (general) probability space $(U,\mathcal{U},\Pi)$, whose probability low Π is independent on the white noise p . The members of the set U will be denoted usually by u .

Our analysis will be carried out on the product probability space

(2.7) $$(S,\mathcal{S},\mathbb{P}) = (U\times E,\mathcal{U}\otimes\mathcal{E},\Pi\otimes p) .$$

The elements of the set S we shall write as pairs (u,ω), $u\in U$, $\omega\in E$. Always $\mathbb{E}\{.\}$ will denote the integral w.r.t. $\mathbb{P}$ but equivalently this integral will be denoted also by

$\int_S \{\ldots\}d\mathbb{P}$, $\int_{U\times E}\{\ldots\}\mathbb{P}(du,d\omega)$, $\int_U\int_E\{\ldots\}p(d\omega)\Pi(du)$, etc.

The integral w.r.t. p on the probability space $(E,\mathcal{E},p)$ will be denoted by $\mathbb{E}_p\{.\}$ and the integral w.r.t. Π on the probability space $(U,\mathcal{U},\Pi)$ will be denoted by $\mathbb{E}_\Pi\{.\}$. By (L^2) we denote the standard space $L^2(S,\mathcal{S},\mathbb{P})$ of square integrable random variables (r.v.'s) provided with the usual structure of a Hilbert space.

On $(S,\mathcal{S})$ define the filtration $(\mathcal{F}_t \equiv \mathcal{U}\otimes\mathcal{E}_t)$, $t\in[0,1]$. By $\bar{\mathcal{S}}$ we denote the completion of $\mathcal{S}$ with all $\mathbb{P}$-null sets and $\bar{\mathcal{F}}_t$ is the corresponding completion of $\mathcal{F}_t$. The completions with p-null sets of $\mathcal{E}$, $\mathcal{E}_t$ are $\bar{\mathcal{E}}$, $\bar{\mathcal{E}}_t$. It is to be kept in mind that in some cases we are dealing with noncompleted σ-fields and for the corresponding results this is essential; so one should not confuse completed and noncompleted σ-fields. Note that filtrations $(\mathcal{E}_t)$, $(\mathcal{F}_t)$, $(\bar{\mathcal{E}}_t)$, $(\bar{\mathcal{F}}_t)$ are all right continuous.

Because of the independence between p and Π with no ambiguity r.v.'s on $(E,\mathcal{E},p)$ or on $(U,\mathcal{U},\Pi)$ might be regarded as r.v.'s on $(S,\mathcal{S},\mathbb{P})$; i.e., in general, all r.v.'s of interest are given on $(S,\bar{\mathcal{S}},\mathbb{P})$. For example, we consider the process (N_t), defined in (2.6), also as being given on $(S,\bar{\mathcal{S}},\mathbb{P})$ by $(u,\omega)\rightarrow N_t(u,\omega) \equiv \langle Z_t,\omega\rangle$, $t\in[0,1]$. In the sequel we shall need to construct stochastic integrals w.r.t. (N_t). In this end we observe the following: (i) (N_t) is mean-square continuous (m.sq.c.) Gaussian process, starting from 0 $(N_0=0)$, having vanishing mean and independent increments; (ii) $(N_t,\bar{\mathcal{F}}_t)$ is martingale, and $N_t-N_{t'} \perp\!\!\!\perp \bar{\mathcal{F}}_s$ for $0\le s\le t'\le t\le 1$; (iii) the quadratic variation (q.v.) of (N_t) is deterministic, nondecreasing function, given by $\langle N\rangle_t=\nu[0,t]\equiv|Z_t|^2$.#

#2.5® In connection with stochastic integrals w.r.t. (N_t) we consider the measurable space

$$([0,1]\times S\ ,\ \mathcal{B}\otimes\bar{\mathcal{S}}) \equiv ([0,1]\times U\times E\ ,\ \mathcal{B}\otimes\overline{\mathcal{U}\otimes\mathcal{E}})$$

and define $\mathcal{P}$ $(\bar{\mathcal{P}})$ to be the σ-field of all $(\mathcal{F}_t)$-predictable ($(\bar{\mathcal{F}}_t)$-predictable) sets; i.e. $\mathcal{P}$ $(\bar{\mathcal{P}})$ is the σ-field generated by the sets $(s,t]\times A$, $A\in\mathcal{F}_s$ $(A\in\bar{\mathcal{F}}_s)$, and the sets $\{0\}\times A$, $A\in\mathcal{F}_0$ $(A\in\bar{\mathcal{F}}_0)$. Since the filtration $(\mathcal{F}_t)$ ($(\bar{\mathcal{F}}_t)$) is right continuous, then it follows that every function $\phi(s;u,\omega)$, $s\in[0,1]$, $u\in U$, $\omega\in E$, which is $\mathcal{P}$-measurable ($\bar{\mathcal{P}}$-measurable), is also adapted in the sense that for every $s>0$ $\phi(s;.,.)$ is $\mathcal{F}_s$-measurable ($\bar{\mathcal{F}}_s$-measurable) function on S (see [3,Theor.3.1.1]).

Define on $[0,1]\times E$ the σ-field $\mathcal{P}_E$ to be those of all $(\mathcal{E}_t)$-predictable sets; i.e. $\mathcal{P}_E$ is generated by the sets $(s,t]\times B$, $B\in\mathcal{E}_s$, and the sets $\{0\}\times B$, $B\in\mathcal{E}_0$. Because $\mathcal{F}_s=\mathcal{U}\otimes\mathcal{E}_s$, then it easily follows that $\mathcal{P}=\mathcal{U}\otimes\mathcal{P}_E$.

By L^2_{loc} we denote the class of all $\bar{\mathcal{P}}$-measurable functions $\phi(s;u,\omega)$ on $[0,1]\times S\equiv[0,1]\times U\times E$ s.t. $\phi(0;.,.)$ is $\bar{\mathcal{F}}_0$-measurable and for $\mathbb{P}$-a.e. $(u,\omega)\in S$

$$\int|\phi(s;u,\omega)|^2\nu(ds) < \infty \ . \tag{2.8}$$

By $\mathcal{L}^2_{loc}$ we denote the class of all functions $\phi(s;u,\omega)$ on

$[0,1]\times S$ s.t. *for every* $(u,\omega)\in S$ $\phi(.;u,\omega)\in H_\nu$ (see #2.2®); i.e. *for every* $(u,\omega)\in S$ $\phi(.;u,\omega)$ is $\mathcal{B}$-measurable function on $[0,1]$ and (2.8) is satisfied. Note that no sort of joint measurability is assumed for the functions of $\mathcal{L}^2_{loc}$, nor it is assumed any sort of measurability w.r.t. the variables $u\in U$ and $\omega\in E$.

For every $\phi\in L^2_{loc}$ one can define the following stochastic integral on the filtered space $(S,\mathcal{T},(\mathcal{F}_t),\mathbb{P})$

(2.9) $$I^\phi_t = \int^t \phi(s;u,\omega)dN_s(\omega) \ , \ t\in[0,1] \ .$$

We convent to write I^ϕ for I^ϕ_1. It is known that $(I^\phi_t,\mathcal{F}_t)$ is local martingale which possesses continuous modification. Its q.v. is given by

$$\langle I^\phi_\cdot\rangle_t = \int^t |\phi(s;u,\omega)|^2\nu(ds) \ , \quad t\in[0,1] \ .$$

If in addition to (2.8) it is required that

$$\mathbb{E}\{\int|\phi(s;u,\omega)|^2\nu(ds)\} < \infty$$

then $(I^\phi_t,\mathcal{F}_t)$ is known to be a martingale. In particular, one has that H_ν is exactly the set of all deterministic elements of L^2_{loc}. Then for every function $f\in H_\nu$ one can define

$$I^f_t = \int^t f(s)dN_s(\omega) \ , \quad t\in[0,1],$$

and $(I^f_t,\mathcal{F}_t)$ is easily seen to be a Gaussian martingale with q.v. $\langle I^f_\cdot\rangle_t = \int^t f^2 d\nu$. For $f,g\in H_\nu$ we have

$$\mathbb{E}\{I^f I^g\} = \int f(s)g(s)\nu(ds)$$

and therefore for every $h,h'\in H$ it holds

$$\mathbb{E}\{I^{\delta h} I^{\delta h'}\} = \int(\delta_s h)(\delta_s h')\nu(ds) = (h|h') \ .$$

Let ϕ be $P \equiv \mathcal{U}\otimes P_E$-measurable function and let $\phi\in\mathcal{L}^2_{loc}$. Then *for every* $u\in U$ and *for every* $\omega\in E$ one has

$$\int|\phi(s;u,\omega)|^2\nu(ds) < \infty \ .$$

Since for every $u\in U$ $\phi(.;u,.)$ is P_E-measurable function on $[0,1]\times E$, it follows that *for every* $u\in U$ one can define the following stochastic integral on the filtered space $(E,\xi,(\xi_t),p)$

(2.10) $${}_uI^\phi_t = \int^t\phi(s;u,\omega)dN_s(\omega) \quad t\in[0,1].$$

For every $u\in U$ we have that $({}_uI^\phi_t,(\xi_t))$ is a local martingale on the probability space (E,ξ,p) , which possesses continuous modification and its q.v. is given by

$$\langle {}_uI^\phi_\cdot\rangle_t = \int^t|\phi(s;u,\omega)|^2\nu(ds) \ , \quad t\in[0,1] \ .$$

We shall write $_uI^\phi$ for the r.v. $_uI_1^\phi$, the last being given on $(E,\mathcal{E},p)$. On the other hand, under the same assumption, that ϕ is $\mathcal{P}$-measurable and $\phi \in \mathcal{L}^2_{loc}$, it holds that $\phi \in L^2_{loc}$ and therefore one can also define as in (2.9) the integral I_t^ϕ, on the filtered space $(S,\mathcal{S},(\mathcal{F}_t),\mathbb{P})$. Note that for every $u\in U$ $_uI^\phi$, as a r.v. on $(E,\mathcal{E},p)$, is defined p-a.e.; i.e. $_uI^\phi$ is a class of p-equivalent functions on E for every $u\in U$. Also, I^ϕ, as a r.v. on $(S,\mathcal{S},\mathbb{P})$, presents a class of $\mathbb{P}$-equivalent functions on S. We have that the mappings $(u,\omega) \longrightarrow {_uI^\phi(\omega)}$ and $(u,\omega) \longrightarrow I^\phi(u,\omega)$ coincide $\mathbb{P}$-a.e. in S, in the sense that if $(u,\omega) \longrightarrow \dot{I}^\phi(u,\omega)$ is a function from the $\mathbb{P}$-class I^ϕ, then $\omega \longrightarrow \dot{I}^\phi(u,\omega)$, as a function on E, is from the p-class $_uI^\phi$ for Π-a.e. $u\in U$; that is to write

(2.11) $\qquad I^\phi(u,\omega) = {_uI^\phi(\omega)}$, for p-a.e. $\omega \in E$, for Π-a.e. $u\in U$.

This coincidence of the integrals $I^\phi(u,\omega)$ and $_uI^\phi(\omega)$ could be verified with some standard arguments, which here we shall omit (when ϕ is simple $\mathcal{P}$-measurable function (2.11) is obvious). #

3. The Girsanov Transformation

In this section we study transformations of $(S,\mathcal{S},\mathbb{P})$ which are of the form $(u,\omega)\longrightarrow(u,T(u,\omega))$, $T(u,\omega) = \omega + h[u,\omega]$, where $h[.,.]$, in its turn, is a mapping from S into $H\subset E$. The h-term will have the form of an integral w.r.t. the curve (Z), as defined in #2.2® . One case of such transformations, which is of a particular interest, is those when $(S,\mathcal{S},\mathbb{P})$ is reduced to the AWS $(E,\mathcal{E},p)$ (one can take for example a set U, having just one element $U = \{u\}$, see (2.7)). Then T above becomes a transformation of E into E (transformation of the white noise), adopting the form $E\ni\omega \longrightarrow T\omega \equiv \omega + h[\omega] \in E$. In the case of the classical AWS this is exactly the transformation given by the r.h.s. of (1.1′), of course if the curve (Z) is defined by (2.4). This transformation of $C_0[0,1]$ is well known in connection with the Girsanov theorem and sometimes it is called the "drift". An important point in what follows it is the invertability of the transformation T (under certain assumptions), which for the purposes of the nonlinear filtering is crucial. It might be of some interest also for the analysis of a class of stochastic equations, having diffusion coefficient 1 (see #3.12 below).

#3.1 Proposition Let for some $t\in[0.1]$ $f:E\to\mathbb{R}$ be $\mathcal{E}_t$-measurable function (note that $\mathcal{E}_t$ is not assumed to be completed with p-null sets). Then *for every* $\omega\in E$ and *for every* $h\in H_t$

(3.1) $$f(\omega + h) = f(\omega) \quad . \#$$

Another form in which (3.1) will be used more frequently is the following

(3.2) $$f(\omega + h) = f(\omega + P_t h) \ , \quad \omega\in E, \quad h\in H \ .$$

#3.2 Definition Let $\phi(s;u,\omega)$ be a real function, defined on $[0,1]\times S \equiv [0,1]\times U\times E$. Then $\phi(s;u,\omega)$ is said to be:

(i) *measurable* - if it is $\mathcal{B}\otimes\mathcal{S}$-measurable;

(ii) *predictable* - if it is $\mathcal{P}$-measurable (not $\overline{\mathcal{P}}$-measurable!).#

If $\phi\in\mathcal{L}^2_{loc}$ then $\phi(.;u,\omega)\in\mathcal{H}_v$ (i.e. (2.8) holds) *for every* $(u,\omega)\in S$, so that

(3.3) $$(u,\omega) \longrightarrow \theta(u,\omega) \equiv Z[\phi(.;u,\omega)] \equiv \int\phi(s;u,\omega)Z(ds) \in H$$

is *everywhere* defined mapping from S into H .

#3.3 Proposition For $\phi\in\mathcal{L}^2_{loc}$ consider the mapping $\theta:S\to H$ given by (3.3). Then if ϕ is measurable, it follows that $\theta(.,.)$ is $\mathcal{S}\backslash\mathcal{B}(H)$-measurable (see #2.4®). #

#3.4 Definition The elements of $\mathcal{L}^2_{loc}$ will be referred to as Girsanov's kernels, or simply kernels. For every $\phi\in\mathcal{L}^2_{loc}$ the transformation $T_\phi:S\longrightarrow E$, given by

$$T_\phi(u,\omega) = \omega - Z[\phi(.;u,\omega)] \equiv \omega - \int \phi(s;u,\omega)Z(ds) \ , \quad (u,\omega)\in S \ ,$$

will be referred to as Girsanov transformation with kernel ϕ .#

Note that the Girsanov transformation T_ϕ is defined *everywhere* in S , not $\mathbb{P}$-a.e.. However, it might occur that two different kernels $\phi,\phi'\in\mathcal{L}^2_{loc}$ define the same Girsanov transformation:

$$T_\phi(u,\omega) = T_{\phi'}(u,\omega) \ , \quad (u,\omega)\in S \ .$$

The last relation is possible if and only if

$$Z[\phi(.;u,\omega)] = Z[\phi'(.;u,\omega)] \ , \quad (u,\omega)\in S \ ,$$

which on the other hand is equivalent to

$$\int |\phi(s;u,\omega)-\phi'(s;u,\omega)|^2 \nu(ds) = 0 \ , \quad (u,\omega)\in S \ .$$

In connection with this we give the following

#3.5 Definition Two kernels $\phi,\phi'\in\mathcal{L}^2_{loc}$ are said to be equivalent (notation $\phi \cong \phi'$), if *for every* $(u,\omega)\in S$ the following equality holds

$$\phi(s;u,\omega) = \phi'(s;u,\omega) \ , \quad \text{for } \nu\text{-a.e. } s\in[0,1] \ .\#$$

#3.6 Proposition Consider the Girsanov transformation $T_\phi : S \longrightarrow E$ for some $\phi\in\mathcal{L}^2_{loc}$. (a) If ϕ is measurable then T_ϕ is $\mathcal{S}\backslash\mathcal{E}$ -measurable. (b) If ϕ is predictable then T_ϕ is $\mathcal{F}_t\backslash\mathcal{E}_t$-measurable for *every* $t\in[0,1]$. #

As a direct consequence of #3.6 we obtain the following

#3.7 Proposition Let $\phi\in\mathcal{L}^2_{loc}$. Consider the transformation $\theta : S \longrightarrow S$, given by $(u,\omega) \longrightarrow \theta(u,\omega) \equiv (u,T_\phi(u,\omega))$. (a) If ϕ is measurable then θ is $\mathcal{S}\backslash\mathcal{S}$-measurable. (b) If ϕ is predictable then θ is $\mathcal{F}_t\backslash\mathcal{F}_t$-measurable for every $t\in[0,1]$. #

#3.8 Proposition Let $\phi\in\mathcal{L}^2_{loc}$. Consider the transformation $\theta : [0,1]\times S \longrightarrow [0,1]\times S$, given by

$$\theta(s;u,\omega) \longrightarrow \theta(s;u,\omega) \equiv (s;u,T_\phi(u,\omega)) \ .$$

(a) If ϕ is measurable then θ is $\mathcal{B}\otimes\mathcal{S} \backslash \mathcal{B}\otimes\mathcal{S}$-measurable.

(b) If ϕ is predictable then θ is $\mathcal{P}\backslash\mathcal{P}$-measurable. #

#3.9 Definition A function $\phi(s;u,\omega)$, defined on $[0,1]\times U\times E$ is called causal if *for every* $(s;u,\omega)\in[0,1]\times S$ and *for every* $h\in H_s$ the following equality holds

$$\phi(s;u,\omega + h) = \phi(s;u,\omega) \ ,$$

or equivalently if

$$\phi(s;u,\omega + h) = \phi(s;u,\omega + P_s h) \ , \quad s\in[0,1],\ u\in U,\ \omega\in E,\ h\in H \ .\ \#$$

Note that if $\phi(s;u,\omega)$ is s.t. $\phi(s;u,.)$ is $\mathcal{E}_s$-measurable for every $s\in[0,1]$ and for every $u\in U$, then it follows from #3.1 that ϕ is causal. Morover, ϕ is causal if $\phi(s;.,.)$ is

$F_s = \mathcal{U}\otimes\mathcal{E}_s$-measurable for all $s\in[0,1]$. If ϕ is predictable, then $\phi(s;.,.)$ is F_s-measurable for every $s\in[0,1]$ (see #2.5®), and therefore ϕ is causal. Note that usually the term "causal" is meant to imply measurability plus (F_s)-adaptivity (see [3,Def.5.1.1]).

For a real function $\phi(s;u,\omega)$, defined on $[0,1]\times S$ let us now consider the following set of conditions:

B1. ϕ is causal in the sense of #3.9 ;

B2. *for every* $(u,\omega)\in S$ $\phi(.;u,\omega)$ is $\mathcal{B}$-measurable (i.e. Borel) function on $[0,1]$ and

(B) $$A_{u,\omega} = \sup_{s\in[0,1]} \{ |\phi(s;u,\omega)| \} < \infty ;$$

B3. *for every* $(u,\omega)\in S$

$$B_{u,\omega} = \sup_{\substack{s\in[0,1] \\ 0\neq h\in H}} \{ \frac{1}{|h|_H} | \phi(s;u,\omega+h) - \phi(s;u,\omega) | \} < \infty .$$

In connection with these conditions we note the following: (i) if ϕ satisfies (B) so does also $1_{[0,t]}(s)\phi(s;u,\omega)$ for every $t\in[0,1]$, with the same $A_{u,\omega}$ and $B_{u,\omega}$, respectively in B2 and B3 ; (ii) the only measurability requirement for ϕ is the one imposed in B2 ; (iii) the constants $A_{u,\omega}$ and $B_{u,\omega}$ are not global, in that they might depend in general on u and ω ; (iv) B2 implies that $\phi(.;u,\omega)\in H_\nu$ *for every* $(u,\omega)\in S$, or which amounts to the same, that $\phi\in\mathcal{L}^2_{loc}$; (v) because of B1 another equivalent form of B3 is this

(3.5) $$|\phi(s;u,\omega+h) - \phi(s;u,\omega)| \leq B_{u,\omega}|P_s|_H , \quad s\in[0,1],\ u\in U,\ h\in H,$$

where $B_{u,\omega}$ is simply a finite constant (independent of s) that exists for every $u\in U$ and every $\omega\in E$; note also that (see #2.2®)

$$|P_t h|^2 = \int^t (\delta_s h)^2 \nu(ds) .$$

As a direct consequence of (3.5) we obtain the following

#3.10 Proposition Let $f,g \in\mathcal{L}^2_{loc}$. Then, if ϕ satisfies conditions (B) , for every $t\in[0,1]$ and every $(u,\omega)\in S$ we have

$$| \phi(t;u,\omega + \int f(s;u,\omega)Z(ds)) - \phi(t;u,\omega + \int g(s;u,\omega)Z(ds)) | \leq$$

$$\leq B_{u,\omega} \int^t |f(s;u,\omega)-g(s;u,\omega)|^2 \nu(ds) \quad .\#$$

Our next result plays the key role in our analysis.

#3.11 Proposition Let $\phi(s;u,\omega)$ be a function on $[0,1]\times S$ which satisfies conditions (B). Then there exists unique up to equivalence (see #3.5) kernel $\phi'(s;u,\omega)$, which also satisfies conditions (B), and which is s.t. the following equality holds *for every* $(u,\omega)\in S$

$$\omega = T_{\phi}[u,T_{\phi'}(u,\omega)] = T_{\phi'}[u,T_{\phi}(u,\omega)] ;$$

that is, for every $u\in U$ the Girsanov transformation $T_{\phi}(u,.):E\to E$ is one-to-one and invertible and its inverse $T_{\phi}^{-1}(u,.)$ is another Girsanov transformation with kernel ϕ'. The kernel ϕ' can be calculated according to the following recursive procedure:

$$\psi_0(t;u,\omega) = 0;$$

$$(3.6) \qquad \psi_{n+1}(t;u,\omega) = \phi(t;u,\omega + \textstyle\int^t \psi_n(s;u,\omega)Z(ds)) \quad , \quad n \geq 0 ;$$

$$\phi'(t;u,\omega) = -\lim_n [\psi_n(t;u,\omega)] .$$

If ϕ is measurable (see #3.2), then ϕ' can be chosen to be also measurable. If ϕ is predictable (see #3.2), then ϕ' can be chosen to be also predictable. #

Let us note, and this is very important for nonlinear filtering, that in the last proposition not only the invertability of the Girsanov transformation is established, but also an algorithm is obtained for the recursive construction of the kernel of the inverse (the one described by (3.6)). Let us also note that kernels ϕ and ϕ' are dual one to another, in the sense that each kernel determines the other one up to equivalence (see #3.5). Indeed, since ϕ' also obeys the assumptions of #3.11 (conditions (B)), with the same recursive procedure (3.6) one can construct the kernel $\phi''\equiv(\phi')'$ of the inverse of T_{ϕ}^{-1}, which is exactly T_{ϕ}; i.e. one has $T_{\phi''}(u,\omega) = T_{\phi}(u,\omega)$, $(u,\omega)\in S$, or which amounts to the same $\phi'' \cong \phi$.

The case when $\phi(t;u,\omega) = \phi(t;\omega)$, i.e. ϕ does not depend on $u\in U$, is particularly interesting, for then T_{ϕ} becomes a transformation of E into E, which, writing simply T for T_{ϕ}, is this

(3.7) $$T\omega = \omega - \int\phi(s;\omega)Z(ds) .$$

The conditions for the kernel $\phi(s;\omega)$, under which this last transformation possesses an inverse are the following

B1'. ϕ is causal in that

$$\phi(s;\omega + h) = \phi(s;\omega + P_s h) , \quad s\in[0,1], \ \omega\in E, \ h\in H ;$$

B2'. *for every* $\omega \in E$ $\phi(.;\omega)$ is $\mathcal{B}$-measurable (i.e. Borel) function on $[0,1]$ and

(B') $$A_\omega = \sup_{s\in[0,1]} \{ |\phi(s;\omega)| \} < \infty ;$$

B3'. *for every* $\omega \in E$ there exists a finite constant $B_\omega \geq 0$ s.t. $|\phi(s;\omega + h) - \phi(s;\omega)| \leq B_\omega \; |P_s|_H$, $s\in[0,1]$, $h\in H$.

When conditions (B') are met, then one can write the inverse of T in the form

$$T^{-1}\omega = \omega + \int\phi'(s;\omega)Z(ds) ,$$

where ϕ' also obeys (B') and can be calculated recurcively, as in (3.6).

#3.12 Remark One can consider (3.7) as a "functional" equation. Namely, let us write in (3.7) ξ for ω and ω for $T\omega$. Then (3.7) can be rewritten as

(3.8) $$\xi = \omega + \int\phi(s;\xi)Z(ds) , \quad \xi,\omega \in E .$$

In (3.8) we regard $\xi = \xi[\omega]$ as an "unknown" functional of the form $\xi:E\to E$. Since (3.8) is equivalent to $\omega = T\xi$, $\omega,\xi\in E$, the solution $\xi = \xi[\omega]$ is then given by

(3.9) $$\xi = T^{-1}\omega = \omega + \int\phi'(s;\omega)Z(ds) .$$

Of course, one can consider $\omega \in E$ as beihg randomly distributed according to some probability low π on the measurable space $(E,\mathcal{E})$. Treated like this, relation (3.8) becomes a stochastic equation and its solution (3.9) is an random element on the probability space $(E,\mathcal{E},\pi)$. Let us note however, that the solution (3.9) is an "everywhere solution", which one can write regardless of the probability low π, just calculating the resolving kernel ϕ' recursively, as in (3.6). In that sense (3.9) presents a solution, which is *stronger* than what is usually referred to as a "strong solution" of a stochastic equation (see [3],[7]). To clarify this, let us consider the classical case when E is taken to be $C_0[0,1]$, the curve (Z) is the one defined in (2.4), and the distribution π of $\omega \in E \equiv C_0[0,1]$ coincides with the Wiener measure μ. In this case the elements $\xi(.)$ and $\omega(.)$ are simply functions from $C_0[0,1]$ and the "functional" equation (3.8) obtains the following form (see #2.3®)

$$(3.8') \qquad \xi(t) = \omega(t) + \int^t \phi(s;\xi)ds , \quad t\in[0,1] .$$

The corresponding solution is then given by

$$(3.9') \qquad \xi[\omega](t) = \omega(t) + \int^t \phi'(s;\omega)ds , \quad t\in[0,1] .$$

Since the coordinate functions $\omega \to W_t(\omega) \equiv \omega(t)$ for $t\in[0,1]$, as a family of r.v. s on the Wiener space $(C_0[0,1],\mu)$, form the standard Wiener process, one can regard (3.8′) as a stochastic equation, which is standardly written as

$$(3.10) \qquad \xi_t(\omega) = W_t(\omega) + \int^t \phi(s;\xi)ds , \quad t\in[0,1] .$$

As all considerations in this section, and mainly #3.11, show, when ϕ satisfies conditions (B′), then one can obtain even an *everywhere solution* of the stochastic equation (3.10), which is $\xi_t(\omega) \equiv \xi[\omega](t)$, and it is given by (3.9′). The obvious interpretation of such a solution is this: *every* realisation (or trajectory) $\omega(.) \equiv W_.(\omega)$ of the Wiener process (W_t) uniquely determines the trajectory of the solution (ξ_t), which can be calculated by (3.9′). It is perhaps worth pointing out that the assumptions (B′) are even less restrictive than the standard assumptions for the diffusion and drift coefficients of the stochastic equations, that are usually made in order to ensure the existence and uniqueness of a strong solution (compare with [3, Sect.5.1] and [7, Theor.4.6], see also [2]). Note however that (3.10) is a special case of the Ito-equation with diffusion coefficient equal to 1. Let us also note that the resolving kernel ϕ' in (3.9′) can be obtained by a "fully deterministic" recursive procedure described in (3.6). Therefore one can consider the stochastic equation (3.10), more generally, in that (W_t) might be assumed to be an *arbitrary* process with continuous realizations, with an *arbitrary* probability distribution π on $C_0[0,1]$. In that case again the solution is given by (3.9′). #

4. The Girsanov Theorem

Let $\phi(s;u,\omega)$ be a measurable function from $\mathcal{L}^2_{loc}$ (see #3.2). From #3.6 we have that $T_\phi: S \longrightarrow E$ is a measurable transformation from $(S,\mathcal{S})$ into (E,ξ). Then $T_\phi^{-1}\circ p$ is a probability measure on $(S,\mathcal{S})$, where p is the canonical Gaussian probabilty distribution on (E,ξ), as described in #2.4® . On the other hand $(S,\mathcal{S},\mathbb{P})$ is a probability space and it is important to describe $T_\phi^{-1}\circ p$ in terms of the probability measure $\mathbb{P}$. This will be our main concern in this section. Namely, for the purposes of nonlinear filtering, we need a certain adaptation of the Kameron-Martin-Girsanov theorem in order to show that under some requirements for ϕ the probability low $T_\phi^{-1}\circ p$ is equivalent to $\mathbb{P}$ and its Radon-Nykodim derivative can be expressed in terms of ϕ. These requirements are described by the following set of conditions:

C1. $\phi \in \mathcal{L}^2_{loc}$ and ϕ is predictable (see #3.2);

(C) C2. for every $u \in U$ it holds that

$$\mathbb{E}_p\left\{\exp[\int\phi(s;u,\omega)dN_s(\omega) - \tfrac{1}{2}\int|\phi(s;u,\omega)|^2\nu(ds)]\right\} < \infty .$$

The process (N_t) that appears in C2 above is the Gaussian martingale, defined by (2.6). The stochastic integral w.r.t. (N_t) is understood as the integral ${}_uI^\phi \equiv {}_uI^\phi_1$ on the filtered space $(E,\xi,(\xi_t),p)$, defined *for every* $u\in U$, as in (2.10). Since in our considerations $(U,\mathcal{U})$ is a *general* measurable space, it makes no difference whether C2 is required to hold for all $u\in U$ or for Π-a.e $u\in U$, for one can always replace the set U with $U' = U\backslash\mathcal{N}$, where $\mathcal{N} \in \mathcal{U}$ and $\Pi(\mathcal{N}) = 0$.

With each function ϕ, which satisfies C1, we associate the following exponential supermartingales

$${}_uR^\phi_t(\omega) = \exp\left\{{}_uI^\phi_t(\omega) - \tfrac{1}{2}\int^t\phi(s;u,\omega)\nu(ds)\right\},\ u\in U,\ t\in[0,1] ;$$

$$R^\phi_t(u,\omega) = \exp\left\{I^\phi_t(u,\omega) - \tfrac{1}{2}\int^t\phi(s;u,\omega)\nu(ds)\right\},\ t\in[0,1] .$$

In the first exponent $({}_uI^\phi_t)$ is regarded as the stochastic integral on the filtered space $(E,\xi,(\xi_t),p)$, given by (2.10), for every $u\in U$; so that $({}_uR_t^\phi,\xi_t)$, $t\in[0,1]$, is an exponential supermartingale on (E,ξ,p) *for every* $u\in U$. In the second exponent (I_t^ϕ) is regarded as the stochastic integral on the filtered space $(S,\mathcal{F},(\mathcal{F}_t),\mathbb{P})$, given by (2.9); so that $(R_t^\phi,\mathcal{F}_t)$, $t\in[0,1]$, is an exponential supermartingale on $(S,\mathcal{F},\mathbb{P})$.

Let us convent to write ${}_uR^\phi$ for ${}_uR_1^\phi$, $u\in[0,1]$, and to write R^ϕ for R_1^ϕ . According to (2.11) we have

(4.1) $\quad {}_uR^\phi(\omega) = R^\phi(u,\omega)$ for p-a.e. $\omega\in E$, for Π-a.e. $u\in U$.

With these notations, condition C2 can be written shortly as

(4.2) $$\mathbb{E}_p\{{}_uR^\phi\} = 1 \ , \quad u\in U \ ,$$

and this condition is known to be equivalent to the requirement that $({}_uR_t^\phi,\ \xi_t)$, $t\in[0,1]$, forms a martingale on the probability space (E,ξ,p) for every $u\in U$ (see [3, Lemma 7.1.1]). There is a variaty of conditions for the function ϕ , under which (4.2) holds. For their thorough description we refer to [3, Chap.7]. Here we mention only the most important of these conditions, namely

(4.3) $\quad \int|\phi(s;u,\omega)|^2\nu(ds) < C_u$ for p-a.e $\omega\in E$,

where C_u is a finite constant assumed to exist *for every* $u\in U$ (see [3,Lemma 7.1.2]).

For a function ϕ , which satisfies conditions (C), because of (4.1), we have $\mathbb{E}\{R^\phi\} = 1$. Therefore one can define on $(S,\mathcal{S})$ the probability measure $\mathbb{P}^\phi$ by

$$d\mathbb{P}^\phi = R^\phi d\mathbb{P} \ .$$

Since $R^\phi > 0$, we have that $\mathbb{P}^\phi \equiv \mathbb{P}$, i.e. $\mathbb{P}^\phi$ and $\mathbb{P}$ are equivalent probability lows on $(S,\mathcal{S})$. Also, under the assumptions (C), one can define for every $u\in U$ the probability measure ${}_uP^\phi$ on the measurable space (E,ξ) by

$$d({}_uP^\phi) = {}_uR^\phi dp \ ,$$

and again we have ${}_uP^\phi \equiv p$, $u\in U$, because ${}_uR^\phi > 0$ for all $u\in U$. Because of (4.1) , the measure $\mathbb{P}^\phi$ can be disintegrated as

$$\mathbb{P}^\phi(.) = \int_U {}_uP^\phi(.)\Pi(du) \ ,$$

which is nothing but saying that for every $\mathbb{P}^\phi$-integrable function $f(u,\omega)$ on S, it holds that

$$\int_S f(u,\omega)\mathbb{P}^{\phi}(du,d\omega) = \int_U(\int_E f(u,\omega)\,{}_uP^{\phi}(d\omega))\Pi(du) .$$

The last relation shows that ${}_uP^{\phi}(B)$, $u\in U$, $B\in\xi$, is in fact a transition probability low on the product $(U,\mathcal{U})\otimes(E,\xi)$ (see [9, 35.4]).

The main result of the present section is our next proposition, which presents the desired version of the Kameron-Martin-Girsanov theorem. By $T_{\phi}:S\to E$ we denote the Girsanov transformation, as defined in Section 3.

#4.1 Proposition Let $\phi(s;u,\omega)$ be a function on $[0,1]\times S$, which satisfies conditions (C) . Then for every Borel set $B\in\xi$ and *for every* $u\in U$ we have

$$p(B) = {}_u\mathbb{P}^{\phi}\{ \omega \in E : T_{\phi}(u,\omega)\in B \} . \#$$

For $\phi \in \mathcal{L}^2_{loc}$ we now define the transformation $\theta_{\phi}:S\longrightarrow S$ (see #3.7) by

$$\theta_{\phi}(u,\omega) = (u,T_{\phi}(u,\omega)) , \quad (u,\omega) \in S .$$

Also for $A \in \mathcal{S}$ and for $u\in U$ we put

$${}_uA = \{ \omega \in E : (u,\omega)\in A \} .$$

Then ${}_uA \in \xi$, for every $u\in U$, and for the indicator function of the set ${}_uA$ it obviously holds $1_{({}_uA)}(.) = 1_A(u,.)$.

#4.2 Proposition Let $\phi(s;u,\omega)$ be a real function on $[0,1]\times S$, which satisfies conditions (C) and let $f(u,\omega)$ be a $\mathcal{S}$-measurable function on S . Then the following equality holds

$$\mathbb{E}\{f(u,\omega)\} = \mathbb{E}\{f(\theta_{\phi}(u,\omega))R^{\phi}(u,\omega)\} , \tag{4.4}$$

in the sense that the existence of the either side implies the existence of the other one and the equality. #

Note that it follows from the last proposition that if ϕ obeys conditions (C) then

$$\mathbb{P}(A) = \mathbb{P}^{\phi}\{ (u,\omega)\in S : \theta_{\phi}(u,\omega)\in A \}$$

for every $A \in \mathcal{S}$. Since $\mathbb{P}^{\phi} \equiv \mathbb{P}$, we then have the following implication, when (C) is satisfied :

$$A \in \mathcal{S} \text{ and } \mathbb{P}(A)=0 \longrightarrow \mathbb{P}\{ (u,\omega)\in S : \theta_{\phi}(u,\omega)\in A \} = 0 . \tag{4.5}$$

Our next proposition is, in fact, a reformulation of

[3, Theor.7.3.1]. It follows from the Girsanov theorem, as formulated in #4.1. The proof involves stopping time arguments and essentially uses the fact that (4.3) implies C2, as well as the implication (4.5) .

#4.3 Proposition Let $\phi(s;u,\omega)$ be a function on $[0,1]\times S$, which satisfies condition C1; that is ϕ is predictable ($\mathcal{P}$-measurable) and *for every* $(u,\omega)\in S$

$$\int|\phi(s;u,\omega)|^2\nu(ds) < \infty .$$

Then the following implication holds

$$A \in \mathcal{S} \text{ and } \mathbb{P}(A)=0 \longrightarrow \mathbb{P}\{ (u,\omega)\in S : \theta_\phi(u,\omega)\in A \} = 0 . \text{ \#}$$

In other words, #4.3 claims that (4.5), which was obtained under the assumptions (C) , in fact holds only under the assumption C1 .

Now we are going to apply the above results under the assumptions that ϕ is predictable ($\mathcal{P}$-measurable), $\phi \in \mathcal{L}^2_{loc}$ and satisfies conditions (B). Note that these assumptions are not independent. Indeed, predictability of ϕ implies B1, because of #3.1. Also B2 implies that $\phi \in \mathcal{L}^2_{loc}$. Thus, practically, we assume predictability + B2 + B3. If these assumptions are satisfied, then #3.11 is in force and one can construct the kernel $\phi' \in \mathcal{L}^2_{loc}$ of the inverse Girsanov transform T_ϕ^{-1} in such a way that ϕ' is also predictable and obeys (B) . Obviously, the transformations $\theta_\phi:S\longrightarrow S$ and $\theta_{\phi'}:S\longrightarrow S$ are inverse to each other . For every set $A \in \mathcal{S}$ we write

$$\theta_\phi A = \{ \theta_\phi(u,\omega) : (u,\omega)\in A \} ;$$

$$\theta_{\phi'} A = \{ \theta_{\phi'}(u,\omega) : (u,\omega)\in A \} .$$

Since $\theta_\phi = \theta_{\phi'}^{-1}$ and $\theta_{\phi'} = \theta_\phi^{-1}$, then we have

$$\theta_\phi A = \{ (u,\omega)\in S : \theta_{\phi'}(u,\omega)\in A \} ;$$

$$\theta_{\phi'} A = \{ (u,\omega)\in S : \theta_\phi(u,\omega)\in A \} .$$

#4.4 Proposition Let $\phi(s;u,\omega)$ be a predictable function on $[0,1]\times S$, which satisfies conditions (B) and let $\phi'(s;u,\omega)$ be a kernel of the inverse Girsanov transform T_ϕ^{-1}, which is also predictable and satisfies (B) . Then for every set $A\in\mathcal{S}$ each one of the following three equalities implies the other two:

$$\text{(i) } \mathbb{P}(A)=0; \quad \text{(ii) } \mathbb{P}(\theta_\phi A)=0; \quad \text{(iii) } \mathbb{P}(\theta_{\phi'} A)=0 \text{ . } \#$$

The last proposition shows that as long as ϕ is predictable and obeys (B) one can define

$$\text{(4.6)} \qquad \rho^\phi(u,\omega) = \frac{d(\mathbb{P}\cdot\theta_\phi)}{d\mathbb{P}}(u,\omega) \qquad \mathbb{P}\text{-a.s. .}$$

Since $\mathbb{P}\cdot\theta_\phi \equiv \mathbb{P}$ it follows that $\rho^\phi(u,\omega) > 0$ $\mathbb{P}$-a.e. in S.

For every $A \in \mathcal{S}$ we have

$$\mathbb{P}(A) = \mathbb{P}(\theta_\phi\theta_{\phi'} A) = \mathbb{P}\cdot\theta_\phi\{ (u,\omega)\in S : \theta_\phi(u,\omega)\in A \} .$$

#4.5 Proposition Let $\phi(s;u,\omega)$ be as in #4.4 and let $f(u,\omega)$ be $\mathcal{S}$-measurable function on S . Then the following equality holds

$$\mathbb{E}\{f(u,\omega)\} = \mathbb{E}\{f(\theta_\phi(u,\omega))\rho^\phi(u,\omega)\} ,$$

in the sense that the existence of the either side implies the existence of the other one and the equality. #

If in #4.5 one substitutes $f(\theta_{\phi'}(u,\omega))$ for $f(u,\omega)$, then it is obtained the following

#4.6 Proposition Let the functions ϕ and f be as in #4.5 and let ϕ' be a predictable kernel of the inverse T_ϕ^{-1}, which satisfies conditions (B) . Then the following equality holds

$$\mathbb{E}\{f(\theta_{\phi'}(u,\omega))\} = \mathbb{E}\{f(u,\omega)\rho^\phi(u,\omega)\}$$

in the same sense as in #4.5 . #

The calculus obtained in #4.5 and #4.6 is all we need for the derivation of the nonlinear filter in the next section. The calculation of the Radon-Nykodim derivative $\rho^\phi(u,\omega)$ is therefore very important. This we shall not discuss here, for it is described in [3, Remark 7.3.2]. If except that ϕ is predictable and obeys (B) , it is additionally assumed that ϕ obeys C2 (C1 is satisfied automatically); that is, we additionally assume that $\mathbb{E}_p\{_u R^\phi\}= 1$ for every $u\in U$, then one can always take

$$\rho^\phi(u,\omega) = R^\phi(u,\omega) \quad \mathbb{P}\text{-a.e. .}$$

For example C2 is satisfied if Novikov's condition

$$\mathbb{E}_p\{\exp\, [\tfrac{1}{2}\int|\phi(s;u,\omega)|^2\nu(ds)]\} < \infty$$

is satisfied for every $u\in U$. In particular C2 is satisfied if the kernel ϕ is glabally bounded.

5. The Problem of Nonlinear Filtering

This section contains rigorous description of the problem of nonlinear filtering, scetched in Section 1. Again we consider the basic probability space

$$(S,\mathcal{S},\mathbb{P}) \equiv (U\times E,\mathfrak{U}\otimes\mathcal{E},\Pi\otimes p)\ ,$$

on which all the three components - signal, noise and observation are assumed to be defined. An *arbitrary* orthogonal curve (Z) that satisfies conditions (C) is fixed. The filtrations $(\mathcal{F}_t\subset\mathcal{S})$ and $(\mathcal{E}_t\subset\mathcal{E})$ are defined relative to (Z) , as described in Section 2 (see #2.4®). All notions "adapted", "predictable", "causal", etc. are therefore considered relative to (Z). Also given is a real function $\phi(s;u,\omega)$ on $[0,1]\times S$, which is assumed to be predictable and to satisfy conditions (B) (i.e. for ϕ we assume $\mathcal{P}$-measurability + B2 + B3). Then we can construct the kernel $\phi'(s;u,\omega)$ of the inverse Girsanov transform T_ϕ^{-1}, according to the recursive procedure (3.6) . As it follows from #3.11, $\phi'(s;u,\omega)$ is also predictable and also obeys (B) . *The signal* $h[u,\omega]$ is considered as an element of $H\subset E$, which depends on $u\in U$ and on $\omega\in E$, and which has the form of an integral w.r.t. (Z)

$$h[u,\omega] = \int\phi'(s;u,\omega)Z(ds) \equiv Z[\phi'(.;u,\omega)]\ .$$

We regard every $\omega\in E$ as a "realization" or a "sample" of the noise. Then the observation $y[u,\omega]$ is an element of E , which is the sum of the noise component and the signal component; that is

$$y[u,\omega] = \omega + h[u,\omega] \equiv T_{\phi'}(u,\omega)\ ,\quad (u,\omega)\in S\ . \tag{5.1}$$

We assume that $\rho^\phi(u,\omega)$ is the Radon-Nykodim derivative, given by (4.6) and we put

$$\alpha(\omega) = \int_U \rho^\phi(u,\omega)\Pi(du) \quad \text{for p-a.e. } \omega\in E\ .$$

Since $\rho^\phi > 0$ $\mathbb{P}$-a.e., it follows that $\alpha > 0$ p-a.e. on E .

For every Borel function $f:H\longrightarrow\mathbb{R}$ (by Borel we mean

$\mathcal{B}(H)$-measurable – see #2.4®) for which

$$\mathbb{E}\{|f(Z[\phi'(.;u,\omega)])|\} \equiv \mathbb{E}\{|f(h[u,\omega])|\} < \infty ,$$

we define the following r.v. on $(E,\mathcal{E},p)$

$$(5.2) \qquad [\mathfrak{F}f](\omega) = \frac{1}{\alpha(\omega)} \int_U f(Z[\phi'(.;u,T_\phi(u,\omega))])\rho^\phi(u,\omega)\Pi(du) .$$

Note that #3.3 and #3.8 imply the $\mathcal{S}$-measurability of the mappings

$$(u,\omega) \longrightarrow f(Z[\phi'(.;u,T_\phi(u,\omega))]) ;$$

$$(u,\omega) \longrightarrow f(Z[\phi'(.;u,\omega)]) \equiv f(h[u,\omega]) .$$

On the other hand from #4.5 we have

$$\int_E p(d\omega)\int_U |f(Z[\phi'(.;u,T_\phi(u,\omega))])|\rho^\phi(u,\omega)\Pi(du) =$$

$$= \mathbb{E}\{|f(Z[\phi'(.;u,\omega)])|\} < \infty ,$$

so that the integral in the r.h.s. of (5.2) is correctly defined r.v. on $(E,\mathcal{E},p)$ (i.e. a class of p-equivalent functions that are finite p-a.e.).

The main result of this paper is the following

#5.1 Proposition Let $f:H\longrightarrow\mathbb{R}$ be $\mathcal{B}(H)$-measurable (i.e. Borel w.r.t. the Hilbertian norm) function, which is s.t.

$$\mathbb{E}\{|f(h[u,\omega])|\} < \infty .$$

Then for every bounded and $\mathcal{E}$-measurable function $\varphi:E\longrightarrow\mathbb{R}$ the following equality holds

$$(5.3) \qquad \mathbb{E}\{\varphi(y[u,\omega])f(h[u,\omega])\} = \mathbb{E}\{\varphi(y[u,\omega])[\mathfrak{F}f](y[u,\omega])\} .$$

Proof According to #4.5, the l.h.s. of (5.3) equals

$$\mathbb{E}\{\varphi(T_{\phi'}(u,\omega))f(h[u,\omega])\} =$$

$$= \mathbb{E}\{\varphi(T_{\phi'}(u,T_\phi(u,\omega)))f(h[u,T_\phi(u,\omega)])\rho^\phi(u,\omega)\} =$$

$$= \int_E \varphi(\omega)\Big\{ \int_U f(h[u,T_\phi(u,\omega)])\rho^\phi(u,\omega)\}\Pi(du)\Big\}p(d\omega) =$$

$$= \int_E \varphi(\omega)\alpha(\omega)[\mathfrak{F}f](\omega)p(d\omega) =$$

$$= \int_E \varphi(\omega)[\mathfrak{F}f](\omega)\Big(\int_U \rho^\phi(u,\omega)\Pi(du)\Big)p(d\omega) =$$

$$= \mathbb{E}\{\varphi(\omega)[\mathfrak{F}f](\omega)\rho^\phi(u,\omega)\} .$$

According to #4.6 the last quantity equals

$$\mathbb{E}\{\varphi(T_{\phi'}(u,\omega))[\mathfrak{F}f](T_{\phi'}(u,\omega))\} = \mathbb{E}\{\varphi(y[u,\omega])[\mathfrak{F}f](y[u,\omega])\} . \quad \#$$

The last assertion means nothing but

$$\mathbb{E}\{ f(h[u,\omega]) \mid y[u,\omega]=\omega' \} = [\mathfrak{F}f](\omega') \quad \text{for p-a.e. } \omega'\in E .$$

Note that the nonlinear filter $[\mathfrak{F}f](.)$, as a function on E, is defined only p-a.e.. At the same time if for some $B \in \mathcal{E}$ one has $p(B)=0$, then from #4.4 it follows that

$$\mathbb{P}\{ (u,\omega)\in S : y[u,\omega]\in B \} = \mathbb{P}\{ (u,\omega)\in S : (u,T_{\phi'}(u,\omega))\in U\times B \} =$$
$$= \mathbb{P}\{\theta_{\phi'}(U\times B)\} = 0 .$$

Therefore $[\mathfrak{F}f](y[u,\omega])$ is correctly defined r.v. ($\mathbb{P}$-equivalence class) on $(S,\mathcal{S},\mathbb{P})$.

Note that in the expression for the nonlinear filter (5.2) both kernels ϕ and ϕ' appear. Thus the above strategy requires the calculation of ϕ , if it is assumed that only ϕ', which describes the signal, is given. We have from #3.11 a recursive procedure, namely those of (3.6), by which each of the kernels ϕ and ϕ' can be restored from the other one.

If a kernel $\phi'(s;u,\omega)\equiv\phi'(s;u)$, which does not depend on ω is taken, then one can easily recognize in (5.2) the Kallianpur-Striebel formula (see [5]). In that particular case, when the signal $h[u,\omega]\equiv h[u]$ is no more a function of ω , one simply has

$$T_{\phi}(u,\omega) = \omega - h[u] , \quad T_{\phi}^{-1}(u,\omega) = \omega + h[u] ;$$

i.e. in that case ϕ and ϕ' are related to each other as simply as $\phi = -\phi'$ and there is no need of the recursive calculations in (3.6).

References

[1] Hida, T., *Brownian Motion*. Springer-Verlag, New York, 1980.

[2] Ikeda, N. and Watanabe, Sh., *Stochastic Differential Equations and Diffusion Processes*. North-Holland, Amsterdam, 1981.

[3] Kallianpur, G., *Stochastic Filtering Theory*. Springer-Verlag, New-York, 1980.

[4] Kallianpur, G. and Karandikar, R. L., White noise calculus and nonlinear filtering. *Ann. Probab.* 13, 1033-1107,1985.

[5] Kallianpur, G. and Striebel, C., Estimation of stochastic process with additive white noise observation error. *Ann. Math. Stat.* 39, 785-801, 1968.

[6] Kuo, H.-H., *Gaussian Measures on Banach Spaces*. Springer-Verlag, Berlin, 1975.

[7] Liptser, R. Sh. and Shiryayev, A., *Statistics of Random Processes I.* Springer-Verlag, New-York, 1977.

[8] Potthoff, J., White noise approach to Malliavin's calculus. *J. Func. Analysis* **71**, 207-217, 1987.

[9] Parthasarathy, K.R, *Introduction to Probability and Measure.* Springer-Verlag, New-York, 1978.

CONTINUOUS LOCAL MARTINGALES: STRONG MARKOV PROPERTY, SOLUTIONS OF STOCHASTIC EQUATIONS, AND THE INTERPLAY BETWEEN THEM

H.J. Engelbert and W. Schmidt
Friedrich-Schiller-Universität
Sektion Mathematik, 6900 Jena, GDR

1. INTRODUCTION

The present paper deals with continuous local martingales as the underlying class of stochastic processes and its stochastic calculus as the basic tool.

There are two main subclasses of continuous local martingales. The first subclass consists of continuous local martingales which, additionally, satisfy the strong Markov property. The second subclass is the set of all solutions of one-dimensional stochastic differential equations without drift.

The main purpose of the present paper is to investigate the connection between these two subclasses of continuous local martingales.

Thus this paper contributes to the general problem of the structure of Markov processes describing them as a solution of a certain stochastic equation.

Historically, this was the motivation of K. Itô [13] as he founded his stochastic calculus at the beginning of the 40's. But only after a half century it became possible to give this description, in the frame of (one-dimensional) continuous local martingales, the complete and final shape.

We believe that strong Markov continuous semimartingales on the real line can be treated similarly but we do not deal with this question in the present paper.

The paper is organized as follows.

In Section 2, we give an introduction to strong Markov continuous local martingales.

In Section 3, we are concerned with additive functionals and random time change of strong Markov continuous local martingales. The basic theorem characterizes those random time changes which again lead to a strong Markov continuous local martingale. The proof is essentially based on the generalized Itô formula for continuous local martingales.

In Section 4, we shall apply the results of Section 3 and derive the representation theorem for perfect additive functionals and investigate the structure of terminal times. We discuss the associated Brownian motion and purity and representation property of strong Markov continuous local martingales.

In Section 5, we show that there is a bijection between strong Markov continuous local martingales and "admissible" measures on the real line called speed measure. The approach to the existence and uniqueness is purely probabilistic and does not use analytical tools as infinitesimal generators. Finally, we state two martingale problems which possess a unique solution in the class of strong Markov continuous local martingales. The existence and uniqueness of the second martingale problem generalizes the theorem of P. Lévy on the martingale characterization of the Brownian motion to arbitrary strong Markov continuous local martingales.

In Section 6, we investigate existence, uniqueness, and structure of solutions of one-dimensional stochastic differential equations without drift

$$X_t = X_o + \int_o^t b(X_s)\, dB_s \ , \qquad t \geq 0,$$

with a Brownian motion B. In particular, we give necessary and sufficient conditions for existence as well as for uniqueness.

The last section (Section 7) was the starting point of the present paper. We ask for conditions to describe a given strong Markov continuous local martingale as a solution of a stochastic equation of the above type. This problem will be solved completely by stating a necessary and sufficient condition. Uniqueness will be enforced by an additional stochastic equation which controls the sojourn time in the set of zeros of b.

Finally, we develop the concept of a stochastic equation with diffusion coefficient b which may take infinite values. We then show that an arbitrary strong Markov continuous local martingale can be described as a solution of a stochastic equation of this type.

Only a few proofs will be sketched in this paper giving some basic ideas, most proofs will be omitted. For details the reader is referred to the forthcoming paper [13].

2. STRONG MARKOV CONTINUOUS LOCAL MARTINGALES

Let $\mathbb{R}$ denote the real line, $B(\mathbb{R})$ the σ-algebra of Borel subsets, and $B^u(\mathbb{R})$ its universal completion.

We consider a family $(\Omega, F^o, P_x,\ x\in\mathbb{R})$ of probability spaces. We always assume that, for every $A\in F^o$, $P_x(A)$ is $B^u(\mathbb{R})$-measurable. For every probability measure μ on $(\mathbb{R}, B(\mathbb{R}))$, let

$$P_\mu(A) = \int_{\mathbb{R}} P_x(A)\ \mu(dx), \qquad A\in F^o.$$

By F we denote the completion of F^o relative to the family (P_μ).

Now let $\mathbb{F}=(F_t)_{t\geq 0}$ be an increasing family of sub-σ-algebras of F (called filtration). Unless otherwise stated we do not assume that $\mathbb{F}$ satisfies the usual conditions. We only assume that F_t is complete relative to the family (P_μ).

Let $X=(X_t)_{t\geq 0}$ be a (real-valued) process. We write $(X, \mathbb{F})$ to indicate that X is $\mathbb{F}$-adapted. By $\mathbb{F}^o=(F_t^o)_{t\geq 0}$ we denote the smallest filtration such that X is adapted to it, F_∞^o denotes the σ-algebra generated by $\bigcup_{t\geq 0} F_t^o$. We say that $(X, \mathbb{F})$ is a stochastic process on the family $(\Omega, F, P_x,\ x\in\mathbb{R})$.

By $\mathbb{F}^X=(F_t^X)_{t\geq 0}$ we denote the completion of the filtration $\mathbb{F}^o$ in F relative to the family (P_μ).

The abbreviation "a.s." means "P_μ-a.s. for every initial distribution μ".

If Z is a process and $B\in B(\mathbb{R})$ then $D_B(Z)$ denotes the first entry time of Z into B. If $Z=X$ then we always write D_B instead of $D_B(X)$.

For any filtration $\mathbb{H}$, $\mathbb{H}_+$ denotes the smallest right continuous filtration containing $\mathbb{H}$.

DEFINITION 1. A stochastic process $(X, \mathbb{F})$ on $(\Omega, F, P_x,\ x\in\mathbb{R})$ is called a <u>strong Markov continuous local martingale</u> if the following conditions hold:

(i) $P_x(X_o=x) = 1$, $x\in\mathbb{R}$.

(ii) $(X, \mathbb{F})$ is a continuous local martingale with respect to P_x for every $x\in\mathbb{R}$.

(iii) $(X, \mathbb{F})$ is a (homogeneous) Markov process, i.e., for all $s,t\geq 0$, $x\in\mathbb{R}$, and $A\in B(\mathbb{R})$

$$P_x(X_{s+t}\in A|F_s) = P_{X_s}(X_t\in A) \qquad P_x\text{-a.s.}$$

(iv) $(X, \mathbb{F}^X)$ possesses the strong Markov property, i.e., for every $\mathbb{F}^X$-stopping time S, $t\geq 0$, $x\in\mathbb{R}$, and $A\in B(\mathbb{R})$

$$P_x(X_{S+t}\in A|F_S^X) = P_{X_S}(X_t\in A) \qquad \text{on } \{S<+\infty\}\ P_x\text{-a.s.}$$

Note that the strong Markov property is only required with respect to the filtration $\mathbb{F}^X$.

In this section, $(X, \mathbb{F})$ always denotes a strong Markov continuous local martingale.

ABSORBING POINTS AND THE FINE TOPOLOGY

DEFINITION 2. A state $x \in \mathbb{R}$ is called <u>absorbing</u> if $P_x(X_t=x) = 1$ for all $t \geq 0$. By E we denote the set of absorbing points. Every state $x \in \mathbb{R} \setminus E$ is said to be <u>regular</u>.

The following proposition is the key for the investigation of the basic properties of $(X, \mathbb{F})$. For the definition and properties of finely continuous functions see, for example, [6] (No.23) and [7]. To avoid ambiguity note that the fine topology C_o is introduced with respect to the state space $(\mathbb{R}, B(\mathbb{R}))$.

PROPOSITION 1. If x is regular then $E_y f(X_t)$ is continuous at x for every bounded finely continuous measurable function f and $t \geq 0$.

COROLLARY. The set E of absorbing points is closed.

Indeed, let (x_n) be a sequence from E converging to x. Suppose that x is regular. From Proposition 1 easily follows $f(x) = E_x f(X_t)$ for every bounded continuous function f and $t \geq 0$. This implies that x is absorbing, contradicting the assumption. □

Now we are able to give an explicit characterization of the fine topology C_o.

THEOREM 1. <u>The fine topology</u> C_o <u>of</u> $(X, \mathbb{F})$ <u>consists of all</u> $G \subseteq \mathbb{R}$ <u>such that</u> $G \setminus E$ <u>is open</u>.

The next proposition collects several important properties.

PROPOSITION 2. (i) $X_{D_E} \in E$ on $\{D_E < +\infty\}$ a.s.

(ii) $X_t = X_{t \wedge D_E}$ for all $t \geq 0$ a.s.

(iii) The semigroup $(P_t)_{t \geq 0}$ of $(X, \mathbb{F})$ leaves the space $\mathcal{C}_o$ of bounded finely continuous functions invariant.

(iv) For every $f \in \mathcal{C}_o$, the process $f(X)$ is right continuous a.s.

P r o o f . Statement (i) follows from the corollary above, (ii) from the strong Markov property, (iii) can easily be obtained from Proposition 1 and its Corollary and Theorem 1. Finally, (iv) is a consequence of Theorem 1 and part (ii). □

BASIC PROPERTIES AND THE ASSOCIATED INCREASING PROCESS

In view of [7] (Theorem 33) and [8] (Theorem 33) we now conclude

THEOREM 2. $(X, \mathbb{F})$ <u>is a perfect Markov class in the state space</u> $(\mathbb{R}, B(\mathbb{R}))$ (see [8], Definition 2).

As X is continuous, $(X, \mathbb{F})$ is even a previsible perfect Markov class in $(\mathbb{R}, B(\mathbb{R}))$. This yields the following beautiful properties of $(X, \mathbb{F})$.

COROLLARY. (i) $(X, \mathbb{F})$ is a Feller process in the fine topology C_o.

(ii) $(X, \mathbb{F})$ has a Borel semigroup of transition probabilities.

(iii) $\mathbb{F}^X = \mathbb{F}^X_+$.

(iv) $(X, \mathbb{F}_+)$ is also a strong Markov continuous local martingale satisfying the strong Markov property with respect to $\mathbb{F}_+$ (and not only with respect to $\mathbb{F}^X$).

(v) Every $\mathbb{F}^X$-stopping time T is $\mathbb{F}^X$-previsible and $F^X_{T-} = F^X_T = F^X_{T+}$.

Note that in (i) the fine topology C_o cannot be replaced by the usual topology. Properties (ii) and (iii) are rather surprising since we started from weaker hypotheses. (For this observe the well-known fact that the Markov property with respect to $\mathbb{F}^X_+$ is equivalent to (iii).)

Now we are going to study the associated increasing process $(\langle X\rangle, \mathbb{F}^X)$. This is the unique (up to indistinguishability) a.s. continuous increasing process $\langle X\rangle$ such that $\langle X\rangle_o = 0$ and $(X^2-\langle X\rangle, \mathbb{F})$ is a local martingale with respect to P_μ for every μ. Such a universal version can be chosen in view of [4], (3.39). Note that $\langle X\rangle$ can also be chosen $\mathbb{F}^o$-adapted.

Here are a few basic properties. We define $X_\infty = \limsup_{t\to\infty} X_t$.

PROPOSITION 3. (i) $\langle X\rangle_{D_E} = \langle X\rangle_\infty$ a.s.

(ii) $\langle X\rangle$ is strictly increasing on $[0,D_E)$ a.s.

(iii) $X_\infty = \lim_{t\to\infty} X_t$ exists on $\{\langle X\rangle_\infty < +\infty\}$ and is finite a.s.

(iv) $\liminf_{t\to\infty} X_t = -\infty$ and $\limsup_{t\to\infty} X_t = +\infty$ on $\{\langle X\rangle_\infty = +\infty\}$ a.s.

(v) $X_\infty \in E$ on $\{\langle X\rangle_\infty < +\infty\}$ a.s.

The associated increasing process $\langle X\rangle$ satisfies two 0-1 laws, the first of which characterizes its behaviour for small t and the second for large t.

PROPOSITION 4. (i) We have either $P_x(\langle X\rangle_t > 0) = 0$ for all $t>0$ or $P_x(\langle X\rangle_t > 0) = 1$ for all $t>0$ in dependence of x is absorbing or regular.

(ii) We have either $P_x(\langle X\rangle_\infty<+\infty)=0$ for all $x\in\mathbb{R}$ or $P_x(\langle X\rangle_\infty<+\infty)=1$ for all $x\in\mathbb{R}$ in dependence of $E=\emptyset$ or $E\neq\emptyset$.

STRUCTURAL FUNCTION AND SPEED MEASURE

The main purpose of the remainder of this section is to introduce the speed measure of $(X, \mathbb{F})$. The first step is the following theorem.

THEOREM 3. There exists a Borel function f on $\mathbb{R}$ such that the process $(Y, \mathbb{F}^X)$ defined by

$$Y_t = f(X_t) + t\ , \qquad t\geq 0,$$

is a continuous local martingale up to time D_E. This function is unique on $\mathbb{R}\setminus E$ up to affinity on every component of $\mathbb{R}\setminus E$.

Note that f is (strictly) superharmonic on $\mathbb{R}\setminus E$ and hence strictly concave on every component of $\mathbb{R}\setminus E$. The function f is called structural function of the strong Markov continuous local martingale $(X, \mathbb{F})$.

Proof. If $[a,b]\subseteq\mathbb{R}\setminus E$ then

$$f(x) = E_x\ T(a,b) + cx + d\ , \qquad x\in\mathbb{R},$$

where $T(a,b)$ is the first exit time from (a,b) is such that $Y_{t\wedge T(a,b)}$ is a continuous martingale. Now the result follows by blowing up (a,b) to a component of $\mathbb{R}\setminus E$ and pasting together the corresponding $f = f_{(a,b)}$. □

Before we introduce the speed measure of $(X, \mathbb{F})$ we need some notation. Let m be an arbitrary measure on $(\mathbb{R}, B(\mathbb{R}))$. By E_m we denote the set of all $x\in\mathbb{R}$ such that $m(G)=+\infty$ for all open sets G containing x. Obviously, E_m is a closed subset of $\mathbb{R}$ and m is locally finite on $\mathbb{R}\setminus E_m$, i.e., for every compact subset K of $\mathbb{R}\setminus E_m$ it follows $m(K)<+\infty$.

For our purpose there is no reason two distinguish two measures which agree on all open sets.

Let m_1 and m_2 be two measures on $(\mathbb{R}, B(\mathbb{R}))$. We say that m_1 and m_2 are essentially equal ($m_1 \cong m_2$) if $m_1(G)=m_2(G)$ for every open subset G of $\mathbb{R}$. It can easily be checked that $m_1\cong m_2$ if and only if $E_{m_1}=E_{m_2}$ and $m_1=m_2$ on $\mathbb{R}\setminus E_{m_1}$.

DEFINITION 3. Let f be a structural function of $(X, \mathbb{F})$. Let m_0 be the nonnegative measure on $\mathbb{R}\setminus E$ which is the second derivative (in the sense of distributions) of the function $-\frac{1}{2}f$ on $\mathbb{R}\setminus E$. The equivalence class with respect to "$\cong$" of all measures m on $(\mathbb{R}, B(\mathbb{R}))$ satisfying $m=m_0$ on $\mathbb{R}\setminus E$ and $E_m=E$ is called speed measure of the strong Markov continuous local martingale $(X, \mathbb{F})$.

In the following, unless otherwise stated we do not distinguish between measures and their equivalence classes. Obviously, the speed measure m is then uniquely determined and we have $m(G)>0$ for all non-empty open subsets G of $\mathbb{R}$.

3. ADDITIVE FUNCTIONALS AND RANDOM TIME CHANGE

Let $(X, \mathbb{F})$ be a strong Markov continuous local martingale on $(\Omega, F, P_x, x\in\mathbb{R})$, admitting shift operators $(\theta_s)_{s\geq 0}$. In addition, we suppose that there is also a shift operator θ_∞ satisfying

$$X_t \circ \theta_\infty = X_\infty \quad \text{for all } t\geq 0 \quad \text{on } \{\langle X\rangle_\infty < \infty\} \quad \text{a.s.}$$

In view of Proposition 2.2(ii), without loss of generality we can, and always do, assume that

$$\theta_s = \theta_{s\wedge D_E} \qquad \text{for all } 0\leq s\leq +\infty .$$

By the Corollary to Theorem 2.2, $(X, \mathbb{F}_+)$ is also a strong Markov continuous local martingale. Thus we may assume that $\mathbb{F}=\mathbb{F}_+$.

ADDITIVE FUNCTIONALS AND TERMINAL TIMES

To begin with we recall a few definitions.

Let $(A, \mathbb{F})$ be a right continuous increasing process taking values in $[0,+\infty]$ and U an $\mathbb{F}^X$-stopping time. We say that $(A, \mathbb{F})$ is <u>an additive functional on</u> $[0,U)$ <u>of</u> $(X, \mathbb{F})$ if $I_{\{t<U\}}\cdot A_t$ is F^X_∞-measurable [1)] and

$$A_s + A_t\circ\theta_s = A_{s+t} \qquad \text{on } \{s+t<U\} \qquad \text{a.s.}$$

for all $s,t\geq 0$. If the exceptional set does not depend on s and t then $(A, \mathbb{F})$ is said to be <u>perfect</u>. If in the definition of an additive functional s can be replaced by an arbitrary $\mathbb{F}^X$-stopping time S then $(A, \mathbb{F})$ is called <u>a strong Markov additive functional on</u> $[0,U)$ <u>of</u> $(X, \mathbb{F})$.

If $U=+\infty$ then we simply speak of additive functionals, perfect additive functionals, and strong Markov additive functionals of $(X, \mathbb{F})$, respectively.

An $\mathbb{F}$-stopping time U is called <u>a terminal time</u> (resp., <u>perfect terminal time</u>) of $(X, \mathbb{F})$ if U is F^X_∞-measurable and

$$s + U\circ\theta_s = U \qquad \text{on } \{s<U\}$$

a.s. for all $s\geq 0$ (resp., for all $s\geq 0$ a.s.). If s can be replaced by an arbitrary $\mathbb{F}^X$-stopping time S then U is said to be <u>a strong Markov terminal time</u>.

1) I_A denotes the indicator function of the set A.

LOCAL TIME

As a basic tool we now come to the investigation of the local time of $(X, \mathbb{F})$. We recall that the local time of a continuous local martingale $(X, \mathbb{F})$ is a measurable mapping L^X defined on $[0,+\infty)\times\mathbb{R}\times\Omega$ such that for every nonnegative Borel function g

(1) $$\int_0^t g(X_s)\, d\langle X\rangle_s = \int_{\mathbb{R}} g(y)\, L^X(t,y)\, dy\ , \qquad t\geq 0, \qquad \text{a.s.}$$

(cf. J. Azéma and M. Yor [2]). The generalized Itô formula states that for every concave function f

(2) $$f(X_t) = f(X_o) + \int_0^t Df(X_s)\, dX_s + \frac{1}{2}\int_{\mathbb{R}} L^X(t,y)\, m(dy) \quad \text{a.s.}$$

where Df is the right (or left) derivative of f, the measure m is the second derivative of f in the sense of distributions, and the first integral is the Itô integral with respect to the continuous local martingale $(X, \mathbb{F})$ (cf. P.A. Meyer [17], J. Azéma and M. Yor [2], A.T. Wang [19]). We notice that there always exists a version which is the stochastic Itô integral with respect to X simultaneously for every P_μ (cf. [4], (3.36)).

Setting $f(x)=|x-y|$ for $x\in\mathbb{R}$ where $y\in\mathbb{R}$ is fixed, from (2) we get the Tanaka formula which is, perhaps, the most convenient way to introduce the lokal time L^X:

(3) $$|X_t-y| = |X_o-y| + \int_0^t \operatorname{sgn}(X_s-y)\, dX_s + L^X(t,y) \quad \text{a.s.}$$

where $\operatorname{sgn}(x) = 1$ if $x\geq 0$ and -1 otherwise. From the Tanaka formula we can recover the basic properties of the local time.

PROPOSITION 1. There exists a version L^X which is the local time of $(X, \mathbb{F})$ with respect to every P_μ and satisfies the following conditions:

(i) For every $y\in\mathbb{R}$, $L^X(\cdot,y)$ is (a.s.) increasing.

(ii) For every $t\geq 0$, $L^X(t,\cdot)$ is $B(\mathbb{R})\times F_t^o$-measurable.

(iii) L^X is continuous on $[0,+\infty)\times\mathbb{R}$ a.s.

(iv) For every $y\in\mathbb{R}$, $L^X(\cdot,y)$ is a perfect additive functional.

COROLLARY 1. Let n be a σ-finite measure on $(\mathbb{R}, B(\mathbb{R}))$ and define

$$B_t = \int_{\mathbb{R}} L^X(t,y)\, n(dy)\ , \qquad t\geq 0.$$

Then $(B, \mathbb{F})$ is a perfect additive functional which is, moreover, $\mathbb{F}^o$-adapted and a.s. continuous on $[0,C_\infty)$ with

$$C_\infty = \inf\{\, t\geq 0 \ :\ B_t=+\infty\}\ .$$

To achieve right continuity we make the convention in defining $\int_{\mathbb{R}} L^X(t,y)\, n(dy)$ that this integral is $+\infty$ if $\int_{\mathbb{R}} L^X(s,y)\, n(dy)=+\infty$

for all $s>t$.

COROLLARY 2. The associated increasing process $(\langle X\rangle, \mathbb{F})$ is a continuous perfect additive functional which, moreover, can be chosen $\mathbb{F}^o$-adapted.

For the proof use Corollary 1 for $n=\ell$ where ℓ is the Lebesgue measure on $(\mathbb{R},\mathcal{B}(\mathbb{R}))$ and (1) for $g=1$.

Here are a few further properties of L^X.

PROPOSITION 2. (i) $L^X(t,y) = L^X(t\wedge D_E,y)$, $(t,y)\in[0,+\infty)\times\mathbb{R}$, a.s.

(ii) $L^X(t,y) = 0$, $(t,y)\in[0,+\infty)\times E$, a.s.

(iii) $L^X(t,X_o)>0$ for all $t>0$ on $\{X_o\in\mathbb{R}\setminus E\}$ a.s.

(iv) For every $\mathbb{F}$-stopping time S

$L^X(S+t,X_S)>0$ for all $t>0$ on $\{X_S\in\mathbb{R}\setminus E,\ S<+\infty\}$ a.s.

RANDOM TIME CHANGE

Now we come to the main subject of the present section: Random time change of strong Markov continuous local martingales. Let us consider a right continuous increasing process $(A, \mathbb{F})$ taking values in $[0,+\infty]$. Let T be the right inverse of A:

(4) $$T_t = \inf\{s\geq 0 : A_s>t\}, \qquad t\geq 0.$$

Unless otherwise stated we shall assume that $(A, \mathbb{F})$ satisfies the following conditions:

(5) A is strictly increasing on $[0,D_E\wedge T_\infty)$ a.s.

(6) $A_\infty = +\infty$ on $\{\langle X\rangle_\infty=+\infty\}$ a.s.

Clearly, $T=(T_t)_{t\geq o}$ is an $\mathbb{F}$-time change, i.e., a right continuous increasing process with values in $[0,+\infty]$ such that T_t is an $\mathbb{F}$-stopping time for every $t\geq 0$. We define the time changed process $(Y, \mathbb{G})$ by

(7) $$Y_t = X_{T_t}, \qquad G_t = F_{T_t}, \qquad t\geq 0,$$

where $X_\infty=\limsup_{t\to\infty} X_t$.

PROPOSITION 3. (i) The $\mathbb{F}$-time change T is X-continuous, i.e., X is constant on the intervals $[T_{t-},T_t]$ for all $t\geq 0$ a.s.

(ii) The time changed process $(Y, \mathbb{G})$ defined by (7) is a continuous local martingale on $(\Omega,F,P_x,\ x\in\mathbb{R})$ such that $P_x(Y_o=x)=1$, $x\in\mathbb{R}$.

(iii) The associated increasing process $(\langle Y\rangle, \mathbb{G})$ is given by

$$\langle Y\rangle_t = \langle X\rangle_{T_t}, \qquad t\geq 0, \qquad \text{a.s.}$$

(iv) There exists a version L^Y which is the local time of $(Y, \mathbb{G})$ for every P_μ and is a.s. continuous on $[0,+\infty)\times\mathbb{R}$. We then have

$$L^Y(t,y) = L^X(T_t,y), \qquad (t,y)\in[0,+\infty)\times\mathbb{R} \qquad \text{a.s.}$$

THE BASIC THEOREM

Now we are able to state the main result of this section.

THEOREM 1. <u>The following conditions are equivalent:</u>

(i) $(Y, \mathbb{G})$ <u>is a strong Markov continuous local martingale.</u>

(ii) <u>There exixts a nonnegative measure</u> m <u>on</u> $(\mathbb{R}, B(\mathbb{R}))$ <u>such that</u>

$$A_t = \int_{\mathbb{R}} L^X(t,y)\, m(dy) \qquad \text{on } \{t < D_E\} \qquad \text{a.s.}$$

(iii) (a) $(A, \mathbb{F})$ <u>is a perfect additive functional on</u> $[0,D_E)$ <u>of</u> $(X, \mathbb{F})$.

(b) A <u>is a.s. continuous on</u> $[0,D_E\wedge T_\infty)$.

(iv) (a) $(A, \mathbb{F})$ <u>is a strong Markov additive functional on</u> $[0,D_E)$ <u>of</u> $(X, \mathbb{F})$.

(b) A <u>is a.s. continuous on</u> $[0,D_E\wedge T_\infty)$.

(v) (a) <u>For every</u> $t\geq 0$, $I_{\{t<D_E\}}\cdot A_t$ <u>is</u> F^X_∞<u>-measurable.</u>

(b) $(\tilde{\theta}_s)_{0\leq s\leq+\infty}$ <u>defined by</u> $\tilde{\theta}_s=\theta_{T_s}$ <u>for</u> $0\leq s\leq+\infty$ <u>are perfect shift operators for</u> Y.

(vi) (a) <u>For every</u> $t\geq 0$, $I_{\{t<D_E\}}\cdot A_t$ <u>is</u> F^X_∞<u>-measurable.</u>

(b) $(\tilde{\theta}_s)_{0\leq s\leq+\infty}$ <u>are strong Markov shift operators for</u> Y.

The definition of perfect and strong Markov shift operators is analogous to the corresponding notions for additive functionals.

It is surprising that each of the conditions (v) and (vi) is equivalent to the remaining conditions (i)-(iv).

P r o o f . We only sketch the equivalence of (i)-(iv). The implication (ii)$\Rightarrow$(iii) can be derived from Corollary 1 to Proposition 1, (iii)$\Rightarrow$(iv) is trivial. The implication (iv)$\Rightarrow$(i) belongs to the mathematical folklore if A is, moreover, continuous and finite everywhere. Here the proof is much more involved. The problem is to handle the strong Markov terminal time $T^0_\infty=T_\infty\wedge D_E$. We omit the details. But as we shall see later this implication includes new results on strong Markov terminal times.

The heart of the proof is the implication (i)$\Rightarrow$(ii) which is based on the generalized Itô formula (2). We give the idea in the special case $E=\emptyset$ and $T_\infty=+\infty$. Let f be the structural function and m the speed measure of $(Y,\mathbb{G})$. By the generalized Itô formula (2) we get

$$f(X_t) = f(X_o) + \int_o^t Df(X_s)\, dX_s - \int_{\mathbb{R}} L^X(t,y)\, m(dy) \qquad \text{a.s.}$$

Substituting the time change T we obtain

$$f(Y_t) = f(Y_o) + \int_o^t Df(Y_s)\, dY_s - \int_{\mathbb{R}} L^X(T_t,y)\, m(dy) \qquad \text{a.s.}$$

But $f(Y_t)+t$ is a continuous local martingale and, consequently,

$$t - \int_{\mathbb{R}} L^X(T_t,y)\, m(dy)$$

is a continuous local martingale, too. On the other side, this process is of finite variation on every finite interval. This yields

$$t = \int_{\mathbb{R}} L^X(T_t,y)\, m(dy) \qquad \text{a.s.}$$

Substituting A and using $T_{A_t}=t$ the representation in (ii) is verified. □

The following corollary collects a few necessary conditions for $(Y,\mathbb{G})$ being a strong Markov continuous local martingale.

COROLLARY. Suppose that A satisfies (6) and that the process $(Y,\mathbb{G})$ obtained from $(X,\mathbb{F})$ by the $\mathbb{F}$-time change T is a strong Markov continuous local martingale with the associated increasing process $\langle Y\rangle_t = \langle X\rangle_{T_t}$. Let T^o be defined by $T^o_t=T_t\wedge D_E$. Then the following properties hold:

(i) T is a.s. continuous on $[0,A_{D_E})$.

(ii) T is strictly increasing on $[0,D_M(Y))$ and $D_M(Y)=A_{D_E}\wedge T_{\infty-}$ where M is the set of absorbing points for $(Y,\mathbb{G})$.

(iii) T^o is an $\mathbb{F}^X$-time change.

(iv) $F^Y_t = F^X_{T^o_t}$, $\quad t\geq 0$.

(v) $(T,\mathbb{G})$ is a perfect additive functional on $[0,D_M(Y))$ of $(Y,\mathbb{G})$.

4. SOME APPLICATIONS

REPRESENTATION OF ADDITIVE FUNCTIONALS

Our first application is the representation theorem for strong Markov additive functionals.

THEOREM 1. *Let* $(V, \mathbb{F})$ *be a strong Markov additive functional on* $[0,D_E)$ *of* $(X, \mathbb{F})$*. Suppose that* V *is a.s. continuous on* $[0,D_E \wedge T_\infty)$ *where*

$$T_\infty = \inf\{ t \geq 0 : V_t = +\infty \}.$$

Then there exists a nonnegative measure n *such that*

$$V_t = \int_{\mathbb{R}} L^X(t,y)\, n(dy) \qquad \text{on } \{T_\infty \wedge t < D_E\} \qquad \text{a.s.}$$

P r o o f . We set $A = V + \langle X \rangle$ and use Corollary 2 to Proposition 3.1. By Theorem 3.1 there is a nonnegative measure m representing A via the local time. It is now easy to establish that $n=m-\ell$ is a nonnegative measure such that the conclusion of the theorem holds.□

The proof of this theorem is mainly based on Theorem 3.1 (implications (iii)⇒(i)⇒(ii)) and hence makes use of the generalized Itô formula. A.T. Wang [19] also used the generalized Itô formula to give a proof for the representation of continuous perfect additive functionals of the Brownian motion. However, his proof involves a deep result of H. Tanaka [17] on the structure of such functionals.

Note that the representation in Theorem 1 holds not only on the set $\{T_\infty \wedge t < D_E\}$ but everywhere if and only if

$$V_t = V_{D_E -} \qquad \text{for all} \quad t \geq D_E .$$

STRUCTURE OF TERMINAL TIMES

Next we investigate the structure of strong Markov terminal times. As we know every first entry time into a Borel subset is even a perfect terminal time. It is most surprising that, in a certain sense, the converse is also true.

THEOREM 2. *Let* T_∞ *be any strong Markov terminal time such that*

$$T_\infty \leq D_E .$$

Then there exixts a closed subset M *of* $\mathbb{R}$ *with* $E \subseteq M$ *and* $T_\infty = D_M$ *a.s.*

P r o o f . We define $(A, \mathbb{F})$ by

$$A_t = \begin{cases} t & \text{if} \quad t < T_\infty \\ +\infty & \text{otherwise.} \end{cases}$$

Then $(A, \mathbb{F})$ satisfies the condition (iv) of Theorem 3.1. By this theorem, $(Y, \mathbb{G})$ obtained from $(X, \mathbb{F})$ by the time change T (the inverse of A) is a strong Markov continuous local martingale. Let M be the set of absorbing points for $(Y, \mathbb{G})$. Then M is closed and it can be shown that $E \subseteq M$ and $T_\infty = D_M$ a.s. □

Note that without the condition $T_\infty \leq D_E$ Theorem 2 is not true. The general result is as follows.

THEOREM 3. Let T_∞ be an $\mathbb{F}^X$-stopping time. Then the following conditions are equivalent:

(i) T_∞ is a strong Markov terminal time such that

$$T_\infty \circ \theta_{T_\infty} = 0 \qquad \text{on} \quad \{T_\infty < +\infty\} \qquad \text{a.s.}$$

(ii) T_∞ is indistinguishable from a perfect terminal time S_∞ for which, moreover,

$$u + S_\infty \circ \theta_u = S_\infty \qquad \text{on} \quad \{u \leq S_\infty\} \text{ for all } u \geq 0 \quad \text{a.s.}$$

(iii) There exixts a set $B \subseteq \mathbb{R}$ such that $T_\infty = D_B$ a.s.

In this case, B can be chosen in such a way that B is finely closed and $B^u(\mathbb{R})$-measurable.

This theorem can be seen as a counterpart to the representation theorem for (continuous) additive functionals, namely, it deals with the representation of terminal times as first entry times.

THE ASSOCIATED BROWNIAN MOTION

Let $A = \langle X \rangle$ and T the right inverse of A. We define $(W, \mathbb{G})$ by

$$W_t = X_{T_t}, \qquad G_t = F_{T_t}, \qquad t \geq 0.$$

For A the conditions (3.5) and (3.6) are satisfied (cf. Proposition 2.3(ii)). Moreover, $(A, \mathbb{F})$ is a continuous perfect additive functional of $(X, \mathbb{F})$ (see Corollary 2 to Proposition 3.1). By Theorem 3.1, $(W, \mathbb{G})$ is a strong Markov continuous local martingale. In view of Proposition 3.3(iii) the associated increasing process $(\langle W \rangle, \mathbb{G})$ has the form

$$\langle W \rangle_t = A_{T_t} = t \wedge A_\infty = t \wedge \langle X \rangle_{D_E} \qquad \text{a.s.}$$

and since $\langle X \rangle_{D_E} = D_E(W)$ it follows

$$\langle W \rangle_t = t \wedge D_E(W) \qquad \text{for all} \quad t \geq 0 \qquad \text{a.s.}$$

The set of absorbing points for $(W, \mathbb{G})$ is E, the same as for $(X, \mathbb{F})$. In the terminology of [10] (p. 266), $(W, \mathbb{G})$ is a Wiener process stopped at $D_E(W)$. More precisely, we shall say that $(W, \mathbb{G})$ is a Wiener process with absorbing points E. We call $(W, \mathbb{G})$ the Brownian motion associated to the strong Markov continuous local martingale $(X, \mathbb{F})$.

From Theorem 3.1 follows that $(\tilde{\theta}_s)$ defined by $\tilde{\theta}_s = \theta_{T_s}$ are shift operators for $(W, \mathbb{G})$. Moreover, in view of the Corollary to this theorem

$(T, \mathbb{G})$ is a perfect additive functional on $[0, D_E(W))$ of $(W, \mathbb{G})$. Because of $T_t = +\infty$ for $t \geq D_E(W)$ this implies that $(T, \mathbb{G})$ is a perfect additive functional (on $[0, +\infty)$) of $(W, \mathbb{G})$, and $(T, \mathbb{G})$ is a.s. continuous on $[0, D_E(W))$.

Conversely, we obtain that $(T, \mathbb{G})$ satisfies (3.5) and (3.6) and

$$X_t = W_{A_t} \quad , \qquad F_t = G_{A_t} \quad , \qquad t \geq 0.$$

Thus we have seen that every strong Markov continuous local martingale $(X, \mathbb{F})$ with absorbing points E can be obtained by random time change from a Wiener process $(W, \mathbb{G})$ with absorbing points E, admitting shift operators (θ_s).

Theorem 3.1 applied to the strong Markov continuous local martingale $(W, \mathbb{G})$ and the perfect additive functional $(T, \mathbb{G})$ of it lead us to the following theorem.

THEOREM 4. (i) *There exists a nonnegative measure* m *on* $(\mathbb{R}, B(\mathbb{R}))$ *such that* $m(G) > 0$ *for every nonempty open subset* G *of* $\mathbb{R}$, m *is* σ-*finite on* $\mathbb{R} \setminus E$, *and*

$$T_t^o = \int_{\mathbb{R}} L^W(t,y)\, m(dy) \quad , \qquad t \geq 0, \qquad \text{a.s.}$$

where $T_t^o = T_t \wedge D_E$.

(ii) *If we assume* $E_m = E$ *then* m *is* (up to equivalence " $\cong$ ", cf. before Definition 2.3) *uniquely determined and is just the speed measure of* $(X, \mathbb{F})$.

(iii) *If* m *is the speed measure of* $(X, \mathbb{F})$ *then*

$$\int_0^{t \wedge D_E} g(X_s)\, ds = \int_{\mathbb{R}} g(y)\, L^X(t,y)\, m(dy) \quad , \qquad t \geq 0, \quad \text{a.s.}$$

for every nonnegative Borel function g.

Note that (iii) is obtained from (i) by time change. It turns out that (iii) is the basic formula for strong Markov continuous local martingales (cf. Section 5 - 7).

PURITY AND REPRESENTATION PROPERTY

Now we introduce pure continuous local martingales which were originally studied by L.E. Dubins and G. Schwarz [5] in case of $\langle X \rangle_\infty = +\infty$.

Let $(X, \mathbb{F})$ be a continuous local martingale on $(\Omega, F, P_x, x \in \mathbb{R})$ and $(W, \mathbb{F}^W)$ be the associated Brownian motion (cf. above).

DEFINITION 1. $(X, \mathbb{F})$ is called *pure* if the following conditions are satisfied:

(i) $\langle X\rangle_\infty$ is an $\mathbb{F}^W$-previsible stopping time.

(ii) $F^X_\infty = F^W_\infty$.

THEOREM 5. Let $(X, \mathbb{F})$ be a strong Markov continuous local martingale We then have:

(i) $F^W_t = F^X_{T_t}$, $t\geq 0$,

where T is the right inverse of $A=\langle X\rangle$.

(ii) A is an $\mathbb{F}^W$-time change.

(iii) $(X, \mathbb{F})$ is pure.

P r o o f . The associated Brownian motion $(W, \mathbb{F}^W)$ is a strong Markov continuous local martingale with the associated increasing process $\langle W\rangle_t = \langle X\rangle_{T_t}$, $t\geq 0$. Thus we can apply the Corollary to Theorem 3.1 and obtain that T^o is an $\mathbb{F}^X$-time change and $F^W_t = F^X_{T^o_t}$, $t\geq 0$, where $T^o_t=T_t\wedge D_E$. But then T is also an $\mathbb{F}^X$-time change and statement (i) follows from $F^X_{T_t} = F^X_{T^o_t}$. Obviously, A is an $(F^X_{T_t})_{t\geq o}$-time change and hence (ii) follows from (i). Finally, to verify (iii) we observe that $A_\infty=D_E(W)$ is $\mathbb{F}^W$-previsible and since $T_t\uparrow\infty$ as $t\uparrow\infty$, from (i) we get $F^W_\infty = F^X_\infty$. Thus $(X, \mathbb{F})$ is pure. □

Now we come to the previsible representation property.

DEFINITION 2. We say that $(X, \mathbb{F}^X)$ possesses the previsible representation property if for every $(Z, \mathbb{F}^X)$ which is a P_μ-local martingale for all μ there exists an $\mathbb{F}^X$-previsible process f such that

$$\int_0^t f^2(s)\, d\langle X\rangle_s < +\infty \qquad \text{for all } t\geq 0 \qquad \text{a.s.}$$

and

$$Z_t = Z_o + \int_0^t f(s)\, dX_s \qquad \text{for all } t\geq 0 \qquad \text{a.s.}$$

By $M^\mu_{loc}(X, \mathbb{F}^X)$ we denote the convex set of all probability measures Q on F^X_∞ such that $(X, \mathbb{F}^X)$ is a continuous Q-local martingale with $Q(A)=P_\mu(A)$ for all $A\in F^X_o$.

PROPOSITION 1. (i) If $(X, \mathbb{F}^X)$ is pure then $(X, \mathbb{F}^X)$ possesses the previsible representation property.

(ii) $(X, \mathbb{F}^X)$ possesses the previsible representation property if and only if P_μ is extremal in $M^\mu_{loc}(X, \mathbb{F}^X)$ for every μ.

P r o o f . For the proof of (i), cf. [9] (Theorem 2 and Proposition 7). In case of a single probability space, statement (ii) was proven by R.S. Liptzer [16] and J. Jacod and M. Yor [15]. □

Now we can conclude

THEOREM 6. Let $(X, \mathbb{F})$ be a strong Markov continuous local martingale. We then have:

(i) $(X, \mathbb{F}^X)$ possesses the previsible representation property.

(ii) P_μ is extremal in $M^\mu_{loc}(X, \mathbb{F}^X)$ for every μ.

5. EXISTENCE AND UNIQUENESS OF STRONG MARKOV CONTINUOUS LOCAL MARTINGALES AND THE SOLUTION OF TWO MARTINGALE PROBLEMS

Now we are going to establish the correspondence between strong Markov continuos local martingales and their speed measure.

DEFINITION 1. A measure m on $(\mathbb{R}, B(\mathbb{R}))$ is called admissible if $m(G) > 0$ for every nonempty open subset G of $\mathbb{R}$.

THEOREM 1. (Existence and Uniqueness)
Let m be an arbitrary admissible measure on $(\mathbb{R}, B(\mathbb{R}))$. Then there exists a unique (in law) strong Markov continuous local martingale $(X, \mathbb{F})$ such that m is the speed measure of it.

P r o o f of Existence. Let $(W, \mathbb{G})$ be a Brownian motion on a family $(\Omega, F, P_x, x \in \mathbb{R})$. We define

(1) $$T_t = \int_{\mathbb{R}} L^W(t,y)\, m(dy) \quad , \qquad t \geq 0.$$

Let A be the right inverse of T. The increasing process $(T, \mathbb{G})$ satisfies the conditions and statement (ii) of Theorem 3.1. According to this theorem the process $(X, \mathbb{F})$ defined by

$$X_t = W_{A_t} \quad , \qquad F_t = G_{A_t} \quad , \qquad t \geq 0,$$

is a strong Markov continuous local martingale. From (1), the property $E = E_m$, and Theorem 4.4 now follows that m is, indeed, the speed measure of $(X, \mathbb{F})$. □

We return later to the proof of the uniqueness.

FIRST MARTINGALE PROBLEM

Now we consider the following martingale problem.

Suppose that we are given an admissible measure m on $(\mathbb{R}, B(\mathbb{R}))$. Let $E = E_m$. We look for a process $(X, \mathbb{F})$ defined on a family $(\Omega, F, P_x, x \in \mathbb{R})$ and satisfying the following conditions:

(i) $(X, \mathbb{F})$ is a continuous local martingale such that

(MP1) $P_x(X_0=x) = 1$, $x \in \mathbb{R}$.

(ii) $t \wedge D_E = \int_{\mathbb{R}} L^X(t,y)\, m(dy)$, $t \geq 0$, a.s.

Note that, because of (i), the local time L^X of $(X, \mathbb{F})$ is well-defined.

THEOREM 2. The martingale problem (MP1) possesses a unique solution. This solution is a strong Markov continuous local martingale admitting m as speed measure.

P r o o f . An application of Theorem 4.4(iii) for the function g=1 shows that the strong Markov continuous local martingale with speed measure m is a solution to (MP1). The proof of the uniqueness is a little more delicate. It is a consequence of Proposition 4.1 and the following two ingredients: 1. Every solution $(X, \mathbb{F})$ to (MP1) is pure. 2. If (Q^1_x) and (Q^2_x) are the families of probability measures of two solutions of (MP1) (without loss of generality, defined on the space of continuous functions) then (Q_x) defined by $Q_x = \frac{1}{2}(Q^1_x + Q^2_x)$, $x \in \mathbb{R}$, again corresponds to a solution of (MP1). □

P r o o f of Uniqueness of Theorem 1. Because of Theorem 4.4, every strong Markov continuous local martingale with speed measure m solves (MP1). It now suffices to apply Theorem 2. □

THE SECOND MARTINGALE PROBLEM

Now we arrive at a remarkable generalization of the theorem of P. Lévy on the martingale characterization of the Brownian motion. We pose the following second martingale problem.

Suppose that we are given a closed subset E of $\mathbb{R}$ and a function f defined on $\mathbb{R} \setminus E$ and strictly concave on every component of $\mathbb{R} \setminus E$. For convenience let us extend f to a Borel function on $\mathbb{R}$. We look for a process $(X, \mathbb{F})$ defined on a family $(\Omega, F, P_x, x \in \mathbb{R})$ and satisfying the following conditions:

(i) $(X, \mathbb{F})$ is a continuous local martingale such that $P_x(X_0=x) = 1$, $x \in \mathbb{R}$.

(MP2) (ii) $X_t = X_{t \wedge D_E}$, $t \geq 0$, a.s.

(iii) $(f(X_t) + t, F^X_t, t \geq 0)$ is a (continuous) local martingale up to D_E.

THEOREM 3. <u>The martingale problem</u> (MP2) <u>possesses a unique solution</u> $(X, \mathbb{F})$. <u>This solution is a strong Markov continuous local martingale with</u> E <u>as set of absorbing points and with</u> f <u>as structural function.</u>

P r o o f . Let m_o be the locally finite nonnegative measure on $\mathbb{R}\setminus E$ which is the second derivative of $-\frac{1}{2}f$ (in the sense of distributions). Extend m_o to a nonnegative measure on $\mathbb{R}$ such that $E_m=E$. Using the generalized Itô formula it is now easy to establish that (MP2) is equivalent to the martingale problem (MP1) for this m. □

This theorem generalizes a result of M.A. Arbib [1]. Setting $E=\emptyset$ and $f(x)=-x^2$, $x\in\mathbb{R}$, from Theorem 3 we recover the theorem of P. Lévy on the martingale characterization of the Brownian motion.

Note that this section gives a purely probabilistic approach to the existence and uniqueness of strong Markov continuous local martingales with given speed measure or structural function as characteristic.

6. STOCHASTIC EQUATIONS

In this section we investigate the stochastic equation

(1) $$X_t = X_o + \int_o^t b(X_s)\, dB_s , \qquad t\geq 0,$$

where b is a universally measurable real function and B is a Brownian motion.

A stochastic process $(X, \mathbb{F})$ on a family $(\Omega, F, P_x, x\in\mathbb{R})$ is called a solution to Eq. (1) if $P_x(X_o=x) = 1$, $x\in\mathbb{R}$, and if there is a Brownian motion $(B, \mathbb{F})$ such that (1) holds.

Let us introduce the following sets:

(2) $$N_b = \{x\in\mathbb{R} : b(x)=0\} .$$

(3) $$E_b = \{x\in\mathbb{R} : \int_{x-a}^{x+a} b^{-2}(y)\, dy = +\infty \text{ for all } a>0\}$$

where $b^{-2}(x)=+\infty$ if $b(x)=0$.

EXISTENCE AND UNIQUENESS

THEOREM 1. <u>There exists a solution</u> $(X, \mathbb{F})$ <u>to Eq.</u> (1) <u>if and only if</u> $E_b\subseteq N_b$.

THEOREM 2. <u>The solution</u> $(X, \mathbb{F})$ <u>to Eq.</u> (1) <u>is unique if and only if</u>

<u>the following condition holds: If</u> $E_b \subseteq N_b$ <u>then</u> $E_b = N_b$.

For the proofs of these theorems, cf. [11].

COROLLARY. Suppose that b has right and left limits $b(x+)$ and $b(x-)$ at every point $x \in \mathbb{R}$ and that $b(x)=0$ whenever $|b(x+)| \wedge |b(x-)| = 0$. Then there is a solution $(X, \mathbb{F})$ to Eq. (1).

This corollary recovers a result of M.T. Barlow and E. Perkins [3] (Theorem 2.1 for V=0).

FUNDAMENTAL SOLUTION AND TIME DELAY

Now we are going to investigate the structure of the general solution to Eq. (1). For this we always assume that the existence condition $E_b \subseteq N_b$ is satisfied.

It turns out that for any solution $(X, \mathbb{F})$ to Eq. (1) the points $x \in E_b$ are absorbing, i.e.,

$$X_t = X_{t \wedge D_{E_b}} \quad , \qquad t \geq 0, \qquad \text{a.s.}$$

We now analyze the behaviour of $(X, \mathbb{F})$ up to D_{E_b}. We are looking for particular solutions which spend no time in the zeros of b up to D_{E_b}.

DEFINITION 1. A solution $(X, \mathbb{F})$ to Eq. (1) is called a <u>fundamental solution</u> if

$$\int_0^\infty I_{N_b \cap E_b^c}(X_s)\, ds = 0 \qquad \text{a.s.}^{\,2)}$$

THEOREM 3. <u>There exists a unique (in law) fundamental solution</u> $(X, \mathbb{F})$. <u>This solution is a strong Markov continuous local martingale with the speed measure</u> m <u>given by</u>

$$m(A) = \int_A b^{-2}(x)\, dx \quad , \qquad A \in \mathcal{B}(\mathbb{R}) .$$

For the proof, cf. [12]. Another simple proof can be given using Theorem 4.4.

In the following we assume that $N_b \cap E_b^c \neq \emptyset$. By Theorem 2 and 3 there exist solutions which spend a positive amount of time in $N_b \cap E_b^c$. As we shall see these solutions can be obtained from the fundamental solution by "delaying" it in the set $N_b \cap E_b^c$. To make this precise we develop the notion of time delay.

DEFINITION 2. Let $(X, \mathbb{F})$ be a solution to Eq. (1). An $\mathbb{F}$-adapted right continuous increasing process V with values in $[0,+\infty]$ is called an

2) A^c denotes the complement of the set A.

$(X, \mathbb{F})$-<u>delay</u> if

(4) $$\int_0^\infty I_{M^c}(X_s)\, dV_s = 0 \qquad \text{a.s.}$$

where $M=N_b \cap E_b^c$. We then define the increasing process $(A, \mathbb{F})$ by $A_t = t+V_t$, $t \geq 0$. Let T be the right inverse of A. Then T is a continuous $\mathbb{F}$-time change with $T_0=0$. We define the time changed process $(Y, \mathbb{G})$ by (3.7). We say that the process $(Y, \mathbb{G})$ <u>is obtained from</u> $(X, \mathbb{F})$ <u>by the delay</u> V.

The following results generalize [12] where it was assumed that $E_b=\emptyset$.

PROPOSITION 1. The process $(Y, \mathbb{G})$ obtained from $(X, \mathbb{F})$ by the delay V is again a solution to Eq. (1) on a, possibly, enlarged family of probability spaces.

Let us look at a few typical examples.

EXAMPLES. Let $(X, \mathbb{F})$ be a fundamental solution to Eq. (1).

(i) Let F be a closed subset of M and define $V_t=0$ if $t<D_F$ and $+\infty$ otherwise. Then V is an $(X, \mathbb{F})$-delay. The process $(Y, \mathbb{G})$ obtained from $(X, \mathbb{F})$ by the delay V is just $(X, \mathbb{F})$ stopped at D_F. Note that $(Y, \mathbb{G})$ is again a strong Markov solution.

(ii) Let U be an $\mathbb{F}$-stopping time such that $b(X_U)=0$ on $\{U<+\infty\}$. Furthermore, let S be a nonnegative F_U-measurable random variable. Now we set $V_t=0$ if $t<U$ and S otherwise. Then the delayed process $(Y, \mathbb{G})$ remains constant on $[U,U+S]$ and behaves like $(X, \mathbb{F})$ outside of this interval. Note that $(Y, \mathbb{G})$ is not strong Markov and, if S is not continuously distributed, even not Markov.

(iii) Let $x_0 \in M$ and $p>0$. We define the $(X, \mathbb{F})$-delay V by $V_t = p \cdot L^X(t,x_0)$, $t>0$. The delayed processes $(Y^p, \mathbb{G}^p)$ are strong Markov with different distributions for different p. This example is due to H.P. McKean. As $(X, \mathbb{F})$ he considered, as we call it, the fundamental solution to Eq. (1) with $b(x)=|x|^a$, $x \in \mathbb{R}$, where $0<a<1/2$ and $x_0=0$. This is just Girsanov's example for nonuniqueness.

The next theorem reveals the structure of the general solution to Eq. (1).

THEOREM 4. <u>Let</u> $(Y, \mathbb{G})$ <u>be an arbitrary solution to Eq.</u> (1). <u>Then there exist a fundamental solution</u> $(X, \mathbb{F})$, <u>defined on a, possibly, enlarged family of probability spaces, an</u> $(X, \mathbb{F})$-<u>delay</u> V, <u>and a filtration</u> $\overline{\mathbb{F}}$ <u>with</u> $G_t \subseteq \overline{F}_t \subseteq G_\infty$ <u>such that</u> $(Y, \overline{\mathbb{F}})$ <u>is obtained from</u> $(X, \mathbb{F})$ <u>by the delay</u> V.

STRONG MARKOV SOLUTIONS

As we have seen any solution to Eq. (1) can be obtained by time delay in the zeros of b from the fundamental solution. We now characterize those time delays which transform the fundamental solution into a strong Markov solution.

Let $(X, \mathbb{F})$ be a fundamental solution to Eq. (1), admitting a family of strong Markov shift operators $(\theta_s)_{0\leq s\leq+\infty}$. Let V be an $(X, \mathbb{F})$-delay. By $(Y, \mathbb{G})$ we denote the solution obtained from $(X, \mathbb{F})$ by the delay V.

THEOREM 5. The following conditions are equivalent:

(i) $(Y, \mathbb{G})$ is a strong Markov solution.

(ii) V is a strong Markov additive functional of $(X, \mathbb{F})$ that is, moreover, a.s. continuous on $[0,T_\infty)$ where

$$T_\infty = \inf\{ s\geq 0 : V_s=+\infty\}.$$

(iii) There exixts a nonnegative measure n on $(\mathbb{R},\mathcal{B}(\mathbb{R}))$ carried by $M=N_b\cap E_b^c$ such that $E_n\cap E_b^c\subseteq N_b$ and

$$V_t = \int_{\mathbb{R}} L^X(t,y)\, n(dy) \quad , \qquad t\geq 0, \qquad \text{a.s.}$$

If one (and therefore all) of the conditions (i)-(iii) is satisfied then $(Y, \mathbb{G})$ is a strong Markov continuous local martingale with

(5) $$m(A) = \int_A b^{-2}(x)\, dx + n(A) \quad , \qquad A\in\mathcal{B}(\mathbb{R}),$$

as speed measure.

The proof of this theorem is based on Theorem 3.1. The last statement is verified by establishing that $(Y, \mathbb{G})$ is a solution to the martingale problem (MP1) for m defined in (5) and by using Theorem 5.2.

7. STRONG MARKOV CONTINUOUS LOCAL MARTINGALES AS SOLUTIONS TO STOCHASTIC EQUATIONS

In this last section we start from a strong Markov continuous local martingale $(X, \mathbb{F})$ with speed measure m defined on a family $(\Omega,\mathcal{F}, P_x, x\in\mathbb{R})$ and admitting shift operators (θ_s).

We pose the problem to find conditions under which $(X, \mathbb{F})$ is a solution to the stochastic equation (6.1) without drift.

In a certain sense this is the converse problem to that treated in the last part of Section 6 where we started from a solution to Eq. (6.1) and asked for conditions ensuring that it is a strong Markov continuous local martingale.

CANONICAL DECOMPOSITION

The speed measure m can be decomposed into its absolutely continuous and singular parts

(1) $$m(A) = \int_A h(x)\,dx + n(A), \qquad A \in \mathcal{B}(\mathbb{R}).$$

Here h is a nonnegative Borel function on $\mathbb{R}$ and n is a measure singular to the Lebesgue measure ℓ. Let m^a be defined by

$$m^a(A) = \int_A h(x)\,dx, \qquad A \in \mathcal{B}(\mathbb{R}).$$

Let L be a Borel subset of $\mathbb{R} \setminus E$ such that

(2) $$n(L) = \ell(L^c \cap E^c) = 0.$$

Without loss of generality, we make the following conventions:

(3) $$h(x) = +\infty \quad \text{if and only if} \quad x \in L^c.$$

(4) $$n(E_{m^a}) = 0, \qquad n(\{x\}) = +\infty, \quad x \in E_m \setminus E_{m^a}.$$

Recall that, for any measure μ, E_μ denotes the set of all $x \in \mathbb{R}$ such that $\mu(G) = +\infty$ for every open set G containing x. Also note that $E_m = E$ where E is the set of absorbing points for $(X, \mathbb{F})$.

The decomposition (1) is unique up to equivalence "$\cong$" (see before Definition 2.3) and is called _the canonical decomposition of_ m into its absolutely continuous and singular parts.

STRUCTURE OF THE ASSOCIATED INCREASING PROCESS

The following proposition gives the sojourn time of X in the set L. It will basically be used to reveal the structure of $\langle X \rangle$ and can easily be derived from Theorem 4.4. We use the notation $A = \langle X \rangle$.

PROPOSITION 1. We have

$$\int_0^t I_L(X_s)\,ds = \int_0^t h(X_s)\,dA_s, \qquad t \geq 0, \qquad \text{a.s.}$$

We now introduce the notation

(5) $$H = \{x \in \mathbb{R} : h(x) = 0\}.$$

THEOREM 1. (i) A _is carried by_ $X^{-1}(L)$, _i.e._,

$$A_t = \int_0^t I_L(X_s)\,dA_s, \qquad t \geq 0, \qquad \text{a.s.}$$

(ii) A _admits the decomposition_ $A = A^a + A^s$ _into the sum of the_ continuous increasing processes A^a and A^s where

$$A_t^a = \int_0^t I_{L\cap H^c}(X_s)\, h^{-1}(X_s)\, ds \quad , \qquad t \geq 0,$$

is absolutely continuous with respect to the Lebesgue measure ℓ_+ on $[0,+\infty)$ and

$$A_t^s = \int_0^t I_H(X_s)\, dA_s \quad , \qquad t \geq 0,$$

is singular to ℓ_+.

(iii) $\ell_+(\{s \geq 0 : X_s \in H\}) = 0$ a.s.

COROLLARY. The following conditions are equivalent:

(i) A is absolutely continuous.

(ii) For all $t \geq 0$

$$\int_0^t I_H(X_s)\, dA_s = 0 \qquad \text{a.s.}$$

(iii) $\ell(H) = 0$.

(iv) $A_t = \int_0^t I_{L\cap H^c}(X_s)\, h^{-1}(X_s)\, ds \quad , \qquad t \geq 0.$

CHARACTERIZATION OF SOLUTIONS

Now we are able to state the following result. For this we assume that $(\Omega, F, P_x,\ x \in \mathbb{R})$ is rich enough to carry a Brownian motion $(B^o, \mathbb{F})$.

THEOREM 2. $(X, \mathbb{F})$ is a solution to Eq. (6.1) for some universally measurable diffusion coefficient b if and only if

$$\ell(H) = 0 .$$

We then have that b^{-2} is locally integrable on $\mathbb{R} \setminus E$. Moreover, $(X, \mathbb{F})$ is then a solution to Eq. (6.1) for every universally measurable b satisfying

$$b^2(x) = h^{-1}(x) \quad , \qquad x \in \mathbb{R}.$$

P r o o f . By a theorem of Doob, $(X, \mathbb{F})$ is a solution to Eq. (6.1) if and only if

$$A_t = \int_0^t b^2(X_s)\, ds \quad , \qquad t \geq 0, \qquad \text{a.s.}$$

Using the Corollary to Theorem 1 the result easily follows. □

UNIQUENESS

If the condition $\ell(H) = 0$ of Theorem 2 holds then $(X, \mathbb{F})$ is a solution to Eq. (6.1) but not the unique solution. For example, we could take another strong Markov continuous local martingale $(\widetilde{X}, \widetilde{\mathbb{F}})$ with speed measure $\widetilde{m}$ the canonical decomposition of which is $\widetilde{m} = m^a + \widetilde{n}$ where $\widetilde{n}(L)=0$. If $n \not\equiv \widetilde{n}$ then $(X, \mathbb{F})$ and $(\widetilde{X}, \widetilde{\mathbb{F}})$ are different solutions to Eq. (6.1). The reason is that Eq. (6.1) does not include any information on the sojourn time of the solution in the zeros of b, except for the zeros of b from E_b.

This gives rise to introducing an additional stochastic equation which precises Eq. (6.1).

Let b be a universally measurable function and n a nonnegative measure on $(\mathbb{R}, B(\mathbb{R}))$ carried by $M=N_b \cap E_b^c$. We denote $E=E_b \cup E_n$.

We look for a solution $(X, \mathbb{F})$ of the stochastic equation

$$X_t = X_o + \int_o^t b(X_s)\, dB_s$$
(6)
$$\ell_+(\{0 \le s \le t : X_s \in N_b \cap E^c\}) = \int_{\mathbb{R}} L^X(t,y)\, n(dy)$$

for all $t \geq 0$ a.s. where $(B, \mathbb{F})$ is a Brownian motion.

Note that the first equation of (6) ensures that $(X, \mathbb{F})$ is a continuous local martingale. Thus the local time L^X is well-defined and the second equation of (6) becomes meaningful. The measure n is called the <u>delay measure</u> of the stochastic equation (6).

THEOREM 3. <u>There exists a unique solution</u> $(X, \mathbb{F})$ <u>to Eq.</u> (6) <u>if and only if</u> $E \subseteq N_b$. <u>This solution is a strong Markov continuous local martingale with speed measure</u> m <u>the canonical decomposition of which is given by</u>

$$m(A) = \int_A b^{-2}(x)\, dx + n(A), \qquad A \in B(\mathbb{R}). \tag{7}$$

P r o o f . Suppose $E \subseteq N_b$. Let $(Y, \mathbb{G})$ be a fundamental solution to Eq. (6.1) and $(X, \mathbb{F})$ the solution to Eq. (6.1) obtained from $(Y, \mathbb{G})$ by the delay

$$V_t = \int_{\mathbb{R}} L^Y(t,y)\, n(dy), \qquad t \geq 0.$$

By Theorem 6.5, $(X, \mathbb{F})$ is a strong Markov continuous local martingale with speed measure (7). The second equation of (6) can be derived from Theorem 4.4. To prove the uniqueness, one verifies that every solution to Eq. (6) is also a solution to the martingale problem (MP1) and the assertion follows from Theorem 5.2. The necessity of the condition can be deduced from the property $X_t = X_{t \wedge D_E}$, $t \geq 0$, which can be established for an arbitrary solution to Eq. (6).

Now we return to a strong Markov continuous local martingale $(X, \mathbb{F})$ on $(\Omega, F, P_x, x\in\mathbb{R})$ admitting shift operators (θ_s). Let m be the speed measure with the canonical decomposition (1). Also we assume that there is a Brownian motion $(B^0, \mathbb{F})$ on $(\Omega, F, P_x, x\in\mathbb{R})$.

THEOREM 4. Suppose $\ell(H)=0$. Let b be a universally measurable function such that

(8) $$b^2(x) = h^{-1}(x) \quad , \quad x\in\mathbb{R}.$$

Then $(X, \mathbb{F})$ is the unique solution to Eq. (6) with diffusion coefficient b and delay measure n.

Note that the condition $\ell(H)=0$ is also necessary for $(X, \mathbb{F})$ being a solution to Eq. (6) for some diffusion coefficient b satisfying (8) and for the delay measure n.

STOCHASTIC EQUATIONS WITH INFINITE DIFFUSION

We now consider the case that the condition $\ell(H)=0$ is not necessarily satisfied. If we formally introduce the diffusion coefficient b as before by

$$b(x) = \pm\, h^{-1/2}(x) \quad , \quad x\in\mathbb{R},$$

it may take the values $-\infty$ and $+\infty$ on a set with, possibly, strictly positive Lebesgue measure. Intuitively speaking, a process with infinite diffusion at some point should leave an "infinitesimal neighbourhood" of it immediately.

In the following we will precise this idea. We introduce the notion of a stochastic equation with infinite diffusion. Let b be an arbitrary universally measurable function defined on $\mathbb{R}$ and taking values in $[-\infty, +\infty]$. Furthermore we consider a nonnegative measure n on $(\mathbb{R}, B(\mathbb{R}))$ carried by $N_b \cap E_b^c$ called delay measure. We set $E=E_b\cup E_n$. We look for a solution $(X, \mathbb{F})$ to the following stochastic equation:

(i) $(X, \mathbb{F})$ is a continuous local martingale.

(9)
(ii) $$\int_0^t I_{\{|b|<\infty\}}(X_s)\, dX_s = \int_0^t I_{\{|b|<\infty\}}(X_s) b(X_s)\, dB_s, \quad t\geq 0,$$

(iii) $$\ell_+(\{0\leq s\leq t : X_s\in N_b\cap E^c\}) = \int_{\mathbb{R}} L^X(t,y)\, n(dy) \quad , \quad t\geq 0,$$

(iv) $$\ell_+(\{0\leq s : |b(X_s)| = +\infty\}) = 0$$

a.s. where $(B, \mathbb{F})$ is a Brownian motion.

THEOREM 5. Suppose that the following conditions are satisfied:

(i) $E \subseteq N_b$.

(ii) $n(G) > 0$ *for all open neighbourhoods* G *of* $x \in \mathbb{R} \setminus E$ *such that* $\ell(G \cap \{|b| < +\infty\}) = 0$.

Then there exists a unique solution $(X, \mathbb{F})$ *to Eq.* (9). *This solution is a strong Markov continuous local martingale with speed measure* m *given by* (7).

We notice that the conditions of Theorem 5 are also necessary for the existence of a solution to Eq. (9).

Here is the final result: Every strong Markov continuous local martingale can be described as the unique solution of some stochastic equation.

THEOREM 6. *Let* $(X, \mathbb{F})$ *be an arbitrary strong Markov continuous local martingale. Let* m *be its speed measure with the canonical decomposition* (1). *Let* b *be a function with*

$$b^2(x) = h^{-1}(x) \quad , \quad x \in \mathbb{R}.$$

Then $(X, \mathbb{F})$ *is the unique solution to the stochastic equation* (9) *with diffusion coefficient* b *and delay measure* n.

REFERENCES

1. M.A. Arbib: Hitting and Martingale Characterizations of One-Dimensional Diffusions. Z. Wahrscheinlichkeitstheorie verw. Gebiete 4, 232-247 (1965)
2. J. Azéma et M. Yor: Temps locaux. Asterisque 52-53 (1978)
3. M.T. Barlow and E. Perkins: Strong Existence, Uniqueness and Non-uniqueness in an Equation Involving Local Time. Seminairé de Probabilités XVII, 32-61, Lecture Notes in Mathematics 986, Springer-Verlag, Berlin Heidelberg New York, 1983
4. E. Çinlar, J. Jacod, P. Protter, and M.J. Sharpe: Semimartingales and Markov Processes. Z. Wahrscheinlichkeitstheorie verw. Gebiete 54, 161-219 (1980)
5. L.E. Dubins and G. Schwarz: Extremal Martingale Distributions. Proceedings 5th Berkeley Symp.Math.Statist.Probab., University of California II, Part I, 295-299 (1967)
6. H.J. Engelbert: Markov Processes in General State Spaces (Part I). Math.Nachr.80, 19-36 (1977)
7. ——————: Markov Processes in General State Spaces (Part IV). Math.Nachr. 84, 277-300 (1978)
8. ——————: Markov Processes in General State Spaces (Part V). Math.Nachr. 85, 111-130 (1978)
9. H.J. Engelbert and Juliane Hess: Stochastic Integrals of Continuous Local Martingales, I. Math.Nachr. 97, 325-343 (1980)

10. ——————————: Stochastic Integrals of Continuous Local Martingales, II. Math.Nachr. 100, 249-269 (1981)

11. H.J. Engelbert and W. Schmidt: On One-Dimensional Stochastic Differential Equations with Generalized Drift. Proceedings 4th IFIP-WG 7/1 Working Conference, Marseille-Luminy, 143-155, Lecture Notes in Control and Information Sciences 69, Springer-Verlag, Berlin Heidelberg New York, 1985

12. ——————————: On Solutions of One-Dimensional Stochastic Differential Equations without Drift. Z. Wahrscheinlichkeitstheorie verw. Gebiete 68, 287-314 (1985)

13. ——————————: Strong Markov Continuous Local Martingales and Solutions of One-Dimensional Stochastic Differential Equations (I-III). To appear in Math.Nachr.

14. K. Itô: Stochastic Integral. Proc.Jap.Acad. 20, 519-529 (1944)

15. J. Jacod et M. Yor: Études des solutions extrémales et représentation intégrale des solutions pour certains problèmes de martingales. Z. Wahrscheinlichkeitstheorie verw. Gebiete 38, 83-125 (1977)

16. R.Š. Liptzer: On the Representation of Local Martingales. Teorija Veroyatn. i ejo Primenen. XXI, No.4, 718-726 (1976) (in Russian)

17. P.A. Meyer: Un cours sur les intégrales stochastiques. Séminaire de Probabilités X, 245-400, Lecture Notes in Mathematics 511, Springer-Verlag, Berlin Heidelberg New York 1976

18. H. Tanaka: Note on Continuous Additive Functionals of the 1-Dimensional Brownian Path. Z. Wahrscheinlichkeitstheorie verw. Gebiete 1, 251-257 (1963)

19. A.T. Wang: Generalized Itô's Formula and Additive Functionals of a Brownian Motion. Z. Wahrscheinlichkeitstheorie verw. Gebiete 41, 153-159 (1977)

TRACKING ALMOST PERIODIC SIGNALS UNDER WHITE NOISE PERTURBATIONS

A. Halanay
University of Bucharest, Faculty of Mathematics
Str. Academiei 14, 70109 Bucharest, Romania

T. Morozan
Department of Mathematics, INCREST, Bd. Pacii 220
79622 Bucharest, Romania

1. INTRODUCTION

The results we report have been obtained jointly with our colleagues Constantin Tudor and Vasile Dragan ([1], [2], [3], [4]). The general idea is the following: assume we have a linear control system, an almost periodic signal, an affine feedback control aiming at tracking the signal and a natural performance, penalizing the tracking errors and the control energy, to measure the quality of the control procedure. We ask the natural question: how does the control procedure behave if the system or the realization of the control are perturbed randomly? There are many ways to model random perturbations; we have considered the case of white noise, which physically may be viewed as being the most unpleasant, but for which the mathematical theory is the most convenient.

In fact we will show that the essential qualitative properties concerning the performance are not affected by the noise, there will only be an additional cost. We shall consider several cases: the main one concerns the situation of a continuous signal tracked by a continuous almost periodic system with the dynamics perturbed by white noise; then we analyse the situation when the control procedure is implemented by measuring the signal at discrete moments and using piece-wise constant controls. Finally, for the case when the control system is stationary we shall consider tracking with high-gain and adaptive tracking. Somehow aside the main problem, we shall discuss briefly a result concerning adaptive stabilization.

2. TRACKING A CONTINUOUS SIGNAL WITH A CONTINUOUS TIME CONTROL

Consider the control system

$$dx(t)=[A(t)x(t)+B(t)u(t)]dt+G(t)dw(t)$$

where A, B, G are almost periodic and w is a d-dimensional standard Wiener process. Assume an almost periodic signal r is given and the performance in tracking it by using a suitable control is measured by

$$\lim_{T\to\infty} \frac{1}{T}\int_0^T \{[x(t)-r(t)]^* Q(t)[x(t)-r(t)]+u^*(t)R(t)u(t)\}dt$$

where star denotes the transpose and Q, R are symmetric almost periodic matrices, $Q(t)\geq 0$, $R(t)\geq 0$.

Assume also we have an affine control of the form

$$u(t)=F(t)x(t)+f(t)$$

with almost periodic F, f, such that the linear system defined by $A(t)+B(t)F(t)$ is exponentially stable.

We want to compute the value of the performance for the control procedure considered. Let us describe the final result (see [4]); denote by I(F,f) the value of the performance. Then almost surely

$$I(F,f)=\lim_{T\to\infty} \frac{1}{T}\int_0^T [r^*(t)Q(t)r(t)+f^*(t)R(t)f(t)+$$

$$+2f^*(t)B^*(t)v(t)+\mathrm{Tr}G^*(t)V(t)G(t)]dt$$

where V is the unique almost periodic solution of the system

$$V'+[A^*(t)+F^*(t)B^*(t)]V+V[A(t)+B(t)F(t)]+Q(t)+F^*(t)R(t)F(t)=0$$

and v is the unique almost periodic solution of the system

$$v'+[A^*(t)+F^*(t)B^*(t)]v+V(t)B(t)f(t)+F^*(t)R(t)f(t)-Q(t)r(t)=0$$

It is seen that V(t) and v(t) do not depend upon the noise and that the effect of the noise consists of the last term in the integral and in the fact that the equality holds almost surely.

A special situation is the one when the control is optimal in the absence of noise: In this case we have to assume

$$Q(t)\geq\gamma I, \quad R(t)\geq\gamma I, \quad \gamma>0;$$

the optimal control, which exists (see [1]) under a uniform controllability assumption, corresponds to

$$F(t)=-R^{-1}(t)B^*(t)K^+(t),$$

$$f(t)=R^{-1}(t)B^*(t)g(t),$$

where K^+ is the almost periodic, positive definite, stabilizing solution of the Riccati equation

$$K'=-KA(t)-A^*(t)K+KB(t)R^{-1}(t)B^*(t)K-Q(t)$$

and g is the unique almost periodic solution of the equation

$$g'=[K^+(t)B(t)R^{-1}(t)B^*(t)-A^*(t)]g-Q(t)r(t) \; .$$

The value of the performance is almost surely

$$\lim_{T\to\infty}\frac{1}{T}\int_0^T[r^*(t)Q(t)r(t)-g^*(t)B(t)R^{-1}(t)B^*(t)g(t)+$$
$$+\mathrm{Tr}\;G^*(t)K^+(t)G(t)]dt \; .$$

3. DISCRETE-TIME IMPLEMENTATION OF THE CONTROL

Assume the above described control procedure is implemented by taking

$$u(t)=F(t_j)x(t_j)+f(t_j) \quad \text{for} \quad t\in[t_j,t_{j+1}) \text{ where } t_j=j\delta,\; \delta>0 \; .$$

We want again to know how the control procedure behaves, what will be the corresponding value of the performance. We prove that for $\delta>0$ small enough the control considered will lead to a performance of the form

$$\lim_{N\to\infty}\frac{1}{N\delta}\sum_{j=0}^{N-1}\{(x_j^*-r_j^*)Q_j(x_j-r_j)+2(x_j-r_j)^*L_ju_j+u_j^*R_ju_j+$$
$$+2(x_j-r_j)^*q_j+2u_j^*p_j\}+\lim_{N\to\infty}\frac{1}{N\delta}\sum_{j=0}^{N-1}(\rho_j+\sigma_j)$$

where $r_j=r(t_j)$, $u_j=F(t_j)x_j+f(t_j)$, $x_j=x(t_j)$.

The sequences Q_j, L_j, R_j, q_j, p_j, ρ_j are effectively described in terms of the data,

$$\sigma_j=\mathrm{Tr}\int_{t_j}^{t_{j+1}}\left(\int_{t_j}^{t} C(t,s)G(s)G^*(s)C^*(t,s)ds\right)\Omega(t)dt$$

is the only one depending upon the noise.

The value of the performance is effectively computed giving almost surely

$$J(F,f)=\lim_{N\to\infty}\frac{1}{N\delta}\sum_{j=0}^{N-1}\{f_j^*B_j^*V_{j+1}B_jf_j+2f_j^*B_j^*v_{j+1}+r_j^*\Omega_jr_j-2r_j^*L_jf_j+$$

$$+f_j^*R_jf_j-2r_j^*q_j+2f_j^*p_j+\mathrm{Tr}V_{j+1}\int_{t_j}^{t_{j+1}}C(t_{j+1},s)G(s)G^*(s)\cdot$$

$$\cdot C^*(t_{j+1},s)ds+\rho_j+\sigma_j\};\quad \text{here } f_j=f(t_j),\quad F_j=F(t_j),$$

V_j is the unique almost periodic solution of the equation

$$V_j=(A_j+B_jF_j)^*V_{j+1}(A_j+B_jF_j)+Q_j+L_jF_j+F_j^*L_j^*+F_j^*R_jF_j$$

and v_j is the unique almost periodic solution of the equation

$$v_j=(A_j+B_jF_j)^*v_{j+1}+L_jf_j+F_j^*R_jf_j+q_j+F_j^*p_j-\Omega_jr_j-$$

$$-F_j^*L_j^*r_j+(A_j+B_jF_j)^*V_{j+1}B_jf_j$$

A detailed analysis shows that the value of the performance corresponding to the discrete-time implementation differs from the one associated to the continuous-time one by terms of order of δ (see [4]).

4. HIGH-GAIN AND ADAPTIVE TRACKING

Consider now the stationary control system

$$x'=Ax+Bu$$

with an output $y=Cx$.

We shall assume that the system is square and CB is invertible; we shall also assume that the system is minimumphase, that is the invariant zeros are stable (the invariant zeros are the roots of the equation $\det \begin{pmatrix} A-\lambda I & B \\ C & 0 \end{pmatrix}=0$)

Take $Q>0$ and consider a control of the form

$$u(t)=-\frac{1}{\varepsilon}[(CB)^{-1}]^*Q[y(t)-r(t)], \quad \varepsilon>0,$$

where r is the almost periodic signal to be tracked; if when using this control, white noise perturbations arise, the corresponding dynamics will be

$$dx(t)=(Ax(t)+Bu(t))dt+\sigma_o(y(t)-r(t))dw(t)$$

with σ_o locally Lipschitz, $|\sigma_o(\eta)|\leq\lambda|\eta|$.

We prove [3] that for ε small enough

$$\overline{\lim_{T\to\infty}} \frac{1}{T}\int_o^T|y(s)-r(s)|^2 ds\leq b\varepsilon^2 \quad \text{almost surely}$$

$$\overline{\lim_{T\to\infty}} \frac{1}{T}\int_o^T E|y(s)-r(s)|^2 ds\leq b\varepsilon^2$$

Instead of using the constant gain

$$K(\varepsilon)=-\frac{1}{\varepsilon}[(CB)^{-1}]^*Q$$

we may use an adaptive procedure

$$u(t)=K^*(t)[y(t)-r(t)]$$

where the columns k_j of K are defined by the adaptation algorithm

$$k_j'(t)=-(g_j^*(y(t)-r(t)))P_j[y(t)-r(t)]-\alpha[k_j(t)-k_j(\varepsilon)]$$

$P_j>0$, $\alpha>0$; g_j^* are the rows of $G=Q(CB)^{-1}$. We prove (see [3]) that for all $\varepsilon>0$, solutions are globally defined on $[0,\infty)$ and that for $\varepsilon>0$ small enough

$$\overline{\lim_{T\to\infty}} \frac{1}{T}\int_o^T(|y(s)-r(s)|^2+|K(s)-K(\varepsilon)|^2)ds\leq b_1 \quad \text{a.s.}$$

$$\overline{\lim_{T\to\infty}} \frac{1}{T}\int_o^T|y(s)-r(s)|^2 ds\leq b_2\varepsilon^2 \quad \text{a.s.}$$

$$\overline{\lim_{T\to\infty}} \frac{1}{T}\int_o^T E(|y(s)-r(s)|^2+|K(s)-K(\varepsilon)|^2)ds\leq b_1\varepsilon$$

$$\overline{\lim_{T\to\infty}}\ \frac{1}{T}\int_0^T E|y(s)-r(s)|^2 ds \leq b_2\varepsilon^2 .$$

Let us finally state a result concerning adaptive stabilization. If we use the adaptation algorithm

$$dk_j(t)=-(g_j^*y(t))P_jy(t)dt+\sigma_j(y(t))dv_j(t)$$

with σ_j with linear growth and Lipschitz, then solutions exist globally on $[0,\infty)$ and $\sup_{t\geq 0}(|x(t)|+|K(t)|)<\infty$ a.s.

$$\sup_{t\geq 0} E(|x(t)|^{2p}+|K(t)|^{2p})<\infty$$

$$\int_0^\infty E|x(t)|^{2p}dt<\infty, \quad \lim_{t\to\infty} E|x(t)|^{2p}=0$$

for all $p\geq 1$; moreover $\lim_{t\to\infty} EK(t)$ exists.

5. LAWS OF LARGE NUMBERS

We have stated above the final results.

We feel most of them are of interest even in the deterministic case, especially the ones related to almost periodicity.

We shall comment in this section some results of "law of large numbers" type, which are essential when the performance is considered under the white noise perturbations.

We state the following result, which is rather simple and at least partially may be considered to be known.

THEOREM. *Let* $F:R_+\times R^n\to R^n$, $G_j:R_+\times R^n\to R^n$, $j=1,\dots,d$ *be measurable and* $S_j:R_+\times R^n\to R$, $j=1,\dots,d$ *be polynomial functions of degree* r *in the second argument, with bounded with respect to the first argument coefficients.*

Assume usual conditions ensuring existence and uniqueness for the stochastic differential equation

$$dx(t)=F(t,x(t))dt+\sum_{j=1}^{d} G_j(t,x(t))dw_j(t)$$

If x is a solution with the property $\sup_{t\geq t_0} E|x(t)|^{4r}<\infty$ then almost surely

$$\lim_{T\to\infty} \frac{1}{T} \sum_{j=1}^{d} \int_0^T S_j(t,x(t))dw_j(t)=0$$

If S_o is a polynomial function of degree r in the second argument with bounded with respect to the first argument coefficients then we have almost surely

$$\lim_{T\to\infty} \frac{1}{T} S_o(T,x(T))=0$$

<u>Proof</u>. Use the standard estimate, the Markov inequality [6] and

$$\sup_{t\geq 0} E|S_j(x(t))|^{2p}=h_p<\infty \ , \quad p=1,2,$$

to deduce

$$P\{\frac{1}{k}|S_o(k,x(k))|>\varepsilon\}\leq\frac{1}{k^2\varepsilon^2}E|S_o(k,x(k))|^2\leq\frac{c_1}{\varepsilon^2 k^2}$$

$$P\{\frac{1}{k}|\int_o^k S_j(t,x(t))dw_j(t)|>\varepsilon\}\leq\frac{c_2}{k^3\varepsilon^4}\int_o^4 E|S_j(t,x(t))|^4 dt\leq\frac{c_3}{\varepsilon^4 k^2}$$

Since $\sum_{k=1}^{\infty} \frac{1}{k^2}<\infty$ we obtain from the Borel-Cantelli lemma

$$\lim_{k\to\infty} \frac{1}{k} S_o(k,x(k))=0, \qquad \lim_{k\to\infty} \frac{1}{k}\int_o^k S_j(t,x(t))dw_j(t)=0$$

almost surely.

We use now the Doob inequality to deduce

$$P\{\frac{1}{k} \max_{k-1\leq t\leq k} |\int_{k-1}^t S_j(s,x(s))dw(s)|>\varepsilon\}\leq$$

$$\leq\frac{L_1}{k^2\varepsilon^2}\int_{k-1}^k E|S_j(s,x(s))|^2 ds\leq\frac{C(\varepsilon)}{k^2}$$

hence

$$\lim_{k\to\infty} \frac{1}{k} \max_{k-1\leq t\leq k} |\int_{k-1}^t S_j(s,x(s))dw_j(s)|=0 \qquad \text{a.s.}$$

Further, since S_o is a polynomial of degree r in the second argument it suffices to prove for a monomial of the form $a(t)x_1^{\alpha_1}\dots x_n^{\alpha_n}$ that

$$\lim_{k\to\infty}\frac{1}{k}\max_{k-1\le t\le k}|a(t)x_1^{\alpha_1}(t)\dots x_n^{\alpha_n}(t)|=0 \qquad \text{a.s.}$$

The Itô formula gets

$$x_1^{2\alpha_1}(t)\dots x_n^{2\alpha_n}(t)=x_1^{2\alpha_1}(k)\dots x_n^{2\alpha_n}(k)+$$

$$+\int_k^t Q(s,x(s))ds+\sum_{j=1}^{d}\int_k^t Q_j(s,x(s))dw_j(s)$$

with Q, Q_j admiting polynomial estimates.

We have already proved that

$$\lim_{h\to\infty}\frac{1}{h^2}x_1^{2\alpha_1}(h)\dots x_n^{2\alpha_n}(h)=0 \qquad \text{a.s.}$$

and

$$\lim_{h\to\infty}\frac{1}{h^r}\int_h^t Q_j(s,x(s))dw_j(s)=0 \qquad \text{a.s.}$$

Since

$$P\{\frac{1}{h^2}\max_{h-1\le t\le h}|\int_{h-1}^t Q(s,x(s))ds|>\varepsilon\}\le$$

$$\le\frac{4}{h^4\varepsilon^2}\int_{h-1}^h E|Q(s,x(s))|^2ds\le\frac{C(\varepsilon)}{h^4}$$

we deduce finally $\lim_{k\to\infty}\frac{1}{k}\max_{k-1\le t\le k}|\int_{k-1}^t Q(s,x(s))ds|=0$ a.s. and

$$\lim_{k\to\infty}\frac{1}{k^2}\max_{k-1\le t\le k}x_1^{2\alpha_1}(t)\dots x_n^{2\alpha_n}(t)=0 \qquad \text{a.s.}$$

Let us remark that the first part of the theorem could be proved assuming only $\sup_{t\ge t_o}E|x(t)|^{2r}<\infty$ or may be considered as a direct application of the strong law of large numbers for general martingales.

To compute the performance in the discretized problem we have to show that almost surely

$$\lim_{N\to\infty} \frac{1}{N} \sum_{j=0}^{N-1} [w_j^* M_j w_j - E w_j^* M_j w_j] = 0, \quad \lim_{N\to\infty} \frac{1}{N} \sum_{j=0}^{N} g_j^* w_j = 0,$$

$$\lim_{N\to\infty} \frac{1}{N} x_N^* H_N x_N = 0, \quad \lim_{N\to\infty} \frac{1}{N} \sum_{j=0}^{N-1} x_j^* L_j w_j = 0$$

where w_j are independent random vectors, $\sup_{j\geq 0} E|w_j|^4<\infty$ and x_j is the solution of $x_{j+1}=f_j(x_j,w_j)$.

The first two limits follows from the known Kolmogorov law of large numbers. The third limit follows under the assumption that $\sup_{j\geq 0} E|x_j|^4<\infty$ by using Cebysev inequality and Borel-Cantelli lemma, and the last one is a consequence of the laws of large numbers for martingales (see [2]).

6. STABILITY, BOUNDEDNESS AND ALMOST PERIODICITY

To get the boundedness condition $\sup_{t\geq t_o} E|x(t)|^{4r}<\infty$ implied in the law of large numbers we use the following result of interest in itself.

Consider the system

$$dx(t)=[A(t)x(t)+f(t)]dt+\sum_{j=1}^{d} [B_j(t)x(t)+h_j(t)]dw_j(t)$$

with almost periodic coefficients.

Let $X(t,s)$, $t\geq s$ be the fundamental solution associated to the linear system obtained if $f=0$, $h_j=0$. If for every $p\geq 1$ there exists $\alpha_p>0$, $K_p>0$ such that

$$E|X(t,s)|^p\leq K_p e^{-\alpha_p(t-s)}, \quad t\geq s, \quad t,s\in R$$

then

(i) the system has a solution $\tilde{x}$ for which all moments exist and are almost periodic and this solution is unique in the sense that for any other solution with such property the difference is almost surely zero;

(ii) for every solution with $E|x(t_o)|^p<\infty$ we have for $t\geq t_o$

$$E|x(t)-\tilde{x}(t)|^p\leq \gamma_p e^{-\alpha_p(t-t_o)}(E|\tilde{x}(t_o)|^p+E|x(t_o)|^p)$$

Let us remark that if the coefficients are just bounded then $\tilde{x}$ has the property that all moments are bounded [5].

The solution $\tilde{x}$ is the uniform on compacts almost everywhere limit of the sequence x_k where x_k is the solution such that $x_k(-k)=0$.

If the stability assumption holds only for $t \geq s \geq 0$, then from the differential equations satisfied by the moments, it follows that every solution has moments bounded on R_+ asymptotic almost periodic respectively, provided the initial condition has finite moments.

Similar results are true for the discrete system

$$x_{j+1}=A_j x_j+f_j+\sum_{l=1}^{d}(B_j^l x_j+h_j^l v_j^l)$$

A_j, f_j, B_j^l, h_j^l being almost periodic and $w_j=(v_j^1,\ldots,w_j^d)$ a sequence of independent random vectors with almost periodic moments. Let X_{ij}, $i \geq j$ be the fundamental matrix associated to the linear system ($f_j=0$, $h_j^l=0$). If for every $p \geq 1$ there exist $\alpha_p>0, K_p>0$ such that $E|X_{ij}|^p \leq K_p e^{-\alpha_p(i-j)}$, $i \geq j$, $i,j \in Z$ then the system admits a solution $\tilde{x}_j$, $j \in Z$ for which all moments exist and are almost periodic; this solution is unique in the sense that for any other solution with the same property the difference is almost surely zero. For any solution x_j for which $E|x_{j_o}|^p<\infty$ we have

$$E|\tilde{x}_j-x_j|^p \leq \gamma_p e^{-\alpha_p(j-j_o)}(E|\tilde{x}_{j_o}|^p+E|x_{j_o}|^p), \quad j \geq j_o$$

Let us finally remark that concerning almost periodicity, the Bochner condition [7] was crucial in the continuous case as well as in the discrete one. We would like to mention preservation of almost periodicity in the process of discretization by using piece-wise constant controls and results concerning existence of almost periodic solutions for the continuous and discrete Riccati equations.

REMARK. Similar problems for performance measured with the mean value were studied by Da Prato and Ichikawa [8].

7. GOOD LIAPUNOV FUNCTIONS

In studying adaptive tracking and stabilization, an essential role is played by the use of good Liapunov function for which the powers are also Liapunov functions; while this fact is always true in the deterministic case, in stochastic situations this was a lucky case. In the case of adaptive stabilization the Liapunov function was

$$V(x,K)=x^*Hx+\sum_{j=1}^{m}(k_j-k_j(\varepsilon))^*P_j^{-1}(k_j-k_j(\varepsilon))$$

and it had for all $p\geq 1$ the property

$$\mathcal{L}V^p(x,K)\leq -\gamma_p|x|^{2p}$$

with $\mathcal{L}$ the differential operator associated to the stochastic system for x, K.

It is such Liapunov function that gives partial stability (with respect to the component x) as well as global existence, although the standard growth conditions are not satisfied.

REFERENCES

1. A. Halanay, T. Morozan, C. Tudor, Tracking almost periodic signals under white noise perturbations, Stochastics 1987, vol. 21, pp. 287-301.
2. A. Halanay, T. Morozan, C. Tudor, Tracking discrete almost periodic signals under random perturbations, Int. J. Control, 1988, vol. 47, no. 1, 381-392.
3. A. Halanay, T. Morozan, Adaptive stabilization and tracking under white noise perturbations, to appear in Int. J. Control.
4. V. Drăgan, A. Halanay, T. Morozan, Performance estimates in discrete-control tracking under white noise perturbations, submitted for publication.
5. T. Morozan, Bounded and periodic solutions of affine stochastic differential equations, Studii şi Cercetări Matematice 38 (1986), 523-527.
6. A. Friedman, Stochastic Differential Equations and Applications, vol. 1, Academic Press 1975.
7. S. Bochner, A new approach to almost periodicity, Proc. Nat. Acad. Sci. U.S. 48 (1962), 2039-2043.
8. G. Da Prato and A. Ichikawa, Optimal control of linear systems with almost periodic inputs, SIAM J. Control and Optimization, vol. 25, no. 4, (1987), 1007-1019.

VARIATIONAL CALCULUS FOR GAUSSIAN RANDOM FIELDS

Takeyuki Hida
Department of Mathematics, Nagoya University
Nagoya 464-01 Japan
and
Si Si
Department of Mathematics, Rangoon University
Rangoon Burma

§0. Introduction

The purpose of this paper is to propose a new method of study of Gaussian random fields using the variational calculus.

For ordinary stochastic processes the Lévy's infinitesimal equation gives us a guiding idea on how to investigate a given stochastic process in an anlytic manner, taking time propagation into account. There, one can see a key role played by the innovation. Interesting results have been obtained in line with what was proposed by Lévy for ordinary stochastic processes with one-dimensional time parameter.

The variational calculus for Gaussian random fields which will be proposed and developed in this paper is one of the generalizations of the time variation of ordinary processes. By using this calculus, we can see way of dependency as the parameter of the field changes, and, in addition, we shall actually form innovation in many favorable cases.

What we shall discuss in this article is, at present, far from a very general theory of stochastic variational calculus, however a few concrete techniques for several particular cases will be presented in what follows as the first step of our approach. They are

1) Fields depending on a plane circle which is wandering around in a two dimensional space R^2. We can use the conformal group to

describe possible deformations of the circle, which means we can see possible change of the field according to the movement of the circle.

2) Use of the Green's function. If a given field is expressed as a white noise integral, then Laplacian applied to this field will take out the original white noise.

3) The Hadamard equation or variational equations arising from electro-magnetic fields may be paraphrased in terms of white noise analysis.

4) Very concrete computation for the R^d-parameter Lévy Brownian motion, when we apply the series expansion in terms of the spherical harmonics, turns the present question into that of the analysis of a system of stochastic processes with one-dimensional time parameter. Thus, we shall be given further suggestions on the stochastic variational calculus in this line.

§1. Background

First of all we introduce white noise with parameter space R^d. Let a characteristic functional

(1.1) $$C(\xi) = \exp[-\frac{1}{2}\|\xi\|^2]$$

be given on a space E ($\subset L^2(R^d)$) of test functions on R^d. Then, by using the Bochner-Minlos theorem (see, e.g. [5]) we can introduce a Gaussian measure μ on the space E^* (= the dual space of E) of generalized functions on R^d in such a way that

$$C(\xi) = \int_{E^*} \exp[i\langle x, \xi\rangle] \, d\mu(x).$$

This measure μ is called a <u>white noise measure</u>.

As soon as the measure space (E^*, μ) is given, a complex Hilbert space $(L^2) \equiv L^2(E^*, \mu)$ is formed. A member $\varphi(x)$ of (L^2) is called a <u>white noise functional</u> or a <u>Brownian functional</u>.

A good representation of white noise functionals may be obtained by the $\mathcal{S}$-transform:

(1.2) $(\mathcal{S}\varphi)(\xi) = \int_{E^*} \varphi(x + \xi)\, d\mu(x), \qquad \xi \in E, \quad \varphi \in (L^2).$

Let $\mathcal{H}_1$ be the subspace of (L^2) spanned by the $\langle x,\xi\rangle$, $\xi \in E$. Then the $\mathcal{S}$-transform establishes an isomorphism

(1.3) $\mathcal{H}_1 \simeq L^2(R^d),$

in such a way that

(1.4) $(\mathcal{S}\varphi)(\xi) = \int_{R^d} F(u)\xi(u)\, du, \qquad \varphi \in \mathcal{H}_1, \; F \in L^2(R^d).$

Using such a representation, we can introduce a class $\mathcal{H}_1^{(-1)}$ of generalized white noise functionals extending the isomorphism (1.3):

(1.5) $\mathcal{H}_1^{(-1)} \simeq H^{-(d+1/2)}(R^d),$

where $H^m(R^d)$ stands for the Sobolev space of order m over R^d. We tacitly assume that the space of test functionals are taken to be the subspace $\mathcal{H}_1^{(1)}$ ($\subset \mathcal{H}_1$) which is isomorphic to $H^{(d+1)/2}(R^d)$. Note that any member of the Soborev space of order (d+1)/2 can be continouous and be evaluated at every point.

Remark 1. What we have so far discussed on $\mathcal{H}_1$ is only a part of the genral theory of generalized nonlinear white noise functionals (see [6], [11]). We have summarized just what we shall use in the following sections.

Proposition 1.1. *Let M be an analytic manifold in R^d and let I_M be the indicator function of M. A functional*

(1.6) $U(\xi) = \int_{R^d} I_M(u)f(u)\xi(u)d\sigma(u), \qquad \xi \in \mathcal{S}(R^d), \quad f \in L^2(M,d\sigma),$

$d\sigma$: volume element on M,

is the $\mathcal{S}$-transform of an $\mathcal{H}_1$-functional φ if M is d-dimensional, while U(ξ) is the $\mathcal{S}$-transform of a generalized functional φ in $\mathcal{H}_1^{(-1)}$ if M is at most (d-1)-dimensional.

Proof is almost obvious, if we observe the expression (1.6) of U(ξ) taking a test function ξ. It is therefore omitted.

A generalized white noise functional whose $\mathcal{S}$-transform is given by (1.6) is often denoted by

(1.7) $\varphi(x) = \int_M f(u)x(u)d\sigma(u).$

We then come to the partial derivative of white noise $x(u)$:

(1.8) $$x_j(u) \equiv \frac{\partial x}{\partial u_j}(u), \qquad u = (u_1, \ldots, u_j, \ldots, u_d),$$

which is defined by

$$\langle x_j, \xi\rangle = - \langle x, \xi_j\rangle, \qquad \xi_j(u) : \text{partial derivative in } u_j.$$

For a smooth $g(u)$, the $\mathcal{S}$-transform of $\langle x_j, g\rangle$ is given by

(1.9) $$U(\xi) = - \int_{R^d} g_j(u)\xi(u)\, d\sigma(u).$$

With this fact in mind, we can easily prove the following proposition.

Proposition 1.2. Let M be a d-dimensional manifold and let g be such that $g_j \in L^2(M, d\sigma)$. Then, a functional of the form

(1.10) $$\varphi(x_j) = \int_M g(u)x_j(u)\, d\sigma(u)$$

is defined and its $\mathcal{S}$-transform is given by (1.9).

Remark.2. If $g_j(u)$ is not in $^2L\ (M, d\sigma)$, then $\varphi(x_j)$ in (1.10) is a generalized white noise functional.

§2 Restriction of parameter.

The parameter u, running through R^d, of the white noise may be restricted to some lower dimensional manifold M, and we are, roughly speaking, still given a white noise with time parameter set M.

We may use the technique developed and used in [7] and [12], but we prefer, in this report, another way of restricting the parameter which is suggested by Proposition 1.1., in order to have consistent restrictions.

Take an anlytic manifold M in R^d, and take a function $F(u)$ in the Sobolev space $H^m(R^d)$ with $m = (d+1)/2$. Then there is a mappimg P_M from $\mathcal{H}_1$ into itself

(2.1) $$\int F(u)x(u)d\sigma(u) \text{ ------> } \int_M F(u)x(u)d\sigma(u)$$

If M is a closure of a d-dimensional domain, then P_M is extended to a projection down to a closed subspace of $\mathcal{H}_1$. However, if M is less than d-dimensional, we need to modify the mapping (2.1) and to give some interpretation. Let the mapping be represented by U-functionals so that the integral of the image is well defined:

(2.2) $\int F(u)\xi(u)d\sigma(u)$ ------- $\int_M F(u)\xi(u)d\sigma_M(u)$, $\xi \in E \subset L^2(R^d)$,

where $d\sigma_M(u)$ is the measure over M induced by $d\sigma(u)$ over R^d. Now we can see that the image under the mapping $\mathcal{S}P_M\mathcal{S}^{-1}$ is continuous on E(M), a nuclear space over M. We can, therefore, define a white noise measure μ_M with parameter set M, and $\int_M F(u)\xi(u)d\sigma_M(u)$ is $\mathcal{S}$-transform of a Gaussian random function expressed as a linear functional of a white noise with parameter set M. This can be shown by using the fact that M is locally Euclidean (see [13]). We can therefore paraphrase (2.2) in terms of white noise:

(2.3) $\int F(u)x(u)d\sigma(u)$ ------> $\int_M F(u)x_M(u)d\sigma_M(u)$.

The above mapping is also denoted by P_M. Summing up, we have proved the following theorem.

Theorem 2.1. The family of mappings

$$\mathcal{P} = \{ P_M ;\ M \text{ closure of domain in } R^d,\ \partial M \text{ analytic}\}$$

satisfies

i) P_M with d-dimensional M is a projection operator on $\mathcal{H}_1$, while P_M maps (L^2)-functionals into the space of generalized functionals, if the dimension of M is less than d.

ii) if $M \supset N$, then $P_M P_N$ is equal to P_N.

iii) $\mathcal{P}$ is a consisten family in the sense that for $M \supset N \supset K$

$$P_M\, P_N\, P_K = P_M\, P_K = P_K$$

Proof of i) comes from Proposition 1.1. Other assertions are obvious.

Remark. We can play the same game in the case of a partial derivative $x_j(u)$ of white noise. Even a normal derivative $\frac{\partial x}{\partial n}$ on the surface of a manifold behaves similary.

§3 Random fields depending on a circle.

Variational calculus for Gaussian random fields depending on a plane circle has been discussed in [8] the Proceedings of the Karpacz

Winter school on Theoretical Physics, held in January, 1988. There circles vary under the action of the conformal group acting on R^2. We have also used the technique of the $\mathcal{S}$-transform, by which random functions can be represented in terms of functionals of C^∞-functions. We can therfore appeal to the classical theory of calculus of variations (cf. for example, P. Lévy [1], [2]).

Some more detailed results can be found in the forthcoming paper by K.-S. Lee [12] so that we do not want to go further in the present report.

The basic idea for this particular case is that the set of all circles is topologized so as to be a 3-dimensional manifold, which is the parameter set of the Gaussian field in question. The conformal group acts on this parameter space as the symmetry gropup so that the irreducible representation of the conformal group can automatically be obtained (see Lee [12] mentioned above . By using this substantial property, we can prove the canonical property, and even we can speak of a generalized notion of innovation.

§4. Green's function method.

What we shall discuss, in this section, is a multi-dimensional parameter generalization of a multiple Markov Gaussian process in the restricted sense. To establish a reasonably systematic development of the Green's function method, we restrict our attention mainly to the case of second order differential operators in two-dimensional variable as well as their powers.

We are now given a general linear partial differential equation of second order over a domain D :

(4.1) $L\varphi = A\varphi_{11} + 2B\varphi_{12} + C\varphi_{22} + D\varphi_1 + E\varphi_2 + F\varphi = f,$

where $A,\dots,F$ and f are given functions of $u = (u_1, u_2)$, and where $\varphi_i = \partial\varphi/\partial_{u_i}$ and $\varphi_{ij} = \partial\varphi^2/\partial_{u_i}\partial_{u_j}$.

The Green's function associated with the differential operator L will be denoted by G(u,v;C), where C stands for the boundary of the the domain D on which (4.1) is defined. To fix the idea we assume, in what follows, that L is formally self-adjoint. The most important example in the present approach is, of course, the Laplacian operator.

Let x(u) be white noise and define X(C) by

(4.2) $$X(u,C) = \int_D G(u,v;C)\, x(v)\, d\sigma(v) \qquad u \in D.$$

Then, applying the operator L, we obtain the white noise x:

(4.3) $$(LX)(u,C) = x(u).$$

Now we understand that the white noise x(u), on which the X(u,C) is based, can be obtained by applying the operator L acting on X(u,C) itself. Thus obtained quantity x(u) may therefore be considered as the <u>innovation</u> for the field X(u,C). Proofs of these facts are given in terms of U-functionals by applying the $\mathcal{S}$-transform.

We then consider Green's functions of higher order. Let G(u,v) u,v $\in R^2$, be a symmetric kernel function, and let $G^i(u,v)$ be defined inductively by

$$G^0(u,v) \equiv 1, \qquad G^1(u,v) = G(u,v),$$
$$G^i(u,v) = \int_D G^{i-1}(u,s)G(s,v)\, d\sigma(s), \qquad i = 1,2,\ldots .$$

The kernel $G^i(u,v)$ is, following V. Volterra, called the i-th power composition of the second kind of G(u,v). Obviously, the equation

$$L\, G^i(u,v) = G^{i-1}(u,v)$$

is obtained, where L should be understood as an operator scting on functions of u.

Now set

(4.4) $$Y(u,C) = \int_D G^N(u,v)\, x(v)\, d\sigma(v).$$

Then we have

(4.5) $$L^N\, Y(u,C) = x(u).$$

Thus, the field Y(u,C) may be thought of as a generalization of N-ple Markov Gaussian process X(t) in the restricted sense which admits us to have an exression

$$X(t) = \int_0^t (t - u)^{N-1} \dot{B}(u) \, du.$$

We are now in a position to note an important property enjoyed by the formula (4.4). If it is viewed as a representation of Y(u,C), then it is a canonical representation (see P. Lévy [3]) of Y(u,C) in terms of the white noise x(u) in the following sense.

Proposition 4.1. We have

$$(4.6) \qquad \mathcal{B}_D(Y) = \mathcal{B}_D(x),$$

where $\mathcal{B}_D(\cdot)$ denotes the σ-field generated by the random variables in the paranthesis with parameter running through D.

Proof. The property (4.6) comes from the equation (4.5).

Thus, we have seen an example of a canonical representation.

§5. Variation of fields.

We now return to X(u,C) given by (4.2). The variable u is now fixed, while C is a variable. Let $\mathbf{C}$ be the class of plane curved introduced in R^2 and be topologized in a usual manner. Again we can appeal to the $\mathcal{S}$-transform technique to give a rigorous expression of the variation $\delta X(u,C)$.

Theorem 5.1. Let $\{X(u,C) \; ; \; C \in \mathcal{C}\}$ be a Gaussian random field such that each X(u,C) is given by (4.2). Then the variation $\delta X(u,C)$ of X(u,C) when C varies in $\mathcal{C}$ is expressed in the form

$$(5.1) \qquad \delta X(u,C) = \int_D \delta G(u,v;C)x(v)d\sigma(v) + \int_C G(u,s;C)x(s)\delta n(s)ds,$$

where $\delta G(\cdot,\cdot;C)$ denotes the variation of $G(\cdot,\cdot;C)$ in C, and where s is the parameter on the curve C and ds is the line element.

Before we come to the proof of the theorem, some interpretations are necessary so that the formula (5.1) can well be understood.

1) We know the exact expression of the variation $\delta G(u,v;C)$. It is given by the so-called Hadamard equation (see [2],[4,§3]) :

$$(5.2) \qquad \delta G(u,v;C) = - \frac{1}{2\pi} \int_C \frac{\partial}{\partial n} G(u,m;C) \frac{\partial}{\partial n} G(m,v;C) \, \delta n(s) ds ,$$

where $m = m(s)$ runs through C and where $\delta n = \delta n(s)$ denotes the normal displacement of the point $m(s)$ when C changes to $C + \delta C$.

Remark 1. We see that it is a great advantage of our analysis that even a restriction of the parameter of white noise to a curve can be rigorously understood as is seen in the second term of (5.1).

Proof of the theorem. The $\mathcal{S}$-transform of $X(u,C)$ is

$$(5.3) \qquad (\mathcal{S}X)(u,C;\xi) = \int_D G(u,v;C)\, \xi(v)\, d\sigma(v), \qquad \xi \in \mathcal{S}(R^d),$$

which will be denoted by $U(u,C;\xi)$. The Lévy's variation technique is now ready to be applied, and we obtain

$$\delta U(u,C)(\xi) = \int_D \delta G(u,v;C)\, \xi(v)\, d\sigma(v) + \int_C G(u,s;C)\xi(s)\delta n(s) ds,$$

Applying the $\mathcal{S}^{-1}$ to each term we have (5.1).

Remark 2. The first and the second terms of the right hand side of the formula (5.1) can be discriminated, since the first one is of order $\|\delta n\|$ while the second one is of order $\|\delta n\|^{1/2}$. This suggests us to deal with the second term separately to obtain the innovation.

§6 Concluding remarks.

1) We shall be able to discuss the variation of $X(C)$, $C \in \mathbf{C}$, of the form

$$(6.1) \qquad X(C) = \int_C F(s;C)\, x(s)\, ds.$$

We obtain

$$(6.2) \qquad \delta X(C) = \int_C \{\delta F(s;C) - \kappa\, F(s;C)\delta n(s)\}\, x(s) ds + \int_C F(s;C) \frac{\partial}{\partial n} x(s)\delta n(s) ds$$

But we have to clarify a very singular random function on the curve, expressed as the normal derivative of $x(s)$ (see also [16]).

Another interesting approach can be seen in [12] again, with C a circle. If F is independent of C, we can think of tranformations that carry C onto itself to determine the value of $x(s)$.

2) Canonical representation, although rigorous definition is not yet given, can be considered for a very important example, i.e. Lévy's Brownian motion. It has been discussed for odd dimensional parameter cases, but it still remains in question for even dimensional parameter cases.

To fix the idea, we shall discuss a two-dimensional parameter Lèvy Brownian motion $\{X(a)\ ;\ a \in R^2\}$. Set $X(a) = X(t,\theta)$, $a \in R^2$, $t \in R_+$, $\theta \in [0,2\pi)$. Following H.P. McKean, we expand $X(t,\theta)$ in a Fourier series for each fixed t. Take a complete orthonormal basis $\{\varphi_n(\theta)\ ;\ n \in N\}$ and set

$$X_n(t) = \int_0^{2\pi} X(t,\theta)\varphi_n(\theta)\, d\theta, \qquad n \geq 0. \tag{6.3}$$

Then, we can prove the existence of the canonical representation of $X_n(t)$ with one dimensional parameter. Denote by $B_n(t)$ the Brownian motion of the representation. The collection $\{\dot{B}_n(t);\ n \in N,\ t \geq 0\}$ is equivalent of white noise.

In this course, particular interest is found in the operations acting on the $X_n(t)$'s to obtain the white noises with R^2-parameter. They, putting together, might play some role of variational operator acting on x(a) itself to obtain the innovation. Since they are not local operators, handy tool is not availabe. However, they seem to be telling us some profound probabilistic properties of Brownian motion.

[REFERENCES]

[1] P. Lévy, Sur la variation de la distribution de l'électricité sur un conducteur dont la surface se déforme. Bull. Société math. de France, 46 (1918), 35-68.

[2] ______ , Problèmes concrets d'analyse fonctionnelle. Gauthier-Villars, Paris, 1951, Part. I, II.

[3] ______ , A special problem of Brownian motion, and a general theory of Gaussian random functions. Proc. 3rd Berkeley Symp. vol.II (1965), 133-175.

[4] K. Aomoto, Formule variationelle d'Hadamard et modèle des variétés différentiables plongées. J. Functional Analysis 34 (1979), 493 -523.

[5] T. Hida, Brownian motion (in Japanese). Iwanami Pub. Co., Tokyo 1975; English Trans. Springer-Verlag, New York Heidelberg Berlin 1980.

[6] ______ , Analysis of Brownian functionals. Carleton Math. Lec. Notes, no.13, Carleton Univ. Ottawa, 1975.

[7] ______ , A note on generalized Gaussian random fields. Jounal Multivariate Analysis 27 (1988), 255-260.

[8] ______ , White noise analysis and Gaussian random fields. Proc. 24th Winter School of Theoretical Physics, Karpacz, 1988, to appear.

[9] T. Hida, K.-S. Lee and S.-S. Lee, Conformal invariance of white noise. Nagoya Math. J. 98 (1985), 87-98.

[10] T. Hida, K.-S. Lee and Si Si, Multidimensional parameter white noise and Gaussian random fields. Balakrishnan volume (1987), 177-183.

[11] H.-H. Kuo, Brownian functionals and applications, Acta Appl. Math. 1 (1983), 175-188.

[12] K.-S. Lee, White noise approach to Gaussian random fields. to appear, 1989.

[13] J.L. Lions and E. Magenes, Non-homogeneous boundary value problems and applications, vol.I, Springer Verlag, New York Heidelberg Berlin, 1972.

[14] Si Si, A note on Lévy's Brownian motion. Nagoya Math. J. 108 (1987), 121-130.

[15] ____ , Gaussian processes and conditional expectations. BiBoS Notes 292/87, Univ. Bielefeld, 1987.

[16] ____ , Topics on Gaussian random fields. RIMS Kokyuroku, #672 (1988), Gaussian Random Fields - Stochastic Variational Calculus and Related Topics, ed. A. Noda.

[17] Y. Yokoi, Positive generalized white noise functionals. Preprint, Kumamoto, 1987; to appear 1989.

Local uniqueness of Feller processes with integrodifferential generators

Jan Kisyński

Abstract. Take into account a Markov process on a compact C^∞ manifold M with boundary, generated by a second order elliptic integrodifferential operator W, whose domain is determined by non-local integrodifferential conditions of Ventcel. If U is an open subset of $M \setminus \partial M$, then the behaviour of the process before its first exit time from U depends only on $\mathbb{1}_U W|_{C_c^\infty(U)}$.

1. Infinitesimal generators determined by integrodifferential systems of Ventcel.

Let M be a compact C^∞-manifold with boundary ∂M, or without boundary ($\partial M = \emptyset$). Every restriction of infinitesimal generator of any Feller submarkov semigroup on M satisfies the principle of maximum. If $\partial M = \emptyset$ then each linear operator $W : C^2(M) \longrightarrow C^2(M)$ satisfying principle of maximum is an operator of Waldenfels, i.e. has the form $W = P + S$, where P is a differential operator of diffusion and S is an integrodifferential operator of Lévy (decomposition need not be smooth). If $\partial M \neq 0$ then, in the same context, together with operators of Waldenfels the "boundary" (in fact global) conditions of Ventcel appear. See [7] and [2].

Hölderian operators of Waldenfels. By a hölderian operator of Waldenfels we mean an operator $W : C^2(M) \longrightarrow C^2(M)$ of the form $W = P + S$, where P is a differential operator of diffusion with coefficients (in local coordinates) satisfying Hölder condition with some exponent $\gamma \in (0,1)$, and S is a hölderian integrodifferential operator of Lévy. The former means that there are: a non-negative function $a \in C^{\gamma}(M)$, a vector field u on M of class C^{γ}, a C^{∞} map $v : M \times M \longrightarrow T(M)$ satisfying for every x and y in M the conditions

$$v(x,y) \in T_x(M), \qquad v(x,x) = 0, \qquad d_y v(x,y)\Big|_{x=y} = id_{T_x(M)},$$

and a hölderian Lévy's kernel s on $M \times \mathcal{B}(M)$, such that

$$(Sf)(x) = -a(x)f(x)+df(x)\cdot u(x)+\int_M \left[f(y)-f(x)-df(x)\cdot v(x,y)\right] s(x,dy)$$

for every $f \in C^2(M)$ and $x \in M$. We say that s is a hölderian Lévy's kernel on M if

$$s(x,B) = \int_B \overline{x,y}^{\,\beta-2} k(x,dy), \qquad x \in M, \qquad B \in \mathcal{B}(M),$$

where

(i) $\overline{x,y}$ is a geodesical distance from x to y in the sense of some C^{∞} riemannian metric on M,

(ii) β is a positive constant, and

(iii) k is a bounded non-negative kernel on $M \times \mathcal{B}(M)$ satisfying the Lipschitz condition

$$d(k(x,\cdot),k(y,\cdot)) \leqslant \mathrm{const}\cdot\overline{x,y}, \qquad x,y\in M,$$

with respect to the Prochorov distance d in the space of bounded Borel measures on M, corresponding to the geodesical distance on M.

Approximation of hölderian operators of Lévy. Let S be a hölderian operator of Lévy and let γ and β be as above. Denote by $d\nu$ an arbitrarily fixed C^∞ density on M, for instance the riemannian volume. Let U and V be open, such that $\overline{V}\subset U\subset M\setminus\partial M$, and that ∂V is a C^∞ submanifold of M. Then for every $\alpha\in(0,\gamma\wedge\frac{1}{3}\beta)$ and every $\varepsilon>0$ it is possible to construct a non-negative C^∞ function Φ on $M\times M$ such that

(i) $\Phi|_{\overline{V}\times\overline{V}}$ depens only on the restriction of s to $\overline{V}\times\mathcal{B}(U)$,
(ii) $\Phi = 0$ on $M\times\mathcal{O}$, where $\mathcal{O}$ is a neighbourhood of $\partial M\cup\partial V$,
(iii) $\| S^* - S \|_{L(C^{2+\alpha}(M),C^{\alpha}(M))} \leqslant \varepsilon$, where S^* is the operator of Lévy defined by

$$(S^*f)(x) = -a(x)f(x) + df(x)\cdot u(x) +$$

$$+ \int_M \left[f(y) - f(x) - df(x)\cdot v(x,y)\right]\Phi(x,y)\,d\nu(y)$$

Corollary. If $\alpha\in(0,\gamma\wedge\frac{1}{3}\beta)$, then $P,S,W\in L(C^{2+\alpha}(M),C^{\alpha}(M))$ and S is a compact operator of $C^{2+\alpha}(M)$ into $C^{\alpha}(M)$.

An operator of Ventcel is an operator $\Gamma : C^2(M) \longrightarrow C(\partial M)$ such that, for every $f \in C^2(M)$ and $x \in \partial M$,

$$(\Gamma f)(x) = (Q\gamma_o f)(x) - \tilde{\alpha}(x)f(x) + df(x)\cdot\tilde{u}(x) +$$

$$+ \int_M \left[f(y) - f(x) - df(x)\cdot\pi(x)v(x,y)\right] t(x,dy),$$

where:

(i) $Q : C^2(\partial M) \longrightarrow C(\partial M)$ is a second order homogeneous differential operator of diffusion on ∂M,

(ii) $\gamma_o : C(M) \longrightarrow C(\partial M)$ is the trace operator,

(iii) $\tilde{\alpha} \in C(\partial M)$ is a non-negative function,

(iv) $\tilde{u}$ is a continuous vector field on ∂M, never directed outside of M,

(v) every $\pi(x)$ is a projection of $T_x(M)$ onto $T_x(\partial M)$, and the whole π is a C^∞ section of vector bundle with base ∂M, whose fiber at any $x \in \partial M$ is $L(T_x(M), T_x(\partial M))$,

(vi) t is a non-negative Borel kernel on $\partial M \times \mathcal{B}(M)$, called the Ventcel kernel.

An integrodifferential system of Ventcel on M is a triple (W, Γ, δ) such that W is an operator of Waldenfels, Γ is an operator of Ventcel and $\delta \in C(\partial M)$ is a non-negative function. We say that (W, Γ, δ) is elliptic hölderian if the following collection of conditions is satisfied:

(i) P is second order elliptic and W is hölderian,

(ii) Γ is hölderian, in a sense similar as in the case of W,

(iii) $r \geq \rho$, where r is the differential order of Γ, including differentiation inside the integral, and $\rho = 0,1,2$ is an order of singularity of the kernel t (see [4] for exact definition),

(iv) if $r = 0$ (i.e. if $\rho = 0$ and $\tilde{u}(x) \equiv \pi(x)\int_M v(x,y)t(x,dy)$) then $\delta > 0$ everywhere on ∂M,

(v) if $r = 1$ then $\tilde{u}$ is always directed strictly to the interior of M,

(vi) if $r = 2$, then Q is elliptic and, for each $x \in \partial M$, either $\delta(x) > 0$, or $t(x, M \setminus \partial M) = \infty$, or $\tilde{u}(x) \neq 0$ is directed strictly to the interior of M.

Theorem. [6], [2: p. 480, Th. XVI, p. 492, Th. XIX]. Let W be a hölderian elliptic operator of Waldenfels on M. If $\partial M = \emptyset$ then $W|_{C^2(M)}$ is a pregenerator of a Feller semigroup on M. If $\partial M \neq \emptyset$ and

$$\mathcal{D}_0 = \{ f \in C^2(M) : f = Wf = 0 \quad \text{everywhere on } \partial M \}$$

then $W|_{\mathcal{D}_0}$ is a pregenerator of a Feller semigroup on $\mathring{M} = M \setminus \partial M$. If $\partial M \neq 0$ and (W, Γ, δ) is an elliptic holderian system of Ventcel on M, and

$$\mathcal{D} = \{ f \in C^2(M) : \Gamma f = \delta \gamma_0 W f \},$$

then $W|_{\mathcal{D}}$ is a pregenerator of a Feller semigroup on M.

Here by a Feller semigroup on a locally compact space S (in the above equal to M or to $\mathring{M}$) we mean a one-parameter strongly continuous semigroup of linear non-negative contractions of the space $C_0(S)$ if S is non-compact, and of the space $C(S)$ if S is compact. By a pregenerator we mean a linear operator which is closeable in $C_0(S)\times C_0(S)$ or in $C(S)\times C(S)$, respectively, and whose closure is the strong infinitesimal generator of a semigroup. The equality $\Gamma f = \delta\gamma_0 \mathcal{A} f$ is called the Ventcel condition. It is very important for procfs of results formulated in present paper, that in the elliptic hölderian case the Ventcel condition has an almost boundary character. Namely we have

Proposition. [2; p. 494]. Suppose that $\partial M \neq \emptyset$ and let $(\mathcal{A},\Gamma,\delta)$ be an elliptic hölderian system of Ventcel on M. Let r be the differential order of Γ. If $r = 0$ or $r = 1$ then suppose additionally that

$$(*)\quad \begin{cases} \text{for some } \alpha\in(0,\gamma\wedge\tfrac{1}{3}\beta) \text{ the map } f\longrightarrow \Gamma f - \delta\gamma_0\mathcal{A}f \text{ is} \\ \text{continuous from } C^{4-r+\alpha}(M) \text{ into } C^{2-r+\alpha}(\partial M). \end{cases}$$

Then, for every $\varepsilon > 0$, every compact $K\subset M\setminus\partial M$ and every function f belonging to $C^{4-r+\alpha}(M)$ there is a function $g\in D$ such that $g = f$ on K and $\|g - f\|_{C(M)}\leq\varepsilon$.

2. Canonical Markov processes.

Everywhere in this Section S denotes a separable metric locally compact space and Δ denotes a fixed point outside of S. By S_Δ we denote the sum $S_\Delta = S\cup\{\Delta\}$ equipped with compact metrizable topology which induces on S the original topology of S. Thus if S is compact, then Δ is a separated point of S_Δ, and if S is non-compact, then S_Δ is the one-point compactification of S.

<u>The space of right-continuous functions with left-side limits.</u> By $D(S,\Delta)$, or simply by D, we denote the space of all S_Δ-valued right continuous functions ω on $[0,\infty)$, having left-side limits everywhere on $(0,\infty)$ and such that $\omega^{-1}(\Delta)$ is either empty, or is equal to $[\zeta,\infty)$ for some $\zeta = \zeta(\omega) \in [0,\infty)$.

<u>Canonical filtration on D.</u> Denote by $\mathcal{B}(S_\Delta)$ the Borel σ-field of S_Δ, and by $\mathcal{B}_u(S_\Delta)$ the σ-field of universally measurable sets over $\mathcal{B}(S_\Delta)$. Denote by $\mathring{\mathcal{F}}(S,\Delta)$, or simply by $\mathring{\mathcal{F}}$, the σ-field of cylindric subsets of D, generated by all the sets of the form $\{\omega \in D : \omega(t) \in B\}$, where $t \in [0,\infty)$ and $B \in \mathcal{B}(S_\Delta)$. Let $\mathcal{F}^* = \mathcal{F}^*(S,\Delta)$ be the σ-field of universally measurable sets over $\mathring{\mathcal{F}}$. For any $t \in [0,\infty)$ define the map $\alpha_t : D \longrightarrow D$ of stopping at the time t, so that

$$[\alpha_t(\omega)](s) = \omega(t \wedge s), \qquad \omega \in D, \quad s \in [0,\infty).$$

Evidently, α_t is a measurable map of $(D,\mathring{\mathcal{F}})$ into itself, whence α_t is also a measurable map of $(D,\mathcal{F}^*)$ into itself. For any $t \geqslant 0$ put

$$\mathcal{F}^*_t = \alpha_t^{-1}(\mathcal{F}^*), \qquad \mathcal{F}^*_{t+} = \bigcap_{h>0} \mathcal{F}^*_{t+h}.$$

Following [5], we call $(\mathcal{F}^*_{t+})_{t\geqslant 0}$ the <u>canonical filtration</u> on $(D,\mathcal{F}^*)$.

By canonical process on D we mean the family $(X_t)_{t\geq 0}$ of evaluation maps $X_t : D \ni \omega \longrightarrow \omega(t) \in S_\Delta$, so that $X_t(\omega) = \omega(t)$. Every map X_t is measurable from $(D, \mathcal{F}_t^*)$ into $(S_\Delta, \mathcal{B}_u(S_\Delta))$. In language of general theory of processes the former is expressed by saying that $(X_t)_{t\geq 0}$ is an $(\mathcal{F}_t^*)_{t\geq 0}$ - adapted process with state space $(S_\Delta, \mathcal{B}_u(S_\Delta))$.

The σ-fields K_T^*. Following [5], for any optional time T of the canonical filtration $(\mathcal{F}_{t+}^*)_{t\geq 0}$ put $K_T^* = K_T^*(S,\Delta) = k_T^{-1}(\mathcal{F}^*)$, where $k_T : D \longrightarrow D$ is the map of killing at the time T, i.e. $[k_T(\omega)](t) = \omega(t)$ if $t < T(\omega)$, $= \Delta$ if $t \geq T(\omega)$. Then $\mathcal{F}_{T-}^* \subset K_T^* \subset A_T^* \subset \mathcal{F}_{T+}^*$, where $\mathcal{F}_{T-}^* = \sigma\{A \cap \{t < T\} : t \in [0,\infty), A \in \mathcal{F}_t^*\}$, $A_T^* = a_T^{-1}(\mathcal{F}^*)$, $[a_T(\omega)](t) = \omega(t \wedge T(\omega))$, $\mathcal{F}_{T+}^* = \{A \in \mathcal{F}^* : A \cap \{T \leq t\} \in \mathcal{F}_{t+}^*$ for every $t \in [0,\infty)\}$.

Canonical Markov process corresponding to a Feller semigroup. To any Feller semigroup $(N_t)_{t\geq 0}$ on S with infinitesimal generator G, or with a pregenerator G, there corresponds unique probability kernel $P = P_x(d\omega)$ on $(S_\Delta, \mathcal{B}_u(S_\Delta)) \times \mathcal{F}^*(S,\Delta)$ such that $E_x[f(X_{t+s}) | \mathcal{F}_{t+}^*] = N_s(X_t, f|_S)$ for every $x \in S_\Delta$, $t \geq 0$, $s \geq 0$, $f \in C(S_\Delta)$ such that $f(\Delta) = 0$. See [1; p. 46]. We shall refer to P, or to the whole system $(D, \mathcal{F}^*, (\mathcal{F}_{t+}^*)_{t\geq 0}, P)$ as to canonical Markov process generated by G, or governed by $(N_t)_{t\geq 0}$.

3. Probabilistic interpretation of Lévy's kernel.

The following result is interesting by itself and is also very important for proof of Theorem 5.2.

3.1. Theorem. Let $(\mathring{N}_t)_{t\geq 0}$ be a Feller semigroup on M, or on $\mathring{M} = M \setminus \partial M$, determined according to Theorem of Section 1 by a single elliptic hölderian operator of Waldenfels W, or by elliptic hölderian system of Ventcel (W, Γ, δ). Let $(N_t)_{t\geq 0}$ be the corresponding Markov transition semigroup on M_Δ, or on $\mathring{M}_\Delta$, respectively. Let $A \in \mathcal{B}(\mathring{M})$ and $B \in \mathcal{B}(M_\Delta)$, or $B \in \mathcal{B}(\mathring{M}_\Delta)$, and suppose that $\overline{A} \cap \overline{B} = \emptyset$. Then for the (canonical) Markov process governed by $(N_t)_{t\geq 0}$ we have

$$E_x\Big[\sum_{0<u\leq t} 1\!\!1_A(X_{u-})\, 1\!\!1_B(X_u)\Big] = \int_0^t du \int_A N_u(x,dy)\big[s(y, B\cap M) + a(y)\, 1\!\!1_B(\Delta)\big]$$

for every $x \in M$, or for every $x \in \mathring{M}$, and for every $t \in (0,\infty)$.

3.2. Remarks. Similar interpretation of Lévy's kernel. without restriction that $A \subset \mathring{M}$, i.e. for every $A \in \mathcal{B}(M)$, was proved in [3] under assumptions that for every $f \in C(S_\Delta)$, $S = M$ or $S = \mathring{M}$,

(i) $\int_M f(y)s(x,dy) = \lim_{t\downarrow 0} \frac{1}{t} N_t(x,f)$ for every $x \in S_\Delta \setminus \operatorname{supp} f$, and

(ii) $\sup\{\frac{1}{t} | N_t(x,f)| : t \in (0,\infty),\ x \in K\} < \infty$ for every compact $K \subset S_\Delta \setminus \operatorname{supp} f$. These assumptions ~~need~~ not be satisfied if $\partial M \neq \emptyset$, because if they were, then one could use the method of [3] for proof of Theorem 1 for every $A \in \mathcal{B}(M)$ and $B \in \mathcal{B}(M)$ with disjoint closures. But the assumption that $A \subset \mathring{M}$ is essential in Theorem 1, because a Markov process generated by (W, Γ, δ) with purely differential W, i.e. with $s \equiv 0$, need not be a.s. continuous, but it may jump from ∂M.

4. Auxiliary theorems.

4.1. Assumptions. Let M be a C^∞ manifold and $\mathring{M}_1$, $\mathring{M}_2$ two open precompact subsets of M. Assume that, for $i = 1,2$, the boundary of $\mathring{M}_i$ either is empty, or is a C^∞ submanifold of M. Put $M_i = \overline{\mathring{M}_i}$, so that $\mathring{M}_i = M_i \setminus \partial M_i$. For $i = 1,2$ let S^i be equal either to $\mathring{M}_i$, or to M_i. Let Δ be a point outside of M. Take into account the canonical process on $D(M, \Delta)$. Let T be the first exit time of this process from an open subset U of $\mathring{M}_1 \cap \mathring{M}_2$. For $i = 1,2$, let $T_i = T|_{D(S^i, \Delta)}$.

4.2. Theorem. Admit the assumptions 4.1. For $i = 1,2$ let P^i be a Markov probability kernel on $(S^i_\Delta, \mathcal{B}_u(S^i_\Delta)) \times \mathcal{F}^*(S^i, \Delta)$. Then the three conditions are equivalent:

(i) $$E^1_x\left[\int_0^{T_1} e^{-\lambda u} f(X_u)\,du\right] = E^2_x\left[\int_0^{T_2} e^{-\lambda u} f(X_u)\,du\right] \text{ for every } x \in U,\ \lambda > 0 \text{ and } f \in C_0(U),$$

(ii) $$E^1_x\left[f(X_t);\ t < T_1\right] = E^2_x\left[f(X_t);\ t < T_2\right] \text{ for every } x \in U,\ t \geq 0 \text{ and } f \in C_0(U),$$

(iii) $$P^1_x(k_T^{-1}(A) \cap D(S^1, \Delta)) = P^2_x(k_T^{-1}(A) \cap D(S^2, \Delta)) \text{ for every } x \in U \text{ and } A \in \mathcal{F}^*(M, \Delta).$$

4.3. Remark. The operator $\hat{N}^i_t : b\mathcal{B}_u(U) \longrightarrow b\mathcal{B}_u(U)$ defined by $\hat{N}^i_t(x,f) = E^i_x[f(X_t);\ t<T]$ constitute a semigroup with resolvent composed of operators $\hat{R}^i_\lambda(x,f) = E^i_x[\int_0^{T_i} e^{-\lambda u} f(X_u)du]$.

4.4. Example. Let Q and R be optional times of the filtration $(\mathcal{F}^*_{t+}(M,\Delta))_{t\geq 0}$, $Q_i = Q|_{D(S^i,\Delta)}$, $R_i = R|_{D(S^i,\Delta)}$. Then $A = \{\omega \in D(M,\Delta) : Q \leq R < T\} \in \mathcal{F}^*(M,\Delta)$ and for $i = 1,2$, $k_T^{-1}(A) \cap D(S^i,\Delta) = \{\omega \in D(S^i,\Delta) : Q_i \leq R_i < T_i\}$, by a theorem of Courrège and Priouret [5; p. 177, (27,1)]. So, if one of the conditions (i) - (iii) is satisfied then $P^1_x\{Q_1 \leq R_1 < T_1\} = P^2_x\{Q_2 \leq R_2 < T_2\}$ for every $x \in U$.

4.5. Theorem. Under Assumptions 3.1 suppose that, for every $k = 0,1,2,\ldots$ and $i = 1,2$, $P^{i,k}$ is a Markov kernel on $(S^i_\Delta, \mathcal{B}_u(S^i_\Delta)) \times \mathcal{F}^*(S^i,\Delta)$ corresponding to a Feller semigroup $(N^{i,k}_t)_{t\geq 0}$ on S^i, with resolvent $(R^{ik}_\lambda)_{\lambda > 0}$. Suppose that, for $i = 1,2$ and some $\lambda > 0$, we have $\lim_{k\to\infty} R^{ik}_\lambda(x,f) = R^{i0}_\lambda(x,f)$ uniformly in $x \in S^i$, for every $f \in C_0(\mathring{M}_i)$ if $S^i = \mathring{M}_i$, or for every $x \in C(M_i)$ if $S^i = M_i$. Suppose that the condition

$$E^{1k}_x\Big[\int_0^{T_1} e^{-\lambda u} f(X_u)du\Big] = E^{2k}_x\Big[\int_0^{T_2} e^{-\lambda u} f(X_u)du\Big] \quad \text{for every } x \in U \text{ and } f \in C_0(U)$$

is satisfied for every $k = 1,2,\ldots$. Then the same condition is satisfied also for $k = 0$.

4.6. Theorem. Let S be a locally compact space, Δ a point outside of S, and $(X_t)_{t \geq 0}$ the canonical cadlag process with state space S_Δ and with cemetery Δ. Let $P = P_x(d\omega)$ be a Markov kernel on $(S_\Delta, \mathcal{B}_u(S_\Delta)) \times \mathcal{F}^*(S, \Delta)$ and G the weak infinitesimal generator of the corresponding transition semigroup. Let U be an open subset of S, and G_o a pregenerator of a Feller semigroup $(\overset{o}{N}_t)_{t \geq 0}$ on U. Denote by T the first exit time of the process $(X_t)_{t \geq 0}$ from U. Suppose that

$$(*) \quad \begin{cases} \text{for every } f_o \in \mathcal{D}(G_o) \text{ there is } f \in \mathcal{D}(G) \text{ such that } f = f_o \\ \text{and } Gf = G_o f_o \text{ on } U, \text{ and } P_x\{T < \infty,\ f(X_T) \neq 0\} = 0 \text{ for} \\ \text{every } x \in U. \end{cases}$$

Then

$$E_x\left[f(X_t);\ t < T\right] = \overset{o}{N}_t(x, f)$$

for every $x \in U$, $t \geq 0$ and $f \in C_o(U)$.

4.7. Remark. Applications of Theorem 4.6 to Markov processes generated by systems of Ventcel are possible thanks to: (1) approximation theorem for hölderian Lévy's kernels stated in Section 1, (2) the Proposition of Bony, Courrege and Priouret concluding Section 1, (3) the Theorem 3.1, and (4) the Theorem 4.5.

5. The main result.

5.1. Assumptions. Let M be a C^∞ manifold (not necessarily connected), and $\overset{o}{M}_1$, $\overset{o}{M}_2$ two open precompact subsets of M such that the boundary of each of them either is empty, ot is a C^∞ submanifold of M. Put $M_i = \overline{\mathring{M}_i}$, so that $\mathring{M}_i = M_i \setminus \partial M_i$. For $i = 1,2$ let either $S^i = \mathring{M}_i$, or $S^i = M_i$. Choose a point Δ outside of M. Suppose that, for $i = 1,2$, $P^i = P_x^i(d\omega)$ is a Markov kernel on $(S^i_\Delta, \mathcal{B}_u(S^i_\Delta)) \times \mathcal{F}^*(S^i, \Delta)$ generated, according to Theorem of Section 1, by a single elliptic hölderian operator of Waldenfels $W_i : C^2(M) \longrightarrow C(M)$ if $S^i = \mathring{M}_i$ or if $\partial M_i = \emptyset$, or generated by an elliptic hölderian system of Ventcel $(W_i, \Gamma_i, \delta_i)$ on M_i if $S^i = M_i$ and $\partial M_i \neq \emptyset$.

5.2. Theorem. Admit the Assumptions 5.1. Let U be an open subset of $\mathring{M}_1 \cap \mathring{M}_2$, and T the first exit time from U of the canonical process on $D(M, \Delta)$. If

$$(W_1 f)(x) = (W_2 f)(x) \quad \text{for every } x \in U \text{ and } f \in C_c^\infty(U),$$

then

$$P_x^1(A \cap D(S^1, \Delta)) = P_x^2(A \cap D(S^2, \Delta)) \quad \text{for every } x \in U \text{ and } A \in K_T^*(M, \Delta)$$

5.3. Remarks. The analytical part of reasonigs which lead to above theorem depends essentially on the ellipticity of differential parts of the occuring operators of Waldenfels, and on correctness of boundary problems resulting from the considered systems of Ventcel when the integral terms are ignored. See [2; Sections III.1 and III.2]. The proofs of presented results are similar to that in [4], where however the canonical σ-fields are not used, and where $S^1 = S^2 = M$.

Bibliography

[1] R.M. Blumenthal and R.K. Getoor, Markov Processes and Potential Theory, Academic Press, 1968.

[2] J.-M. Bony, Ph. Courrège et P. Priouret, Semi-groupes de Feller sur une variété a bord compacte et problèmes aux limites intégro-différentielles du second ordre donnant lieu au principe du maximum, Ann. Inst. Fourier 18, 2 (1968), pp. 369-521.

[3] N. Ikeda and S. Watanabe, On some relations between the harmonic measure and the Lévy measure for a certain class of Markov processes, J. Math. Kyoto Univ. 2(1962), pp. 79-95.

[4] J. Kisyński, Localizations of Feller infinitesimal generators and uniqueness of corresponding killed processes, in "Probability Measures on Groups IX", Lecture Notes in Mathematics, Springer-Verlag, to appear.

[5] P.-A. Meyer, Presentation des processus de Markov, pp. 165-198 in J.L. Bretagnolle, S.D. Chatterji, P.-A. Meyer, Ecole d'Eté de Probabilités: Processus Stochastiques, Lecture Notes in Mathematics, Vol. 307, Springer-Verlag, 1973.

[6] K. Sato and T. Ueno, Multi-dimensional diffusion and the Markov process on the boundary, J. Math. Kyoto Univ. 4(1965), pp. 529-605.

[7] W. von Waldenfels, Fast positive Operatoren, Zeitschrift für Warscheinlichkeitstheorie 4(1965), pp. 159-174.

ON GENERAL ARMA MODELS AND REGULARITY CONDITIONS

Aleksander Kowalski, Dominik Szynal
Institute of Mathematics, Maria Curie-Sklodowska University
Pl. Marii Curie-Sklodowskiej 1, 20-031 Lublin, POLAND

SUMMARY

Some regularity conditions for general ARMA models have been introduced in [9]. We now discuss situations when those regularity conditions are fulfilled.

1. INTRODUCTION

Let $(\Omega,\mathcal{F},P)$ be a fixed probability space and let $\mathbb{Z}$ and $\mathbb{C}$ denote the sets of all integers and complex numbers, respectively.

The L^2-stationary process $\{y_k, k\in\mathbb{Z}\}$ is an ARMA(p,q) (autoregressive-moving-average of the order (p,q) process) if there exists a normed orthogonal valued process $\{e_k, k\in\mathbb{Z}\}$ and complex numbers $a_0, \dots, a_p, c_0, \dots, c_q$ such that

$$\sum_{i=0}^{p} a_i y_{k-i} = \sum_{i=0}^{q} c_i e_{k-i}, \quad k\in\mathbb{Z}. \tag{1.1}$$

In an ARMA(p,q) model (1.1) we assume in the sequel that $p=q=n$, as every ARMA(p,q) process admits a representation as an ARMA(n,n)model with $n=\max\{p,q\}$.

The model (1.1) is AR-regular (cf. [11]) or MA-regular if all roots of the polynomial $A(z^{-1}) = 1+a_1z^{-1}+ \dots +a_pz^{-p}$ or the polynomial $C(z^{-1}) = 1+c_1z^{-1}+ \dots +c_qz^{-q}$, respectively, belong to the unit circle. The model AR- and MA-rgular is called regular.

The following theorem is fundamental in the prediction

theory ([4], [11]).

Theorem 1.1. Suppose that the model (1.1) is AR-regular. Then there exists a unique solution $\{y_k, k\in\mathbb{Z}\}$ of (1.1) in the class of stationary processses, given by the L^2-convergent series

$$y_k = \sum_{j=0}^{\infty} b_j e_{k-j}, \quad k\in\mathbb{Z}, \tag{1.2}$$

where

$$b_j = \frac{1}{2\pi}\int_0^{2\pi} e^{ij\lambda} \frac{C(e^{-i\lambda})}{A(e^{-i\lambda})} d\lambda .$$

If additionally the model (1.1) is MA-regular, then the linear spaces spaned by processes $\{e_l, l\le k\}$ and $\{y_l, l\le k\}$ coincide for every $k\in\mathbb{Z}$ (i.e. $\{e_k, k\in\mathbb{Z}\}$ is an innovation process) and the prediction problem can be solved.

H. Niemi [11] has considered the ARMA equation (1.1) with a process $\{e_k, k\in\mathbb{Z}\}$ satisfying the following conditions:

$$Ee_k = 0, \quad E[e_k\bar{e}_l] = \sigma_k^2\delta_{k,l}, \tag{1.3}$$

where $0<m\le\sigma_k^2\le M<\infty$. The process $\{e_k, k\in\mathbb{Z}\}$ satisfying (1.3) is an UBLS process. The classical theory of stationary ARMA processes has been extended to UBLS ARMA processes in [11] as follows:

Theorem 1.2. Suppose that a process $\{e_k, k\in\mathbb{Z}\}$ satisfies the conditions (1.3). Then:

(i) There exists a unique, UBLS, purely nondeterministic, solution given by (1.2) of the ARMA equation (1.1) if the polynomial $A(z^{-1})$ has all roots in the unit circle.

(ii) If additionally all roots of the polynomial $C(z^{-1})$ belong to the unit circle, then the process $\{e_k, k\in\mathbb{Z}\}$ is an innovation process.

(iii) The all prediction formulas derived for stationary ARMA processes are valid in the UBLS case if the assumptions from (i) and (ii) hold true.

The ARMA model

$$\sum_{j=0}^{p} a_k(j)y_{k-j} = \sum_{j=0}^{q} c_k(j)e_{k-j}, \quad k\in\mathbb{Z}, \tag{1.5}$$

with time-dependent coefficients $a_j(k)$, $c_j(k)\in\mathbb{C}$, $j=1,\dots,n$,

$a_k(0) = c_k(0) = 1$, $k \in \mathbb{Z}$, and an orthogonal, stationary or UBLS process $\{e_k, k \in \mathbb{Z}\}$ were discussed in several papers (cf. [5], [12] and [13]).

We quote the basic results given by M. Hallin [5]. Let $G(t,s)$ and $H(t,s)$ be Green functions associated with

$$\sum_{j=0}^{p} a_k(j) y_{k-j} = 0 \quad \text{and} \quad \sum_{j=0}^{q} c_k(j) e_{k-j} = 0 ,$$

respectively.

Suppose that $E|e_k|^2 = \sigma^2 = \text{const}$. If

$$\sum_{s=-\infty}^{k} |G(k,s)| < \infty, \quad k \in \mathbb{Z}, \tag{1.6}$$

and if there exists a constant M such that

$$\sum_{j=0}^{q} |c_k(j)| < M, \quad k \in \mathbb{Z}, \tag{1.7}$$

then the process

$$y_k = e_k + \sum_{r=1}^{\infty} \Big[\sum_{j=1}^{\min(r,n)} [G(k,k+j-r) c_{k+j-r}(j) - G(k,k-r)] \Big] e_{k-r} \tag{1.8}$$

is a second-order, purely nondeterministic, mean zero process which is a solution of (1.5).

Moreover, if additionally

$$\sum_{s=-\infty}^{k} |H(k,s)| < \infty, \quad k \in \mathbb{Z}, \tag{1.9}$$

then the linear spaces generated by $\{e_l, l \le k\}$ and $\{y_l, l \le k\}$ coincide for every $k \in \mathbb{Z}$.

N.Singh and M.S. Peiris [13] have given simpler conditions than the above ones under which the model (1.5) is regular. Namely, it is stated that there exists a purely nondeterministic solution (1.8) of (1.5) if $E|e_k|^2 \le M < \infty$ and all zeros of the polynomials $A_k(z^{-1}) = 1 + a_k(1) z^{-1} + \ldots + a_k(p) z^{-p}$ and $C_k(z^{-1}) = 1 + c_k(1) z^{-1} + \ldots + c_k(q) z^{-q}$ lie in the region $|z| \le \lambda < 1$. We note that then the process (1.8) is UBLS.

In [7] has been considered the time invariant L^2-ARMA model with nonstationary noise. There are given sufficient conditions under which there exists a unique purely nondeterministic L^2-solution of the ARMA equation and the noise is the innovation process. Moreover, it is solved the prediction problem in this case. Similar problems for a L^1-ARMA process are investigated in [8].

2. A GENERAL ARMA MODEL

In this section we quote the basic definitions and properties of a general ARMA model which has been introduced in [9].

Let on a probability space $(\Omega, \mathcal{F}, \{\mathcal{F}_k, k \in \mathbb{Z}\}, P)$ with a filtration $\{\mathcal{F}_k, k \in \mathbb{Z}\}$ be defined a stochastic process $\{e_k, k \in \mathbb{Z}\}$ such that

$$\begin{aligned} &e_k \text{ is } \mathcal{F}_k\text{-measurable} \\ &E[e_k | \mathcal{F}_{k-1}] = 0 \text{ a.s..} \end{aligned} \qquad \text{(MD)}$$

Consider an ARMA(n,n) equation

$$\sum_{j=0}^{n} a_k(j) y_{k-j} = \sum_{j=0}^{n} c_k(j) e_{k-j} \quad \text{a.s.,} \quad k \in \mathbb{Z}, \qquad (2.1)$$

where the process $\{e_k, k \in \mathbb{Z}\}$ satisfies (MD) conditions. This equation can be rewritten in the following equivalent form

$$x_{k+1} = A(k) x_k + K(k) e_k \quad \text{a.s.,} \qquad (2.2)$$

$$y_k = B x_k + e_k, \quad k \in \mathbb{Z}, \qquad (2.3)$$

where $x_k = [x_k^{(1)}, \dots, x_k^{(n)}]'$,

$$x_k^{(n-j)} = \sum_{i=j+1}^{n} [c_i(k+j) e_{k-i+j} - a_i(k+j) y_{k-i+j}], \quad j = 0, 1, \dots, n-1,$$

$$A(k) = \begin{bmatrix} 0 & 0 \dots 0 & -a_n(k+n) \\ 1 & 0 \dots 0 & -a_{n-1}(k+n-1) \\ 0 & 1 \dots 0 & -a_{n-2}(k+n-2) \\ \vdots & & \vdots \\ 0 & 0 \dots 1 & -a_1(k+1) \end{bmatrix}, \quad K(k) = \begin{bmatrix} c_n(k+n) - a_n(k+n) \\ \vdots \\ c_1(k+1) - a_1(k+1) \end{bmatrix}$$

$B = [0 \ \dots \ 0 \ 1]$.

Definition 2.1. (i) The model (2.1) is said to be AR-regular iff the series

$$\sum_{j=0}^{\infty} \alpha_k(j) e_{k-j} \qquad (2.4)$$

converges a.s., where $\alpha_k(j) = \mathbf{B}\mathbf{A}(k,k-j+1)\mathbf{K}(k-j)$, $j \geq 1$, $\alpha_k(0)=1$, $\mathbf{A}(k+1,l) = \mathbf{A}(k)\mathbf{A}(k,l)$, $\mathbf{A}(l,l) = \mathbf{I}$, $l \in \mathbb{Z}$.

(ii) The model (2.1) is MA-regular iff the series

$$\sum_{j=0}^{\infty} \beta_k(j) y_{k-j} \qquad (2.5)$$

converges in probability, where $\beta_k(j) = \mathbf{B}\bar{\mathbf{A}}(k,k-j+1)\mathbf{K}(k-j)$, $j \geq 1$, $\beta_k(0)=1$, and $\bar{\mathbf{A}}(k) = \mathbf{A}(k) - \mathbf{K}(k)\mathbf{B}$.

(iii) The model (2.1) is regular iff it is both AR- and MA-regular.

Under the above regularity conditions it can be proved the following result.

Theorem 2.2. [9] Suppose that (2.4) holds true. Then:

(i) There exists a unique $\{\mathcal{F}_k, k \in \mathbb{Z}\}$ adapted and purely nondeterministic, with respect to the filtration $\{\mathcal{F}_k^e, k \in \mathbb{Z}\}$, solution

$$y_k = (\text{a.s.}) \sum_{j=0}^{\infty} \alpha_k(j) e_{k-j}, \quad k \in \mathbb{Z}, \qquad (2.6)$$

of the ARMA-equation (2.1) iff the model (2.1) is AR-regular,

(ii) $\mathcal{F}_k^e = \mathcal{F}_{-\infty}^e \cup \mathcal{F}_k^y$ if the model (2.1) is regular.

3. L^2-ARMA MODELS SATISFYING THE REGULARITY CONDITIONS

In this section we give sufficient conditions for AR-regularity and regularity of ARMA processes when $\{e_k, k \in \mathbb{Z}\} \subset L^2$.

Consider an ARMA process generated by the equation (2.1), where now we assume that $\{e_k, k \in \mathbb{Z}\}$ is a L^2-stochastic process satisfying the (MD)-conditions with $E[|e_k|^2] = \sigma_k^2 < \infty$, $k \in \mathbb{Z}$. Define

$$u_k(j) = -\sum_{i=1}^{n} a_{k+i}(i) u_k(j-i), \quad u_k(0)=1, \quad u_k(-i)=0, \quad i>0. \qquad (3.1)$$

It can be proved the following lemma

Lemma 3.1. Let $\{\mathbf{A}_j, j \in \mathbb{N}\}$ be a sequence of matrices having the form

$$A_j = \begin{bmatrix} 0 & 0 \ldots 0 & -a_n(j) \\ 1 & 0 \ldots 0 & -a_{n-1}(j) \\ \cdot & \cdot \quad \cdot & \cdot \\ 0 & 0 \ldots 1 & -a_1(j) \end{bmatrix},$$

Then

$$\|BA_1A_2\ldots A_m\| \leq n \max\{|u(j)|,\ m-n+1\leqslant j\leq m\},$$

where

$$u(j) = -\sum_{i=1}^{n} a_i(j)u(j-i),\ u(0)=1,\ u(-i)=0,\ i>0.$$

Proof. Let $M_j = A_1A_2\ldots A_j$. Then

$$BM_{j-1}A_j = B\begin{bmatrix} m^j_{11} & \ldots & m^j_{1n} \\ \vdots & & \vdots \\ m^j_{n1} & & m^j_{nn} \end{bmatrix} A_j = [\ m^j_{n2}\ \ldots m^j_{nn}\ u(j)\ ] =$$

$$= [\ m^j_{n2}\ \ldots\ m^j_{nn-1}\ u(j-1)\ u(j)\ ] = \ldots =$$

$$= [\ u(j-n+1)\ \ldots\ u(j-1)\ u(j)\ ]$$

and $\|BA_1A_2\ldots A_m\| \leq n \max\{|u(j)|,\ m-n+1\leqslant j\leq m\}$, which ends the proof.

We need the following result which is a consequence of the Rademacher-Menchoff theorem ([14], pp.20-21).

Lemma 3.2. Suppose that for every $k\in\mathbb{Z}$

$$\sum_{j=2}^{\infty} (\ln^2 j)|\alpha_k(j)|^2\sigma^2_{k-j} <\infty \tag{3.2}$$

Then the series $\sum_{j=0}^{\infty}\alpha_k(j)e_{k-j}$ converges both a.s and in L^2, and the model (2.1) is AR-regular.

Theorem 3.3. Assume that

$$\limsup_{k\to-\infty}\ \limsup_{m\to\infty}\ [\bar{u}_k(m)\|K(k-m)\|\sigma_{k-m}]^{\frac{1}{m}} =q < 1,$$

$$\|K(t)\| = \max\{\ |c_i(t+i) - a_i(t+i)|,\ 1\leq i\leq n\ \},\ t\in\mathbb{Z}, \tag{3.3}$$

$$\bar{u}_k(m) = \max\{|u_k(j)|,\ m-n+1\leqslant j\leq m\ \}.$$

Then the model (2.1) is AR-regular and the series (2.6) L^2-converges.

Proof. It is enough to verify (3.1). To do this we see that (3.3) implies that there exists $k_o \in \mathbb{Z}$ such that for every $k<k_o$

$$\limsup_{m\to\infty} (\alpha_k(m)\sigma_{k-m})^{\frac{2}{m}} < 1. \tag{3.4}$$

By Lemma 3.1 for every $k\in\mathbb{Z}$, $j>0$, we have

$$\|BA(k-1)A(k-2)\cdots A(k-m)\| \le n \max\{|u_k(j)|,\ m-n<j\le m\}.$$

Hence

$$|\alpha_k(m)\sigma_{k-m}|^2 \le n^2\|BA(k-1)A(k-2)\cdots A(k-m)\|^2\|K(k-m)\|^2\sigma^2_{k-m} \le$$

$$\le (nu_k(m)\|K(k-m)\|)^2\sigma^2_{k-m} .$$

Therefore, by (3.3), for $k<k_o$

$$\limsup_{m\to\infty} (\alpha_k(m)\sigma_{k-m})^{\frac{2}{m}} \le \limsup_{m\to\infty} (n\bar{u}_k(m)\|K(k-m)\|\ )^{\frac{1}{m}}\ \sigma^{\frac{1}{m}}_{k-m} < 1$$

which implies (3.2) for $k<k_o$. The convergence of series (3.2) for $k\ge k_o$ follows from (2.1) (cf. [9]).

Now let us denote $y_k = (L^2)\sum_{j=0}^{\infty}\alpha_k(j)e_{k-j}$ and $\delta_k^2 = E|y_k|^2$.

Theorem 3.4. Supose that the model (2.1) is AR-regular and $\{y_k,\ k\in\mathbb{Z}\}\subset L^2$. If

$$\limsup_{k\to-\infty}\ \limsup_{m\to\infty} [\bar{v}_k(-m)\|K(k-m)\|\delta_{k-m}]^{\frac{1}{m}} < 1, \tag{3.5}$$

where $\bar{v}_k(m) = \max\{|v_k(j)|,\ m-n+1\le j\le m\}$,

$$v_k(j) = -\sum_{i=1}^{n} c_{k+i}(i)v_k(j-i),\ v_k(0)=1,\ v_k(-i)=0,\ i>0.$$

then the series (2.5) L^2-converges and the model (2.1) is regular.

Proof. Note that the matrix $\bar{A}(k) = A(k) - K(k)B$ is exactly the $A(k)$ matrix with $a_{k+i}(i)$ replacing by $c_{k+i}(i)$. Then the proof is similar to the previous one.

We give now the alternative sufficient conditions for AR-regularity and regularity of ARMA processes.

Let

$$\tilde{u}_k(j) = -\sum_{i=1}^{n} a_k(i)\tilde{u}_k(j-i),\ \tilde{u}_k(0)=1,\ \tilde{u}_k(-i)=0,\ i>0,$$

$$\tilde{v}_k(j) = -\sum_{i=1}^{n} c_k(i)\tilde{v}_k(j-i),\ \tilde{v}_k(0)=1,\ \tilde{v}_k(-i)=0,\ i>0$$

and $\bar{\bar{u}}_k(m) = \max\{|\tilde{u}_k(j)|,\ m-n+1\leqslant j\leq m\}$, $\bar{\bar{v}}_k(m) = \max\{|\tilde{v}_k(j)|,\ m-n+1\leqslant j\leq m\}$.

Theorem 5.5. If for some $k_o\in\mathbb{Z}$ and all $k<k_o$

$$\limsup_{m\to\infty} [\bar{\bar{u}}_k(-m)\|\tilde{C}(k-m)\|\sigma'_{k-m}]^{\frac{1}{m}} < 1, \tag{3.6}$$

where $\sigma'_k = \max\{\sigma_k, \ldots, \sigma_{k-n}\}$, then the model (2.1) is AR-regular and the series (2.6) L^2-converges.

If additionally for all $k<k_o$

$$\limsup_{m\to\infty} [\bar{\bar{v}}_k(-m)\|\tilde{A}(k-m)\|\delta'_{k-m}]^{\frac{1}{m}} < 1, \tag{3.7}$$

where $\delta'_k = \max\{\delta_k, \ldots, \delta_{k-n}\}$, then $\mathcal{F}^e_{-\infty}\cup\mathcal{F}^y_k = \mathcal{F}^e_k$.

Proof. Rewrite the ARMA equation (2.1) as follows (cf.[6])

$$\tilde{y}_{k+1} = \tilde{A}(k)\tilde{y}_k+\tilde{K}(k)\tilde{e}_k \quad \text{a.s.}, \tag{3.8}$$

$$y_k = \tilde{B}\tilde{y}_k+e_k, \quad k\in\mathbb{Z}, \tag{3.9}$$

where $\tilde{y}_k = [y_k, \ldots, y_{k-n+1}]'$, $\tilde{e}_k = [e_k, \ldots, e_{k-n+1}]'$,

$$\tilde{A}(k) = \begin{bmatrix} -a_1(k) & -a_2(k) & \ldots & -a_n(k) \\ 1 & 0 & \ldots & 0 \\ 0 & 1 & & 0 \\ \vdots & & & \\ 0 & 0 & & 1 \end{bmatrix}, \quad \tilde{C}(k) = \begin{bmatrix} -c_1(k) & -c_2(k) & \ldots & -c_n(k) \\ 1 & 0 & \ldots & 0 \\ 0 & 1 & & 0 \\ \vdots & & & \\ 0 & 0 & & 1 \end{bmatrix},$$

$B=[1\ 0\ \ldots\ 0]$.

Hence

$$\tilde{y}_k = \tilde{A}(k,k-m)\tilde{y}_m + \sum_{j=1}^{k-m} \tilde{A}(k,k-j+1)\tilde{C}(k-j)\tilde{e}_{k-j}, \quad m<k. \tag{3.10}$$

Suppose that (3.6) holds. Then the Cauchy criterion implies that the series

$$\sum_{j=1}^{k-m} \tilde{A}(k,k-j+1)\tilde{C}(k-j)\tilde{e}_{k-j}$$

absolutely L^2-converges to L^2-vector

$$\tilde{y}^*_k =(L^2) \sum_{j=1}^{\infty} \tilde{A}(k,k-j+1)\tilde{C}(k-j)\tilde{e}_{k-j}$$

being a L^2-solution of the recurrence equation (3.8).

We see that the process $\{y^*_k, k\in\mathbb{Z}\}$, where

$$y^*_k =(L^2) \sum_{j=1}^{\infty} \tilde{B}\tilde{A}(k,k-j+1)\tilde{C}(k-j)\tilde{e}_{k-j} + e_k, \; k\in\mathbb{Z},$$

which can be rewritten as

$$y^*_k =(L^2) \sum_{j=0}^{\infty} \alpha^*_k(j)e_{k-j} \;, \tag{3.11}$$

is a purely nondeterministic process being a L^2-solution of the ARMA equation (2.1). But from the Theorem 2.2.(i) (see [9]) folows that this solution is unique so $\alpha^*_k(j) = \alpha_k(j)$. Then the series (2.6) a.s. and L^2-converges which proves that the model (2.1) is AR-regular.

To prove the second part of Theorem 3.5 we note that (3.8) and (3.9) implies that

$$e_k = \tilde{B}\tilde{C}(k,k-m)\tilde{e}_m + \sum_{j=1}^{k-m} \tilde{B}\tilde{C}(k,k-j+1)\tilde{A}(k-j)\tilde{y}_{k-j} \;, \; m<k.$$

and the series

$$\sum_{j=1}^{\infty} \tilde{C}(k,k-j+1)\tilde{A}(k-j)\tilde{y}_{k-j}$$

absolutelly L^2-converges if (3.7) holds. Then we have

$$e_k = (L^2)\lim_{m\to\infty} \tilde{B}\tilde{C}(k,k-m)\tilde{e}_m +(L^2) \sum_{j=1}^{\infty} \tilde{B}\tilde{C}(k,k-j+1)\tilde{A}(k-j)\tilde{y}_{k-j} \;,$$

which implies that $\mathcal{F}^e_k \subseteq \mathcal{F}^e_{-\infty} \cup \mathcal{F}^y_k$, $k\in\mathbb{Z}$, as the first limit in above is $\mathcal{F}_{-\infty}$-measurable.

The inclusion $\mathcal{F}^y_k \subseteq \mathcal{F}^e_k$ follows from the equation (3.11) which ends the proof.

Now, we consider the asymptotically time-invariant system. Suppose that $\lim_{k\to-\infty} A(k) = A_{-\infty}$ i.e. there exists a matrix A_{∞} such that $\lim_{k\to-\infty} \|A(k) - A_{-\infty}\| = 0$. We see that the matrix $A_{-\infty}$ has the form

$$A_{-\infty} = \begin{bmatrix} 0 & 0 \dots 0 & -a_n \\ 1 & 0 \dots 0 & -a_{n-1} \\ \cdot & \cdot \quad \cdot & \cdot \\ 0 & 0 \dots 1 & -a_1 \end{bmatrix}$$

and $a_k(i) \to a_i$ as $k \to -\infty$, $i=1,2,\dots,n$.

Let $r(A_{-\infty}) = \max\{|z_i| : A_{-\infty}(z^{-1}) = 0,\ i=1,2,\dots,n\}$.
We proove

Theorem 3.5. Let

$$\limsup_{k \to -\infty} \sum_{j=0}^{\infty} \ln^2 j (r(A_{-\infty})+\varepsilon)^{2j} \beta_{k-j}^2 \sigma_{k-j}^2 < \infty \qquad (3.12)$$

for some $\varepsilon > 0$. Then the model (2.1) is AR-regular and the series (2.6) L^2-converges.
If additionally there exists the limit $\lim_{k \to -\infty} \bar{A}(k) = \bar{A}_{-\infty}$, where $\bar{A}(k) = A(k) - K(k)B$, such that for some $\varepsilon' > 0$

$$\limsup_{k \to -\infty} \sum_{j=0}^{\infty} (r(C_{-\infty})+\varepsilon')^j \beta_{k-j} \delta_{k-j} < \infty \qquad (3.13)$$

with

$$\bar{A}_{-\infty} = \begin{bmatrix} 0 & 0 \dots 0 & -c_n \\ 1 & 0 \dots 0 & -c_{n-1} \\ \cdot & \cdot \quad \cdot & \cdot \\ 0 & 0 \dots 1 & -c_1 \end{bmatrix},$$

$r(C_{-\infty}) = \max\{|z_i| : C_{-\infty}(z^{-1}) = 0,\ i=1,2,\dots,n\}$ and

$$\delta_k^2 = \sum_{j=0}^{\infty} (r(A_{-\infty})+\varepsilon)^{2j} \beta_{k-j}^2 \sigma_{k-j}^2$$, then the model (2.1) is regular.

Proof. Suppose that (3.12) holds. Then there exists $k_0 \in \mathbb{Z}$ such that for every $k < k_0$

$$\sum_{j=0}^{\infty} \ln^2 j (r(A_{-\infty})+\varepsilon)^{2j} \beta_{k-j}^2 \sigma_{k-j}^2 < \infty .$$

Since $a_k(i) \to a_i$, $i=1,2,\dots,n$, then $A(k-1)\dots A(k-j) \xrightarrow[k \to -\infty]{} A_{-\infty}^j$, $j \in \mathbb{N}$, and for every $\varepsilon > 0$ there exists $k_1 \in \mathbb{Z}$ such that for every $k < k_1$

$$\|A(k-1)\dots A(k-j)\| \le \|A_{-\infty}^j\| + \frac{\varepsilon}{2} .$$

Moreover, there exists $k_2 \in \mathbb{Z}$ such that for $k<k_2$

$$\|A^j_{-\infty}\| \le \rho^j(A_{-\infty}) + \frac{\varepsilon}{2} = r^j(A_{-\infty}) + \frac{\varepsilon}{2} ,$$

where $\rho(A_{-\infty})$ denote the spectral radius of the matrix $A_{-\infty}$ (cf. [10], pp.15). Now, for $k<\min\{k_1, k_2\} = k_0$ we have

$$\sum_{j=0}^{\infty} \ln^2 j \|A(k-1)A(k-2)\dots A(k-j)\|^2 \beta^2_{k-j}\sigma^2_{k-j} \ll$$

$$\ll \sum_{j=0}^{\infty} \ln^2 j (r(A_{-\infty})+\varepsilon)^{2j}\beta^2_{k-j}\sigma^2_{k-j} < \infty$$

and by the Rademacher-Menchoff Lemma the model (2.1) is AR-regular for $k<k_0$, then, as well, for $k\ge k_0$, by the ARMA-equation (2.1) (cf. [9]).

To prove the second statement of Theorem note that there exists $k_0 \in \mathbb{Z}$ such that for $k<k_0$ we have

$$\sum_{j=0}^{\infty} \|\bar{A}(k-1)\bar{A}(k-2)\dots\bar{A}(k-j)\| \|K(k-j)\| \|y_{k-j}\|_{L^2} \ll$$

$$\sum_{j=0}^{\infty} \|\bar{A}(k-1)\bar{A}(k-2)\dots\bar{A}(k-j)\|\beta_{k-j}\delta_{k-j} \ll \sum_{j=0}^{\infty} (r(C_{-\infty})+\varepsilon')^j\beta_{k-j}\delta_{k-j}<\infty,$$

since

$$\|y_k\|^2_{L^2} \le \sum_{j=0}^{\infty} (r(A_{-\infty})+\varepsilon)^{2j}\beta^2_{k-j}\sigma^2_{k-j} = \delta^2_k .$$

This prooves that the model is regular for $k<k_0$ and in a consequence it is regular for all $k\in\mathbb{Z}$.

Crollary 3.7. Suppose that there exists the limit $\lim_{k\to-\infty} A(k) = A_{-\infty}$. Then the model (2.1) is AR-regular if

$$\limsup_{k\to-\infty} \limsup_{j\to\infty} (\beta_{k-j}\sigma_{k-j})^{\frac{1}{j}} < (r(A_{-\infty})+\varepsilon)^{-1}$$

for some $\varepsilon>0$.

Proof. This follows immediately from the Cauchy criterion.

Example 3.8. (c.f. [13]) Consider the L^2 ARMA model (2.1) satisfying the following regularity conditions from [13]

(i) $\sigma^2_k \le M^2<\infty$

(ii) $\max\{r(\tilde{A}_k), r(\tilde{C}_k)\}<\lambda<1$, $k\in\mathbb{Z}$.

The following simple example shows that the assumption

(ii) can not be replaced by the weaker one

(ii') $\max \{r(\tilde{A}_k), r(\tilde{C}_k)\}<1, k\in\mathbb{Z}$.

Consider the ARMA equation

$$y_k = a(k)y_{k-1} + e_k, \quad k\in\mathbb{Z}, \tag{3.14}$$

where $a(k) = (|k|+0,5)^{0,5}(|k|+1)^{-0,5}$, $k\in\mathbb{Z}$ and $\{e_k, k\in\mathbb{Z}\}$ is an independent-valued L^2-process with zero means and $E|e_k|^2 = 1$.Then the Raabe criterion shows that the model (2.1) is not AR-regular, i.e. it does not exist a purely nondeterministic, and $\{\mathcal{F}_k, k\in\mathbb{Z}\}$ adapted process satisfying the equation (2.1). Indeed, for $k=-2$, the series

$$\sum_{j=1}^{\infty} a(-2)\dots a(-2-j+1)e_{-2-j}$$

a.s. diverges, since the series

$$\sum_{j=1}^{\infty} a^2(-2)\dots a^2(-2-j+1) := \sum_{j=1}^{\infty} \varphi_j$$

diverges which follows from the Raabe criterion:

$$[j(\frac{\varphi_{j+1}}{\varphi_j} -1)+1] \ln j +2 = [j(\frac{j-0,5}{j} -1)+1] \ln j +2 =$$

$$=(j\frac{-0,5}{j} +1)\ln j +2 = 0,5\ln j +2 \to\infty.$$

REFERENCES

[1] B.D.O. Anderson, J.B. Moore (1979). *Optimal Filtering*, Prentice Hall, Englewood Cliffs, NJ.

[2] R.Azencot, D. Dacunha-Castelle (1986). *Series of Irregular Obsevations*, Springer-Verlag, NY.

[3] G.P.E. Box, G.M. Jenkins (1976). *Time Series Analysis. Forecasting and Control*, Holden-Day Inc., California.

[4] C.W.J. Granger, P. Newbold (1977). *Forecasting Economic Time Series*, Academic Press, NY.

[5] M. Hallin (1978). Mixed Autoregresive-Moving Average Multivariate Processes with Time-Dependent Coefficients, *J. Multivar. Anal.*, 8, pp. 567-572.

[6] A. Kowalski, D. Szynal (1988). On State Estimation and

Control in Discrete-Time Systems on Linear Type, *Automatica*, vol **23**.

[7] A. Kowalski and D. Szynal (1988). On Optimal Prediction in Nonstationary ARMA Models, *Scand. J. Stat.*, vol.**15**, No 2, pp. 111-116.

[8] A. Kowalski, D. Szynal (1989). A Simple Characterization of Optimal Predictors for L^1-ARMA Processes, *System and Cntrol Letters*, vol. **12**.

[9] A. Kowalski, D. Szynal (1988). An Optimal Prediction in General ARMA Model, submited to *Jourmal of Multivariate Analysis*.

[10] M.A. Krasnoselski (1969). *An Approximative Solutions of Operator Equations*, Nauka, Moscov.

[11] H. Niemi (1983). On the Effects of a Nonstationary Noise on ARMA Models, *Scand. J. Statist.*, 10, pp. 11-17.

[12] M. S. Peiris (1986). On Prediction in Time Dependent ARMA models, Commun. Statist.-Theory Meth., 15(12), pp. 3659-3668.

[13] N. Singh, M.S. Peiris (1987). A Note of the Properties of Some Nonstationary ARMA Processes, *Stochastic Processes and Their Applications*, 24, pp. 151-155.

[14] W.F. Stout (1974). *Almost Sure Convergence*, Academic Press, NY.

ON LIMIT POINTS OF A SEQUENCE OF WEAK SOLUTIONS OF ONE-DIMENSIONAL STOCHASTIC DIFFERENTIAL EQUATIONS

Andrzej Rozkosz and Leszek Słomiński
Institute of Mathematics, Nicholas Copernicus University
ul. Chopina 12/18, 87-100 Toruń, Poland.

1. INTRODUCTION.

Let $\{X^n\}_{n\in\mathbb{N}}$ be a sequence of solutions to stochastic differential equations (s.d.e.) of the form :

$$(1)\qquad X_t^n = \int_0^t b_n(X_{s-}^n)dM_s^n + \int_0^t a_n(X_{s-}^n)d[M^n]_s , \qquad t\in\mathbb{R}^+,$$

where for every $n\in\mathbb{N}$ M^n is a local martingale, $[M^n]$ is the process of quadratic variation of M^n and $a_n = a_n(x)$, $b_n = b_n(x)$ are measurable functions. Denote by $L(\{X^n\})$ the set of limits of convergent in distribution subsequences of $\{X^n\}_{n\in\mathbb{N}}$. Our purpose is to collect conditions under which each element of $L(\{X^n\})$ is also a solution of the s.d.e. of the form (1) and, secondly, to characterize a, b, M in terms of a_n, b_n, M^n.

In the present paper we generalize results contained in [11]. We have considered there the above problem under the additional assumption, that every limit equation has a weak solution unique in the sense of law and for equations driven by local martingales with continuous trajectories.

Similar problems have been treated in [1], [6], [7] and [9].

We give brief proofs only. For more details we refer to [11].

Without loss of generality we can assume, that all processes appearing in the sequel are defined on the same probability space $(\Omega, \mathcal{F}, P)$ and have trajectories in $C(\mathbb{R})$ or $D(\mathbb{R})$, where $C(\mathbb{R})$ is the space of continuous mappings x, $x: \mathbb{R}^+ \longrightarrow \mathbb{R}$, $x(0)= 0$ with the topology of uniform convergence on compact subsets of $\mathbb{R}^+$, $D(\mathbb{R})$ is the space of cadlag mappings x, $x: \mathbb{R}^+ \longrightarrow \mathbb{R}$, $x(0)= 0$ with the Skorokhod J_1 topology.

2. STOCHASTIC DIFFERENTIAL EQUATIONS DRIVEN BY CONTINUOUS LOCAL MARTINGALES.

Let $\{M^n\}_{n\in\mathbb{N}}$ be a sequence of local martingales with continuous trajectories and let $\{a_n\}_{n\in\mathbb{N}}$, $\{b_n\}_{n\in\mathbb{N}}$ be sequences of measurable functions satisfying the following conditions :

$$(H_1)\quad \begin{cases} a_n^2(x) + b_n^2(x) \leq K(1+x^2), & K>0,\ n\in\mathbb{N},\ x\in\mathbb{R}, \\ \{b_n^{-2}\}_{n\in\mathbb{N}} & \text{is uniformly integrable on all compact subsets of } \mathbb{R}, \\ b_n^2(x) > 0, & n\in\mathbb{N},\ x\in\mathbb{R}. \end{cases}$$

It is shown in [11] (Proposition 2) that, under above conditions there exist weak solutions X^n of the equation (1) . In this section we characterize the set $L(\{X^n\})$.

THEOREM 1. <u>Assume</u> (H_1) <u>holds.</u>

(i) $\{X^n\}_{n\in\mathbb{N}}$ <u>is tight in</u> $C(\mathbb{R})$ <u>iff</u> $\{[M^n]\}_{n\in\mathbb{N}}$ <u>is tight in</u> $C(\mathbb{R})$.

(ii) <u>Let</u> $X\in L(\{X^n\})$. <u>Then there exists a continuous local martingale</u> M <u>such, that</u> $[M]\in L(\{[M^n]\})$ <u>and the following equation holds</u>

$$(2)\qquad X_t = \int_0^t b(X_s)\,dM_s + \int_0^t a(X_s)\,d[M]_s, \qquad t\in\mathbb{R}^+,$$

<u>where</u> ab^{-2}, b^{-2} <u>are limits points of some subsequences of</u> $\{a_nb_n^{-2}\}_{n\in\mathbb{N}}$ <u>and</u> $\{b_n^{-2}\}_{n\in\mathbb{N}}$ <u>respectively in the topology of weak convergence in</u> $\mathbb{L}^1_{loc}$.

REMARK 1. Let $a_nb_n^{-2}\longrightarrow f$, $b_n^{-2}\longrightarrow h$ weakly in $\mathbb{L}^1_{loc}$. Then $b^2=h^{-1}$ and $a=fh^{-1}$ are defined modulo Lebesgue measure only. It is proved in [11] that there exist versions of b^2, a which satisfy the hypothesis :

$$(H_1')\quad \begin{cases} a^2(x) + b^2(x) \leq K(1+x^2), & K>0,\ x\in\mathbb{R}, \\ b^{-2} & \text{is uniformly integrable on all compact subsets of } \mathbb{R}, \\ b^2(x) > 0, & x\in\mathbb{R}. \end{cases}$$

In what follows we will consider just these versions. □

PROOF of Theorem 1. (i) Sufficiency. It is an easy consequence of Aldous' tightness criterion and Rebolledo's inequality (see e.g. [12]). Necessity follows from (ii).

(ii) By (H_1) there exists a subsequence $(n')\subset(n)$ such that $X^n \xrightarrow[\mathcal{D}]{} X$ and $a_n b_n^{-2} \longrightarrow ab^{-2}$, $b_n^{-2} \longrightarrow b^{-2}$ weakly in $\mathbb{L}^1_{loc}$, where ab^{-2}, b^{-2} satisfy (H_1') (for simlicity we will write n instead of n' in the sequel).

Let us fix $n\in\mathbb{N}$. Using Itô's formula to the function

$$\Phi_n(x) = \int_0^x \int_0^y (a_n b_n^{-2} - ab^{-2})(u)du\,dy$$

we get

$$\Phi_n(X_t^n) = \int_0^t \Phi_n(X_s^n)dX_s^n + \frac{1}{2}\int_0^t (a_n b_n^{-2} - ab^{-2})(X_s^n)d[X^n]_s \quad , \quad t\in\mathbb{R}^+.$$

Since $\Phi_n \rightrightarrows 0$, $\Phi_n' \rightrightarrows 0$ (uniformly on compact subsets of $\mathbb{R}$) then

$$(3) \qquad \sup_{t\leq q} \left| \int_0^t (a_n b_n^{-2} - ab^{-2})(X_s^n)d[X^n]_s \right| \xrightarrow[P]{} 0 \;, \quad q\in\mathbb{R}^+ .$$

Due to Krylov's estimates (see [10] , [11])

$$(4) \qquad \left(X^n , \int_0^{\cdot} ab^{-2}(X_s^n)d[X^n]_s\right) \xrightarrow[\mathcal{D}]{} \left(X , \int_0^{\cdot} ab^{-2}(X_s)\,d[X]_s\right) \text{ in } C(\mathbb{R}^2).$$

We have

$[M^n]_t = \int_0^t b_n^{-2}(X_s^n)d[X^n]_s$, hence from (3) , (4) we get the convergence

$$(5) \qquad \left(X^n , \int_0^{\cdot} a_n(X_s^n)\,d[M^n]_s\right) \xrightarrow[\mathcal{D}]{} \left(X , \int_0^{\cdot} ab^{-2}(X_s)\,d[X]_s\right) \text{ in } C(\mathbb{R}^2).$$

In particular, we obtain

$$X_\cdot^{nc} = \int_0^{\cdot} b_n(X_s^n)\,dM_s^n = X_\cdot^n - \int_0^{\cdot} a_n(X_s^n)d[M^n]_s$$

$$\xrightarrow[\mathcal{D}]{} X_\cdot - \int_0^{\cdot} ab^{-2}(X_s)d[X]_s \quad \text{in} \quad C(\mathbb{R}) .$$

Because $\{X^{nc}\}_{n\in\mathbb{N}}$ is a sequence of continuous local martingales, then its weak limit is a continuous local martingale, too. Therefore the continuous martingale part X^c of the semimartingale X is equal $X_t^c = X_t - \int_0^t ab^{-2}(X_s)\, d[X]_s$, and by (5) we get

$$(6) \qquad (X^n, X^{nc}) \xrightarrow[\mathcal{D}]{} (X, X^c) \quad \text{in} \quad C(\mathbb{R}^2).$$

We are ready to define the martingale M.

Let $M_t = \int_0^t b^{-1}(X_s)dX_s^c$. We have $X_t^c = \int_0^t b(X_s)dM_s$ and

$$[X]_t = [X^c]_t = \int_0^t b^2(X_s)\, d[M]_s ,$$

hence

$$X_t = X_t - \int_0^t ab^{-2}(X_s)\, d[X]_s + \int_0^t ab^{-2}(X_s)\, d[X]_s =$$

$$= X_t^c + \int_0^t ab^{-2}(X_s)\, d[X]_s =$$

$$= \int_0^t b(X_s)\, dM_s + \int_0^t a(X_s)\, d[M]_s , \quad t \in \mathbb{R}^+ .$$

Now we show that $[M^n] \xrightarrow[\mathcal{D}]{} [M]$.
By Krylov's estimates

$$\int_0^{\cdot} b^{-2}(X_s^n)\, d[X^n]_s \xrightarrow[\mathcal{D}]{} \int_0^{\cdot} b^{-2}(X_s)\, d[X]_s = [M]. \quad \text{in} \quad C(\mathbb{R}).$$

On the other hand, applying Itô's formula to the function

$\Phi_n(x) = \int_0^x \int_0^y (b_n^{-2} - b^{-2})(u)\, du\, dy$ we get analogously as in (4)

$\sup_{t \leq q} \left| \int_0^t (b_n^{-2} - b^{-2})(X_s^n)\, d[X^n]_s \right| \xrightarrow[P]{} 0, \quad q \in \mathbb{R}^+$ and the proof is finished. □

REMARK 2. Directly from the proof ((6) and Krylov's estimates) we see that, if $X^n \xrightarrow[\mathcal{D}]{} X$, then the martingale M appearing in Theorem 1 is defined uniquely as a weak limit of the sequence

$$\left\{ \int_0^{\cdot} b^{-1}(X_s^n)\, dX_s^{nc} \right\}_{n \in \mathbb{N}} . \ \square$$

COROLLARY 1. ([11]). (i) Assume that the deterministic process $I(t) = t$ is the only possible limit point of $\{[M^n]\}_{n \in \mathbb{N}}$ or equivalently $M^n \xrightarrow[\mathcal{D}]{} W$, where W is a standard Wiener process

Then the following two conditions are equivalent:

(7) $X^n \xrightarrow[\mathfrak{D}]{} X$ in $C(\mathbb{R})$,

(8) $a_n b_n^{-2} \longrightarrow ab^{-2}$ weakly in $\mathbb{L}^1_{loc}$,

$b_n^{-2} \longrightarrow b^{-2}$ weakly in $\mathbb{L}^1_{loc}$.

Moreover, if (7) or (8) holds, then X is the unique in the sense of law weak solution of the equation

(9) $$X_t = \int_0^t b(X_s)dW_s + \int_0^t a(X_s)ds , \qquad t \in \mathbb{R}^+ .$$

(ii) Conversly, assume (7) , (8) and let X be a weak solution of (9) . Then $M^n \xrightarrow[\mathfrak{D}]{} W$ in $C(\mathbb{R})$.

PROOF. Firstly let us note, that uniqueness of solutions of (9) for b^2 , a satisfying condition (H_1') easy follows from Engelbert-Schmidt's theorem [2] (see [11] , Proposition 1).

(i) (8) ⇒ (7) . Let $X^{n'} \xrightarrow[\mathfrak{D}]{} X'$ for some subsequence $(n') \subset (n)$. By Theorem 1 X' satisfies the equation (9) and because of uniqueness $\mathcal{L}(X) = \mathcal{L}(X')$.

(7) ⇒ (8). It is a consequence of a generalized version of Kulinič results [6] , (see [11] , Proposition 5).

(ii) From the proof of Theorem 1 we have

$$[M^n]. \xrightarrow[\mathfrak{D}]{} \int_0^{\cdot} b^{-2}(X_s)\,[X^c]_s = I \qquad \text{in} \quad C(\mathbb{R}) .$$

□

THEOREM 2. Assume (H_1) and additionaly that $\{b_n\}_{n \in \mathbb{N}}$ is relatively compact in $\mathbb{L}^1_{loc}$. Let $X \in L(\{X^n\})$. Then there exists a continuous local martingale M such that $M \in L(\{M^n\})$ and the equation (2) holds, where a is a limit point of some subsequence of $\{a_n\}_{n \in \mathbb{N}}$ in the topology of weak convergence in $\mathbb{L}^1_{loc}$ and b is a limit point of some subsequence of $\{b_n\}_{n \in \mathbb{N}}$ in the topology of $\mathbb{L}^1_{loc}$.

PROOF. Suppose $X^n \xrightarrow[\mathfrak{D}]{} X$, $a_n \longrightarrow a$ weakly in $\mathbb{L}^1_{loc}$, $b_n \longrightarrow b$ in $\mathbb{L}^1_{loc}$. Analogously as in the proof of Theorem 1 we define a local martingale M such that the equation (2) is satisfied.

For $M^n_t = \int_0^t b_n^{-1}(X^n_s)dX^{nc}_s$, therefore , in view of Remark 2

it is sufficient to show

$$\sup_{t \leq q} \left| \int_0^t (b_n^{-1} - b^{-1})(X_s^n) dX_s^{nc} \right| \xrightarrow[P]{} 0 , \qquad q \in \mathbb{R}^+.$$

But by (H_1) $b_n^{-1} \longrightarrow b^{-1}$ in $\mathbb{L}^2_{loc}$, and so it is a consequence of Krylov's estimates (see [11]). □

COROLLARY 2. <u>Assume that the equation (2) has a unique weak solution,</u> $a_n \longrightarrow a$ <u>weakly in</u> $\mathbb{L}^1_{loc}$, $b_n \longrightarrow b$ <u>in</u> $\mathbb{L}^1_{loc}$. <u>Then the following conditions are equivalent:</u>

(10) ≡ (7) $\qquad X^n \xrightarrow[\mathcal{D}]{} X$ <u>in</u> $C(\mathbb{R})$,

(11) $\qquad M^n \xrightarrow[\mathcal{D}]{} M$ <u>in</u> $C(\mathbb{R})$.

<u>Moreover, if (10) or (11) holds, then X is a solution of (2).</u>

PROOF. (11) ⇒ (10). We follow the proof of Corollary 1 (i), (8) ⇒ (7) . (10) ⇒ (11) . From the proof of Theorem 2

$$M^n_\cdot \xrightarrow[\mathcal{D}]{} M_\cdot = \int_0^\cdot b^{-1}(X_s)\, dX_s^c \qquad \text{in} \quad C(\mathbb{R}).$$
□

3. <u>STOCHASTIC DIFFERENTIAL EQUATIONS DRIVEN BY LOCAL MARTINGALES WITH JUMPS.</u>

Let us consider s.d.e. of the form (1) , where for each $n \in \mathbb{N}$ M^n is a local martingale with trajectories in $D(\mathbb{R})$, a_n , b_n are continuous and have at most linear increase at infinity. It is known, that in such a case there exist weak solutions X^n of s.d.e. (1) (see [4]). To characterize $L(\{X^n\})$ we will assume additionaly:

(H_2) $\qquad E \sup_{t \leq q} |\Delta M_t^n| \longrightarrow 0$ as $n \to +\infty$, $q \in \mathbb{R}^+$,

(H_3) $\qquad (H_1)$ with $b_n^2(x) \geq k_N > 0$ for $x \in [-N , N]$, $n \in \mathbb{N}$ instead of the weaker $\{b_n^{-2}\}_{n \in \mathbb{N}}$ locally uniformly integrable.

(H_4) $\qquad \{a_n\}_{n \in \mathbb{N}}$, $\{b_n^2\}_{n \in \mathbb{N}}$ are equicontinuous on compact subsets of $\mathbb{R}$.

We will say that a sequence of processes $\{Y^n\}_{n \in \mathbb{N}}$ with trajectoris in $D(\mathbb{R})$ is C-tight, if $\{Y^n\}_{n \in \mathbb{N}}$ is tight in $D(\mathbb{R})$ and each its limit point has continuous trajectories.

THEOREM 3. Assume (H_2), (H_3), (H_4).

(i) $\{X^n\}_{n\in\mathbb{N}}$ is C-tight iff $\{[M^n]\}_{n\in\mathbb{N}}$ is C-tight.

(ii) Let $X \in L(\{X^n\})$. Then there exists a local martingale M with continuous trajectories such that $[M] \in L(\{[M^n]\})$ and s.d.e. (2) is satisfied, where a, b^2 are limit points of some subsequences of $\{a_n\}_{n\in\mathbb{N}}$, $\{b_n^2\}_{n\in\mathbb{N}}$ respectively in the topology of uniform convergence on compact sets.

PROOF. (i) Exactly the same as for Theorem 1.

(ii) Let, for some subsequence if needed, $X^n \xrightarrow[\mathcal{D}]{} X$ and $a_n \rightrightarrows a$, $b_n^2 \rightrightarrows b^2$ uniformly on compact subsets of $\mathbb{R}$.

Note, that by (H_2) X has continuous trajectories.

Define $$\bar{X}_t^n = \int_0^t b_n(X_{s-}^n)dM_s^n \quad , \quad n\in\mathbb{N}, \quad t\in\mathbb{R}^+ .$$

By (H_2) $\{X^n\}_{n\in\mathbb{N}}$ satisfies the condition U.T. (see [3], [5]), hence $(X^n, [X^n]) \xrightarrow[\mathcal{D}]{} (X, [X])$ in $D(\mathbb{R}^2)$. From (H_2) we have also $\sup_{t\le q} \left| [X^n]_t - [\bar{X}^n]_t \right| \xrightarrow[P]{} 0$, $q\in\mathbb{R}^+$. Therefore, using continuity of stochastic integral (see [5]) we get

$$\left(X^n , \int_0^{\cdot} a_n(X_{s-}^n)d[M^n]_s\right) = \left(X^n , \int_0^{\cdot} a_n b_n^{-2}(X_{s-}^n)d[\bar{X}^n]_s\right)$$

$$\xrightarrow[\mathcal{D}]{} \left(X , \int_0^{\cdot} ab^{-2}(X_s)d[X]_s\right) \quad \text{in} \quad D(\mathbb{R}^2),$$

and consequently

$$\bar{X}_{\cdot}^n = X_{\cdot}^n - \int_0^{\cdot} a_n(X_{s-}^n)d[M^n]_s \xrightarrow[\mathcal{D}]{} X_{\cdot} - \int_0^{\cdot} ab^{-2}(X_s)d[X]_s = Y ,$$

which, in view of (H_2), implies that Y is a continuous local martingale. So $X^c = Y$.

If we define now $M_t = \int_0^t b^{-1}(X_s)dX_s^c$, then it is clear, that the s.d.e. (2) is satisfied.

Besides, because of $(X^n, [\bar{X}^n]) \xrightarrow[\mathcal{D}]{} (X, [X] = [X^c])$ in $D(\mathbb{R}^2)$ we get

$$(12) \quad [M^n]_{\cdot} = \int_0^{\cdot} b_n^{-2}(X_{s-}^n)d[\bar{X}^n]_s \xrightarrow[\mathcal{D}]{} \int_0^{\cdot} b^{-2}(X_s)d[X]_s = [M]_{\cdot}$$

in $D(\mathbb{R})$. □

COROLLARY 3. Assume (H_2), (H_3), (H_4). (i) Let $M^n \xrightarrow[\mathcal{D}]{} W$. Then the following two conditions are equivalent:

(13) ≡ (7) $\quad X^n \xrightarrow[\mathcal{D}]{} X \quad$ in $D(\mathbb{R})$,

(14) $\quad a_n \rightrightarrows a$, $\quad b_n^2 \rightrightarrows b^2$ $\quad$ uniformly on compact subsets of $\mathbb{R}$.

Furthermore, if (13) or (14) holds, then X is the unique in the sense of law weak solution of the equation (9).

(ii) Suppose (13), (14) and let X be a weak solution of (9). Then $M^n \xrightarrow[\mathcal{D}]{} W$.

PROOF. (i) (14) ⇒ (13). By Theorem 3 for every subsequence (n'), $(n') \subset (n)$ the limit process X satisfies equation of the form (9). Therefore we can apply the considerations in the proof of Corollary 1.
(13) ⇒ (14). Like the proof of implication (7) ⇒ (8) in Corollary 1.
(ii) Due to (12) $[M^n]_t \xrightarrow[P]{} t$, $t \in \mathbb{R}^+$, hence by (H_2) $M^n \xrightarrow[\mathcal{D}]{} W$ (see [8]). □

REMARK 3. Under assumptions (H_2), (H_3) and

(H_5) $\quad \{a_n\}_{n \in \mathbb{N}}$, $\{b_n\}_{n \in \mathbb{N}}$ are equicontinuous on compact subsets of $\mathbb{R}$,

the martingale M appearing in Theorem 3 is a weak limit of $M_t^n = \int_0^t b_n^{-1}(X_{s-}^n)\,d\bar{X}_s^n$. Therefore one can easily show results corresponding to Theorem 2 and Corollary 2, too.

□

REFERENCES

1. L.A. Alyushina, N.V. Krylov, On the Passage to the Limit in Ito's Stochastic Equations. Theory Probab. Appl., 33 (1) (1988), 3-13, (in Russian).

2. H.J. Engelbert, W. Schmidt, On Solutions of One-Dimensional Stochastic Differential Equations Without Drift. Z. Wahr. verw. Gebiete., 68 (3) (1985), 287-314.

3. J. Jacod, Convergence en loi de semimartingales et varation quadratique. Lect. Notes in Math. 850, (1980), 547-560.

4. J. Jacod, J. Mémin, Weak and strong solutions of stochastic differential equations: existence and stability. Lect. Notes in Math. 851, (1981), 169-212.

5. A. Jakubowski, J. Mémin, G. Pagés, Convergence en loi des suites

d`integrales stochastiques sur l`espace D^1 de Skorokhod (to appear in Probab. Th. and Rel. Fields).

6. G.L. Kulinič, On necessary and sufficient conditions for the convergence of solutions of one-dimensional diffusion stochastic equations with a non-regular dependence of coefficients on a parameter. Theory Probab. Appl., 27 (4)(1982), 795-801 , (in Russian).

7. G.L. Kulinič, Limit behaviour of solutions of stochastic diffusion equations when the convergence of the coefficients is non-regular. Lect. Notes in Math. 1021,(1983), 352-354.

8. R. Liptser, A. Shiryayev, On a problem of necessary and sufficient conditions in the functional central limit theorem for local martingales. Z. Wahr. verw. Gebiete, 59 (1982), 311-318.

9. S.Ya. Makhno, Sufficient conditions for convergence of solutions of stochastic equations. Teor. Sluchajnych Protsessov, 16 (1988), 66-73, (in Russian).

10. A.V. Mel`nikov, Stochastic Equations and Krylov`s Estimates for Semimartingales, Stochastics 10 (1983) , 81-102.

11. A. Rozkosz, L. Słomiński, On weak convergence of solutions of one-dimensional stochastic differential equations (submitted to Stochastics) (1988).

12. K. Yamada, A Stability Theorem for Stochastic Differential Equations and Application to Stochastic Control Problems, Stochastics 13 (1984), 257-279.

Modelling of random fatigue accumulation

KAZIMIERZ SOBCZYK

Institute of Fundamental Technological Research
Polish Academy of Sciences, Warsaw

1. INTRODUCTION

An important deterioration process which takes place in structures operating under time-varying actions is fatigue. It has been recognized as a frequent cause for failure of engineering structures. On the other hand, fatigue phenomenon itself is still far from being completly understood and remains a challanging task today.

It has been widely accepted that fatigue process takes place via formation and growth of cracks in a material. However, during the course of propagation the crack encounters various types of metellurgical structures and imperfections, so that the rate of growth is variable in time.

Fatigue phenomenon can be studied from various points of view. It is of interest for *physics*, since it is intriguing to recognize the true physical (e.g. atomic and molecular) mechanisms of fatigue deterioration process on microlevel which might be responsible for generating macroscopic fatigue. Fatigue is obviously a subject of *mechanics*, since for mechanical engineer there is of interest why, when and where fatigue might occur and consequences it could have on overal stress distribution in structure. Fatigue is an important issue in *metallurgy*, since the manufacturer always wants to produce structural elements that will be free from fatigue as long as possible. Fatigue creates the problems in design of many structures and they lead to the questions commonly considered in *reliability theory*.

In spite of the above problems there exists an important issue which seriously determines the direction of recent research on fatigue. This is the fact that *our basic knowledge on fatigue comes from experiments.* The experimental data constitute basic source of information on fatigue of various materials subjected to various loading conditions. In such situation it is important to look at fatigue phenomenon from the point of view of mathematical modelling.

However, experimental results as well as field data indicate that fatigue process in real materials involves considerable random variability. This variability varies with respect to many parameters such as material properties, type of load, environmental conditions etc. Because of this inherent randomness in fatigue data a *stochastic modelling* seems to be the most appropriate one. However, the question of how to perform a stochastic modelling and analysis of random fatigue process which would lead to a consistent theory of random fatigue still constitutes an important and intriguing problem of research.

The older studies were mostly concerned with statistics of dispersed fatigue data and application of the results to estimation of fatigue reliability of parts in service. This leads to the estimation of the life-time distribution. Let us denote a random life-time of a structural element by T. A number of papers have appeared in which various probability distributions for random variable T have been proposed. One of the best known in this context is the Weibull distribution (of the extreme-value type) of the form

$$F(t) = 1 - \exp\Big[-\big(\frac{t-\varepsilon}{\nu}\big)^k\Big], \qquad t \geq \varepsilon > 0 \tag{1}$$

where $F(t)$ is the (cumulative) probability distribution of T; the function $1 - F(t)$ is called the survival function. Also such distributions as: exponential, log-normal and the Gumbel distribution (of extreme values, cf. [1]) have also been used.

It is clear however, that a general and mathematically natural approach to formulation of fatigue life disribution is as follows. The accumulation of fatigue (expressed, for example, in terms of the lenght of a dominant crack) in real materials subjected to realistic actions is, as a function of time t, a certain non-decreasing stochastic process $X(t,\gamma)$ defined for $t \in [t_0, \infty)$, $\gamma \in \Gamma$ where $\{\Gamma, \mathcal{F}, \mathcal{P}\}$ is the basic probability space. Fatigue failure occures at such time $t = T(\gamma)$, for which $X(t,\gamma)$ crosses, for the first time, a fixed critical level x^*. Therefore, the fatigue life distribution is the first-passage time distribution for the process $X(t,\gamma)$, that is, the distribution of random variable T defined as

$$T = \sup\{t : \quad X(\tau,\gamma) < x^*\}, \qquad T_0 \leq \tau \leq t. \tag{2}$$

To be able to characterize random variable T, first a stochastic process $X(t,\gamma)$ has to be constructed using the avaliable data.

At present there exists in the literature a number of proposals for $X(t,\gamma)$— cf.[2]. Here we wish to describe briefly a cummulative random compound model for fatigue crack growth proposed recently in paper [3] and to indicate its possible extensions; a detailed elaboration of these generalizations is now under the authors consideration.

2. Description of the Model

Fatigue process is characterized by a stochastic process $L(t,\gamma)$. This process is taken as the lenght of a dominant crack at time t. The letter γ symbolizes the elementary event belonging to Γ, where Γ is a space (sample space) on which probability is defined (for each $\gamma \in \Gamma$, $L(t,\gamma)$ represents a possible sample-function of the crack lenght). Process $L(t,\gamma)$ is represented as

$$L(t,\gamma) = L_0 + \sum_{i=1}^{N(t,\gamma)} Y_i(\gamma) \tag{3}$$

where L_0 is the initial crack lenght of sufficient size to propagate. $Y_i(\gamma) = \Delta L_i(\gamma)$ are the random crack increments and $N(t) = N(t,\gamma)$ is an integer-valued stochastic process characterizing the number of crack increments in the interval $[0,t]$. Here, it is assumed that $Y_i(\gamma)$, $i = 1,2,\dots,$ are independent and identically distributed random variables.

Randomness of the fatigue crack growth process is, therefore, taken into account via the probabilistic mechanism of the transition from one state to another of the process $N(t,\gamma)$ and by the fact that the single crack increments are allowed to be random.

Although the counting process $N(t,\gamma)$ could be taken in a more general form, here we assume that $N(t,\gamma)$ is a Poisson homogeneous process, that is a process for which probability $P_k(t)$, that within time interval $[0,t]$ the number of events (crack's jumps) is equal to k, is

$$P_k(t) = e^{-\lambda_0 t}\frac{(\lambda_0 t)^k}{k!}. \tag{4}$$

3. Probability Distribution of a Life-Time

Let T be a positive random variable which characterizes a random time of reaching by process $L(t,\gamma)$ of a fixed, critical level ξ.

Probability distribution of T is

$$\begin{aligned} \mathrm{P}(T > t) &= P\{L(t,\gamma) < \xi\} = P\{L_0 + \sum_{i=1}^{N(t,\gamma)} Y_i(\gamma) < \xi\} \\ &= \sum_{k=0}^{\infty} P\{L_0 + \sum_{i=1}^{k} Y_i(\gamma) < \xi\}\, P_k(t) = \sum_{k=0}^{\infty} P_k(\xi)\, P_k(t), \end{aligned} \tag{5}$$

where $P_k(\xi)$ is the probability that after k increments the crack's size is less than critical value ξ.

The probability density of T is

$$f_T(t) = -\frac{d}{dt} P\{T > t\}$$
$$= -\sum_{k=0}^{\infty} \frac{p_k(\xi)}{k!} \frac{d}{dt} \left[(\lambda_0 t)^k e^{-\lambda_0 t} \right]. \tag{6}$$

After transformation and under assumption that random variables $Y_i(\gamma)$ are exponentially distributed (with parameter a) one obtains (cf. [3])

$$f_T(t) = \lambda_0 e^{-\lambda_0 t - a(\xi - L_0)} I_0 \left(2\sqrt{\lambda_0 t a(\xi - L_0)} \right) \tag{7}$$

where $I_0(\cdot)$ is a modified Bessel function of order zero.

4. Probability Distribution of a Crack Size

Let us denote by $f_{L(t)}(l)$ the probability density of process $L(t)$ at time t. Using the concept of moment generating function one obtains that $f_{L(t)}(l) = f_{L_1(t)}(l - L_0)$, where

$$f_{L_1(t)}(l) = e^{-\lambda_0 t} \sqrt{\frac{\lambda_0 a t}{l}}\, e^{-\lambda_0 t - a l} I_1 \left(2\sqrt{\lambda_0 a l t} \right) \tag{8}$$

and I_1 is a modified Bessel function of order one.

5. Relation to Empirical Data

The model just described needs approximate experimental verification (by comparison with data from repeated experiments). The parameters of the model (e.g. statistics of a single crack increments (a), properties of the growth intensity (λ_0) should be estimated from fatigue crack growth data. Since the experiments associated with model (3) have not been performed yet, the model parameters have been related to those occuring in empirical fatigue crack growth equations, like Paris-Erdogan equation, and to characteristics of random loading process.

A possible way of characterization of parameter a is the use of the mean-square criterion; namely, we determine a from the condition

$$\int_{t_0}^{\bar{t}} E\left\{ \left[L_{P-E}(t) - L(t,\gamma) \right]^2 \right\} dt = \min_a \tag{9}$$

where $L_{P-E}(t)$ is a solution of the modified Paris-Erdogan equation (cf. [4])

$$\frac{dL}{dt} = \mu_0 \, C \, g(Q) \; (K_{rms})^m \tag{10}$$

where K_{rms} is the the double root mean square of the stress intensity factor (for infinite elastic sheet $K_{rms} = 2S_{rms}\sqrt{\pi L}$; S_{rms}—the root mean square of stationary loading), Q is a counterpart of a stress ratio, μ_0 can be taken as the average number of maxima of $S(t,\gamma)$ in the interval $[t_0, t]$; the constants C and m characterize the material properties (they are known from traditional experiments); $\bar{t}$ is the time in which $L_{P-E}(t)$ reaches the critical level ξ. Condition (9) gives an effective procedure for relating parameter a with data contained in $L_{P-E}(t)$.

Much weaker criterion which can be used for determining a makes use of the entropies of two processes. Namely, we require

$$\int_{t_0}^{\bar{t}} H_{R-PE}(t)\, dt = \int_{t_0}^{t} H_L(t) \tag{11}$$

where $H_{R-PE}(t)$ and $H_L(t)$ are one-dimensional (informational) entropies of the process $L_{R-PE}(t,\gamma)$ defined by the randomized Paris equation (constant C—regarded as a random variable) and process $L(t,\gamma)$ defined by the cumulative model, respectively. One dimensional entropy is defined in usual way

$$H_L(T) = -\int_0^{\infty} f_{L(t)}(l) \, \ln f_{L(t)}(l) \, dl. \tag{12}$$

The above criterion means that global (or integrated) randomness of process $L_{R-PE}(t,\gamma)$ and process $L(t,\gamma)$ represented by the model discussed are required to be the same.

Parameter λ_0, the intensity of crack growth, is assumed to be equal to the average number of maxima above level s_0 of the loading process. To estimate the parameter ξ, the critical crack lenght, a limiting critical condition from fracture mechanics have been adopted (cf. [5]). In the paper cited the results of numerical calculations of the life-time distribution and the distribution of the crack size have been illustrated graphically for real data.

REMARK. Since the intensity of crack growth is expexted to depend on the crack size or, in other words, on the fatigue state, the birth process (with state-dependent intensity) can be used to count a number $N(t)$ of crack increments in interval $[t_0, t]$.

6. Extended Cumulative Models

6.1. Inhomogeneous crack growth intensity

In general, the intensity of jumps in crack growth varies in time. So, the Poisson process $N(t)$ should be assumed to be inhomogeneous, i.e. $\lambda = \lambda(t)$, $t > 0$. In this case, the probability $P_k(t)$ characterizing a number of jumps in interval $[0, t]$ is governed by differential equations with variable coefficient $\lambda(t)$. However, introducing the function η defined as

$$\lambda(t) = \frac{d\eta(t)}{dt}, \qquad \eta(t) = \int_0^t \lambda(\tau)\, d\tau \tag{13}$$

we get

$$P_k(t) = \frac{[\eta(t)]^k}{k!} e^{-\eta(t)}.$$

The probability density of the crack lenght at time t is given by the formula

$$f_{L_1(t)}(l) = e^{-\eta(t)-al} \sum_{k=0}^{\infty} \frac{[a\eta(t)]^{k+1} l^k}{(k+1)!\, k!}. \tag{14}$$

This formula reduces to (8) when $\lambda(t) = \lambda_0$, i.e. $\eta(t) = \lambda_0 t$.

6.2. Dependent elementary crack increments

The assumption that random variables $Y_i(\gamma)$, $i = 1, 2, \ldots,$ characterizing random sizes of elementary increments (in the crack growth) are independent might be not adequate to real situations. Succesive increments may be dependent, since accumulation of fatigue may result in a loss of resistance to further fatigue damage. In such situation it is reasonable to assume that:

$$P\{Y_k \leq u \mid Y_1, \ldots, Y_{k-1}\} = P\{Y_k \leq u \mid Z_{k-1}\}, \tag{15}$$

where $Z_{k-1} = Y_1 + \ldots + Y_{k-1}$.

Then it is natural to state two conditions:

(i) $P\{Y_k \leq u \mid Z_{k-1} = z\}$ is decreasing in $z \geq 0$,

(ii) $P\{Y_k \leq u \mid Z_{k-1} = z\} \geq P\{Y_{k+1} \leq u \mid Z_k = z\}$

for $k = 1, 2, \ldots,$ and Z_0.

Condition *(i)* means that an accumulation of fatigue damage lowers resistance to further damage, whereas condition *(ii)* says that for a given fatigue state the later jumps are more severe.

Taking the above into account we have

$$F_k(\xi) = P\Big\{\sum_{i=1}^{k} Y_i(\gamma) < \xi\Big\} = \int_0^{\xi} P\{Y_k < \xi - z \mid Z_{k-1} = z\}\, dF_{k-1}(z) \qquad (16)$$

and

$$P\{T > t\} = \sum_{k=0}^{\infty} e^{-\lambda_0 t} \frac{(\lambda_0 t)^k}{k!} F_k(\xi). \qquad (17)$$

6.3. Use of counting integral

The model described is based on representation of the crack growth in the form of accumulation of countable number of jumps. More general representation of a cumulative crack growth may be accomplished by use of the counting integral with respect to random Poisson measure, i.e.

$$L(t, I) = L_0 + \int_{t_0}^{t} \int_{I} h(s, y, \gamma)\, M(ds, dy) \qquad (18)$$

where h is a random function of time and magnitude of jump (which can take every non-negative real number belonging to interval I), M is the time-space Poisson Process. The above integral can be evaluated as

$$L(t,\ I) = L_0 + \begin{cases} 0, & N_t = 0 \\ \sum_{k=1}^{N_t} h_{\tau_k}(Y_k), & N_t \geq 1 \end{cases} \qquad (19)$$

where

$$N_t = \int_{t_0}^{t} \int_{I} M(ds, dy) \qquad (20)$$

characterizes a random number of jumps occuring in the interval $[t_0, t)$ with magnitudes y belonging to interval I and

$$E\{N_t\} = P(y \in I) \int_{t_0}^{t} \lambda(s)\, ds \qquad (21)$$

where $\lambda(s)$ is the intensity of the Poisson Process.

More general model which enables us to account for the dependence of $L(t)$ on its history is

$$L(t, I) = L_0 + \int_{t_0}^{t} \int_{I} h(s, L(s), y)\, M(ds, dy). \tag{22}$$

The form of function h should be inferred from the data.

References

1. Gumbel E., *Statistics of Extremes,*Columbia Univ. Press, N.Y., 1958.
2. Sobczyk K., *Stochastic Models for Fatigue Damage Materials,* Adv. Appl. Prob. vol. 19, 652–673, 1987.
3. Sobczyk K., and Trębicki J. *Modelling of Random Fatigue by Cumulative Jump Processes,* Eng. Fracture Mech., (submitted for publication).
4. Sobczyk K., *Modelling of Random Fatigue Crack Growth,* Eng. Fracture Mech., vol. 24, 609–623, 1986.
5. Fuchs H.O. and Stephens R.I. *Metal Fatigue in Engineering,* J. Wiley, N.Y. 1980.

FUNCTIONALS ON STOCHASTIC PROCESSES

K. Urbanik
Institute of Mathematics, Wrocław University
Pl. Grunwaldzki 2/4 50-384 Wrocław, Poland

1. PRELIMINARIES AND NOTATION. We shall be concerned in this paper with the space B of all complex-valued essentially bounded with respect to the Lebesgue measure Borel functions f on the real line R with the norm

$$\|f\| = \text{ess sup}\{|f(x)| : x \in R\} . \qquad (1.1)$$

In the sequel we shall say shortly almost everywhere instead of almost everywhere with respect to the Lebesgue measure. We say that two functions f and g from B are equivalent, in symbols $f \sim g$, if they are equal almost everywhere. Let C be the space of all complex-valued bounded and uniformly continuous functions on R with the norm (1.1). In this paper we shall identify the space C and its natural embedding into the space B. Let γ be the standard Gaussian probability measure on R. Given $f \in B$ we put

$$\|f\|_\gamma = \int_{-\infty}^{\infty} |f(x)|\gamma(dx) .$$

We define the γ-topology on B as follows: given $f, f_n \in B$ $(n=1,2,\dots)$ we say that the sequence f_n is γ-convergent to f, in symbols $f_n \overset{\gamma}{\to} f$, if the norms $\|f_n\|$ $(n=1,2,\dots)$ are bounded in common and $\|f_n - f\|_\gamma \to 0$ as $n \to \infty$. It is clear that the set C is dense in B in the γ-topology.

Let M be the space of all signed Borel measures μ on R with finite total variation $|\mu|(R)$. Given $\mu \in M$ and $f \in B$ we put

$$(f * \mu)(x) = \int_{-\infty}^{\infty} f(x+y)\mu(dy) \qquad (x \in R) .$$

Since

$$\|f * \mu\|_\gamma \le |\mu|(R)\|f\| , \tag{1.2}$$

we conclude that $\|f * \mu\| = 0$ whenever $\|f\| = 0$. Thus the relation $f \sim g$ yields $f * \mu \sim g * \mu$. Moreover,

$$\|f * \mu\| \le |\mu|(R)\|f\| \tag{1.3}$$

and

$$\int_{-\infty}^{\infty} |f(x+y)|\gamma(dx) \le \int_{-\infty}^{\infty} |f(u)|e^{|uy|}\gamma(du) .$$

Setting $g_r(x) = f(x)e^{r|x|}$ and $E_r = \{y : |y| > r\}$ for any $r > 0$ we have the inequality

$$\|f * \mu\|_\gamma \le \int_{-\infty}^{\infty}\int_{-\infty}^{\infty}|f(x+y)|\gamma(dx)|\mu|(dy) \le |\mu|(R)\|g_r\|_\gamma + |\mu|(E_r)\|f\| .$$

Hence in particular it follows that the map $B \ni f \to f * \mu \in B$ is continuous in the γ-topology. Further, it is easy to check the formula

$$(f * \mu) * \nu = f * (\mu * \nu) \tag{1.4}$$

for any pair $\mu,\nu \in M$ and $f \in B$. We note that the absolute continuity of μ with respect to the Lebesgue measure yields $f * \mu \in C$ for all $f \in B$. Denoting by ω_h $(h > 0)$ the uniform probability distribution on the interval $[0,h]$ we have the relations $f * \omega_h \in C$ and

$$f * \omega_h \overset{\gamma}{\to} f \quad \text{as} \quad h \to 0 \tag{1.5}$$

for any $f \in B$.

Throughout this paper $X(t,\omega)$ $(t \ge 0)$ will denote a stochastic process with stationary and independent increments, right-continuous sample functions with $X(0,\omega) = 0$. It is well-known that the probability measure induced by the process in question is uniquely determined by the one-parameter semigroup of the convolution operators on C defined by the formula

$$T_t f = f * \tau_t \quad (t \ge 0) , \tag{1.6}$$

where $\tau_t(E) = P(\{\omega : X(t,\omega) \in E\})$. Of course, operators (1.6) are well defined on B too. Moreover, setting for $\lambda > 0$

$$\rho_\lambda = \int_0^\infty e^{-\lambda t}\tau_t dt$$

we conclude that the resolvent $R_\lambda f = f * \rho_\lambda$ is also well defined on B. Finally, we quote the formula

$$\rho_\lambda * \tau_h = e^{\lambda h}(\rho_\lambda - \int_0^h e^{-\lambda t}\tau_t dt). \tag{1.7}$$

2. THE γ-EXTENSION OF THE INFINITESIMAL GENERATOR. Let A be the infinitesinal generator of the semigroup $\{T_t\}$ on C with the domain $D(A)$. We note that for any $f \in B$ the map $t \to f * \tau_t = T_t f$ is continuous in the γ-topology on B. By $\tilde{A}$ we shall denote the infinitesimal generator of the semigroup $\{T_t\}$ in the γ-topology on B. More precisely a function f from B belongs to the domain $D(\tilde{A})$ if and only if there exists a function $g \in B$ such $\frac{1}{h}(T_h f - f) \overset{\gamma}{\to} g$ as $h \to 0$. Further we put $\tilde{A}f = g$. It is obvious that $D(A) \subset D(\tilde{A})$ and the operator $\tilde{A}$ is an extension, called the γ-extension, of the generator A.

EXAMPLE. Let $X(t,\omega) = W(t,\omega)$ at where $W(t,\omega)$ is the standard Brownian motion and $a \in R$. It is well-known that the domain $D(A)$ consists of all twoce differentiable functions f satisfying the condition $f, f', f'' \in C$. The generator A is given by the formula

$$A = \frac{1}{2}\frac{d^2}{dx^2} + a\frac{d}{dx} . \tag{2.1}$$

It is easy to verify that $f \in D(\tilde{A})$ if and only if there exists a twice differentiable function g such that $f \sim g$, $g, g' \in C$ and $g'' \in B$.

By the well-known formula $D(A) = R_\lambda(C)$ $(\lambda > 0)$ every function f from $D(A)$ has a representation $f = g * \rho_\lambda$ with $g \in C$ which yields the formula

$$Af = \lambda f - g , \tag{2.2}$$

(see [1], Chapter 37). Hence we get the following Lemmas.

LEMMA 2.1. *For any* $f \in D(A)$ *and* $\mu \in M$ *we have the relations* $f * \mu \in D(A)$ *and* $A(f * \mu) = (Af) * \mu$.

LEMMA 2.2. *Suppose that* $f \in B$, $\mu, \nu \in M$ *and* $f * \mu$, $f * \nu \in D(A)$. *Then* $(A(f * \mu)) * \nu = (A(f * \nu)) * \mu$.

In what follows I will denote the unit operator on B.

PROPOSITION 2.1. *For every* $\lambda > 0$ *we have the relations* $D(\tilde{A}) = R_\lambda(B)$,

$$(\lambda I - \tilde{A})R_\lambda = I \tag{2.3}$$

and

$$R_\lambda(\lambda I - \tilde{A})f = f \tag{2.4}$$

for $f \in D(\tilde{A})$.

P r o o f . Suppose first that $f \in D(\tilde{A})$ and $\frac{1}{h}(T_h f - f) \to \tilde{A}f$ as $h \to 0$. Then

$$\frac{1}{h}(R_\lambda T_h f - R_\lambda f) \overset{\gamma}{\to} R_\lambda \tilde{A} f \ .$$

On the other hand, by (1.7), the left-hand side of the above formula tends to $\lambda f * \rho_\lambda - f$. Thus $f = R_\lambda(\lambda I - \tilde{A})f$ which yields the relation $f \in R_\lambda(B)$ and formula (2.4).

Now let us assume that $f \in R_\lambda(B)$. Taking a representation $f = R_\lambda g$ with $g \in B$ and taking into account (1.7) we get the relation

$$\frac{1}{h}(T_h f - f) = \frac{1}{h}(T_h R_\lambda g - R_\lambda g) \overset{\gamma}{\to} \lambda f - g$$

which ahows that $f \in D(\tilde{A})$ and $\tilde{A}f = \lambda f - g$. Hence formula (2.3) follwos which completes the proof.

PROPOSITION 2.2. *Let* $f \in B$. *Then* $f \in D(\tilde{A})$ *if and only if* $f * \omega_h \in D(A)$ *and* $A(f * \omega_h) = g * \omega_h$ *for some* $g \in B$ *and every* $h > 0$.

P r o o f . *The necessity*. Suppose that $f \in D(\tilde{A})$. Then, by Proposition 2.1, the function f can be written in the form $f = g * \rho_\lambda$, where $g \in B$ and $\tilde{A}f = \lambda f - g$. Since $g * \omega_h \in C$ and $f * \omega_h = g * \omega_h * \rho_\lambda = R_\lambda(g * \omega_h)$, we infer that $f * \omega_h \in D(A)$ and by formula (2.2),

$$A(f * \omega_h) = \lambda f * \omega_h - g * \omega_h = \tilde{A}f * \omega_h$$

which completes the proof of the necessity.

The sufficiency. Suppose that $f \in B$, $f * \omega_h \in D(A)$ and $A(f * \omega_h) = g * \omega_h$ for some $g \in B$ and every $h > 0$. Taking into account (2.2) we observe that

$$f * \omega_h = (\lambda f - g) * \rho_\lambda * \omega_h$$

for all $h > 0$. Thus, by formula (1.5), $f = (\lambda f - g) * \rho_\lambda$.

Consequently $f \in R_\lambda(B)$, which, by Proposition 2.1, completes the proof.

As an immediate consequence of the above Proposition and Lemma 2.1 we get the following result.

PROPOSITION 2.3. *For any* $f \in D(\tilde{A})$ *and* $\mu \in M$ *we have the relations* $f * \mu \in D(\tilde{A})$ *and* $\tilde{A}(f * \mu) = (\tilde{A}f) * \mu$.

PROPOSITION 2.4. *The operator* $\tilde{A}$ *is closed in the* γ*-topology.*

P r o o f . Suppose that $f_n \in D(\tilde{A})$ $(n = 1,2,\ldots)$, $f \in B$, $f_n \overset{\gamma}{\to} f$, $g \in B$ and $\tilde{A}f_n \to g$. We have to prove that $f \in D(\tilde{A})$ and $\tilde{A}f = g$. Setting $g_n = (\lambda I - \tilde{A})f_n$ $(n = 1,2,\ldots)$ we have $g_n \overset{\gamma}{\to} \lambda f - g$ which yields $R_\lambda g_n \overset{\gamma}{\to} R_\lambda(\lambda f - g)$. On the other hand, by Proposition 2.1 (formula (2.4)), we have the formula $R_\lambda g_n = f_n$ $(n = 1,2,\ldots)$ which yields $f = R_\lambda(\lambda f - g)$. Thus, by Proposition 2.1, $f \in D(\tilde{A})$ and $\lambda f - \tilde{A}f = (\lambda I - \tilde{A})R_\lambda(\lambda f - g) = \lambda f - g$. Consequently, $\tilde{A}f = g$ which completes the proof.

As a consequence of the above Proposition we get the following Corollaries.

COROLLARY 2.1. *Suppose that* $f(\cdot,y) \in D(\tilde{A})$ $(y \in (0,\infty))$ *and the functions* $y \to f(\cdot,y)$, $y \to \tilde{A}f(\cdot,y)$ *from* $(0,\infty)$ *into* B *are differentiable in the* γ*-topology. Then* $\frac{\partial}{\partial y} f(\cdot,y) \in D(\tilde{A})$ *and* $\tilde{A} \frac{\partial}{\partial y} f(\cdot,y) = \frac{\partial}{\partial y} \tilde{A}f(\cdot,y)$.

COROLLARY 2.2. *Suppose that* $g(\cdot,y) \in D(\tilde{A})$ $(y \in (0,\infty))$ *and the functions* $y \to g(\cdot,y)$, $y \to \tilde{A}g(\cdot,y)$ *from* $(0,\infty)$ *into* B *are continuous in the* γ*-topology. Moreover, assume that both integrals* $\int_0^\infty \|g(\cdot,y)\|dy$, $\int_0^\infty \|\tilde{A}g(\cdot,y)\|dy$ *are finite. Then* $\int_0^\infty g(\cdot,y)dy \in D(\tilde{A})$ *and* $\tilde{A}\int_0^\infty g(\cdot,y)dy = \int_0^\infty \tilde{A}g(\cdot,y)dy$.

3. FUNCTIONALS ON STOCHASTIC PROCESSES. In what follows by $\operatorname{Re} B$ we shall denote the set of all real-valued functions belonging to B . Let $f \in \operatorname{Re} B$, $x \in R$ and $t \geq 0$. Denote by $\mu(f,x,t,\cdot)$ and $\hat{\mu}(f,x,t,\cdot)$ the probability distribution and the characteristic function of the random functional $\int_0^t f(x + X(u,\omega))du$ respectively. Put for $\lambda > 0$

$$L(f,x,s,\lambda) = \int_0^\infty e^{-\lambda t}\hat{\mu}(f,x,t,s)dt . \tag{3.1}$$

It is easy to check that the function $L(f,s,\cdot,\lambda)$ is continuous, the function $L(f,\cdot,s,\lambda)$ belongs to B,

$$\|L(f,\cdot,s,\lambda)\| \leq \lambda^{-1} \tag{3.2}$$

and the function $L(f,x,s,\cdot)$ has an analytic extension in the right half-plane.

LEMMA 3.1. *Let* $s \in R$ *and* $\lambda > 0$. *The map* $\mathrm{Re}\, B \ni f \to$ $\to L(f,\cdot,s,\lambda) \in B$ *is continuous in the* γ*-topology.*

P r o o f . Inequality (3.2) shows that the map $f \to \|L(f,\cdot,s,\lambda)\|$ is bounded on $\mathrm{Re}\, B$. Setting

$$\beta(E) \doteq \int_0^\infty e^{-\lambda t} \int_0^t \tau_y(E)dy\, dt$$

we get the inequality

$$\|L(f,\cdot,s,\lambda) - L(g,\cdot,s,\lambda)\|_\gamma \leq$$

$$\leq |s| \int_{-\infty}^{\infty} \int_{-\infty}^{\infty} |f(x+y) - g(x+y)|\gamma(dx)\beta(dy) .$$

It is evident that the right-hand side of the sbove inequality tends to 0 as $f \overset{\gamma}{\to} g$ which yields the assertion of the Lemma.

The investigation of the function L is based on the method due to A.V. Skorohod ([2], Chapter 19).

THEOREM 3.1. *For any* $f \in \mathrm{Re}\, B$, $s \in R$ *and* $\lambda > 0$ *the function* $L(f,\cdot,s,\lambda)$ *bolobgs to* $D(\tilde{A})$ *and fulfils the equation*

$$\tilde{A}L(f,\cdot,s,\lambda) + (isf(\cdot)-\lambda)L(f,\cdot,s,\lambda) + 1 = 0 \tag{3.3}$$

almost everywhere. The solution of this equation is unique up to the equivalence relation under the condition that for almost all $x \in R$ $L(f,x,s,\cdot)$ *can be extended to an analytic function in the right half--plane.*

P r o o f . Given $s \in R$ and $\lambda > 0$ we denote by $E(s,\lambda)$ the set of all functions f from $\mathrm{Re}\, B$ for which $L(f,\cdot,s,\lambda) \in D(\tilde{A})$ and equation (3.3) is fulfilled almost evervwhere. For any twice differentiable real-valued function g with $g',g'' \in C$ it has been proved in [2] Chapter 19 that the function $\hat{\mu}(g,x,\cdot,s)$ is differentiable, the function $\hat{\mu}(g,\cdot,t,s)$ belongs to $D(A)$ and the equation

$$\frac{\partial}{\partial t} \hat{\mu}(g,\cdot,t,s) = A\hat{\mu}(g,\cdot,t,s) + isg(\cdot)\hat{\mu}(g,\cdot,t,s)$$

with the initial condition $\hat{\mu}(g,x,o,s) = 1$ is fulfilled. Hence, by a standard calculation, we conclude that the Laplace transform $L(g,\cdot,s,\lambda)$ belongs to $D(A)$ and fulfils equation (3.3). Thus $g \in E(s,\lambda)$. Since the subset of Re B consisting of all twice differentiable functions g with $g',g'' \in C$ is dense in Re B in the γ-topology, to prove the first part of the Theorem it suffices to show that the set $E(s,\lambda)$ is closed in this topology. Suppose that $f_n \in E(s,\lambda)$ $(n = 1,2,\ldots)$, $f \in$ Re B and $F_n \overset{\gamma}{\to} f$. Then, by Lemma 3.1, $L(f_n,\cdot,s,\lambda) \overset{\gamma}{\to} L(f,\cdot,s,\lambda)$ and, consequently,

$$(isf_n(\cdot) - \lambda)L(f_n,\cdot,s,\lambda) \overset{\gamma}{\to} (isf(\cdot) - \lambda)L(f,\cdot,s,\lambda) .$$

Taking into account equation (3.3) we obtain the convergence

$$AL(f_n,\cdot,s,\lambda) \overset{\gamma}{\to} (isf(\cdot) - \lambda)L(f,\cdot,s,\lambda)$$

which, by Proposition 2.4, shows that the function $L(f,\cdot,s,\lambda)$ belongs to $D(\tilde{A})$ and fulfils equation (3.3) almost everywhere. Thus the set $E(s,\lambda)$ is closed in the γ-topology which completes the proof that equation (3.3) holds for all $f \in$ Re B.

To prove the uniqueness of the solution of (3.3) it suffices to show that each solution $F(\cdot,\lambda)$ from $D(\tilde{A})$ of the homogebeous equation

$$\tilde{A}F(\cdot,\lambda) + (g(\cdot) - \lambda)F(\cdot,\lambda) = 0$$

with $g \in B$ such that for almost all x the function $F(x,\cdot)$ can be extended to an analytic function in the right half-plane vanishes almost everywhere. Applying Proposition 2.1 (formula (2.4)) we have the equality $F(\cdot,\lambda) = R_\lambda(g(\cdot)F(\cdot,\lambda))$ almost everywhere which yields the inequality $\|F(\cdot,\lambda)\| \leq \lambda^{-1}\|g\| \, \|F(\cdot,\lambda)\|$. Hence it follows that $\|F(\cdot,\lambda)\| = 0$ whenever $\lambda > \|g\|$. Introducing the notation $Z(\lambda) = \{x : F(x,\lambda) \neq 0\}$, we have the formula $\gamma(Z(\lambda)) = 0$ if $\lambda \geq \|g\|$. Taking a sequence $\lambda_1 > \lambda_2 > \ldots > \|g\| + 1$ and setting $Z = \bigcup_{n=1}^{\infty} Z(\lambda_n)$ we have $\gamma(Z) = 0$ and $F(x,\lambda_n) = 0$ for $n = 1,2,\ldots$ and $x \notin Z$. By the analyticity of $F(x,\cdot)$ for almost all x we conclude that $F(x,\lambda) = 0$ for almost all x and all $\lambda > 0$. This completes the proof of the Theorem.

Given $f \in$ Re B, $a \in R$, $x \in R$ and $t \geq 0$ we introduce the notation

$$p(f,x,t,a) = P(\{\omega : \int_0^t f(x + X(u,\omega))du = at\}) .$$

THEOREM 3.2. *For any* $f \in \operatorname{Re} B$, *and* $a \in R$ *the equality*

$$(f(x) - a)p(f,x,t,a) = 0$$

holds for almost all pairs (x,t).

P r o o f . Put for $\lambda > 0$

$$P(f,x,a,\lambda) = \int_0^\infty e^{-\lambda t} p(f,x,t,a)dt .$$

To prove the Theorem it suffices to show that for every $\lambda > 0$ and for almost every x the equality $(f(x) - a)P(f,x,a,\lambda) = 0$ holds. Since $p(f,x,t,a) = \mu(f,x,t,\{at\})$, we have the formula

$$P(f,x,a,\lambda) = \lim_{Z\to\infty} \frac{1}{2z} \int_{-Z}^{Z} L(f,x,s,\lambda+ias)ds .$$

By Theorem 3.1 and the analyticity of the function $L(f,x,s,\cdot)$ in the right half-plane we conclude that the equation

$$\tilde{A}L(f,\cdot,s,\lambda-ias) + (is(f(\cdot)-a) - \lambda)L(f,\cdot,s,\lambda+ias) + 1 = 0$$

holds almost everywhere which, by Proposition 2.1 (formula (2.4)) yields

$$L(f,\cdot,s,\lambda+ias) = isR_\lambda(f(x) - a)L(f,\cdot,s,\lambda+ias) + \lambda^{-1} .$$

Consequently, for $z > 0$

$$\frac{1}{2z} \int_{-z}^{z} (is)^{-1}(L(f,\cdot,s,\lambda+ias) - \lambda^{-1})ds$$
$$\le R_\lambda(f(\cdot)-a) \frac{1}{2z} \int_{-z}^{z} L(f,\cdot,s,\lambda+ias)ds \qquad (3.4)$$

almost everywhere. Taking into account (3.1) we infer that the left-hand side of (3.4) is majorized by the expression

$$\int_0^\infty e^{-\lambda t} \int_{-\infty}^{\infty} \frac{1}{2z} \left| \int_{-z}^{z} s^{-1}(e^{is(u-at)} - 1)ds \right| \mu(f,\cdot,t,du)dt .$$

Since the integrand is less then $|u-at|$ and

$$\int_{-\infty}^{\infty} |u+at|\mu(f,x,t,du) \le at + \|f\|$$

we can change the order of integration and passing to the limit $z \to \infty$. Consequently, the left-hand side of formula (3.4) tends to 0 as $z \to \infty$. The right-hand side of (3.4) tends to

$$R_\lambda(f(\cdot) - a)P(f,\cdot,a,\lambda) \quad \text{as} \quad z \to \infty .$$

Thus $R_\lambda(f(\cdot) - a)P(f,\cdot,a,\lambda) = 0$ for all $\lambda > 0$ and almost all x.

Our assertion is now an immediate consequence of formula (2.3) and the uniqueness Theorem for the Laplace transform.

THEOREM 3.3. *Let* $f \in \operatorname{Re} B$. *Suppose that for almost all* $x \in R$ *the limit distribution* $\nu(f,x,\cdot)$ *of* $\frac{1}{t}\int_0^t f(x+X(u,\omega))du$ *exists as* $t \to \infty$. *Setting for* $z > 0$

$$H(f,x,z) = \int_{-\infty}^{\infty} \frac{\nu(f,x,ds)}{z-is}$$

we have $H(f,\cdot,z) \in D(\tilde{A})$ *and* $\tilde{A}H(f,\cdot,z) = 0$ *almost everywhere.*

P r o o f . Given $f \in \operatorname{Re} B$. For any triple $y,z,\lambda \in (0,\infty)$, by Theorem 3.1, the function $L(f,\cdot,y,\lambda+yz)$ belongs to $D(\tilde{A})$ and

$$\tilde{A}L(f,\cdot,y,\lambda+yz) = (\lambda+yz-iyf(\cdot))L(f,\cdot,y,\lambda+yz) - 1 . \tag{3.5}$$

Hence, by a standard reasoning, we conclude that both functions $\lambda \to L(f,\cdot,y,\lambda+yz)$, $\lambda \to \tilde{A}L(f,\cdot,y,\lambda+yz)$ from $(0,\infty)$ into B are infinitely differentiable in the γ-topology. Applying Corollary 2.1 we infer that $\frac{\partial^k}{\partial\lambda^k} L(f,\cdot,y,\lambda+yz) \in D(\tilde{A})$ and

$$\tilde{A}\frac{\partial^k}{\partial\lambda^k} L(f,\cdot,y,\lambda+yz) = \frac{\partial^k}{\partial\lambda^k}\tilde{A}L(f,\cdot,y,\lambda+yz) \qquad (k = 1,2,\ldots) .$$

Consequently, formula (3.5) yields

$$\tilde{A}\frac{\partial^2}{\partial\lambda^2} L(f,\cdot,y,\lambda+yz) = 2\frac{\partial}{\partial\lambda} L(f,\cdot,y,\lambda+yz) + \\ + (\lambda+yz-iyf(\cdot))\frac{\partial^2}{\partial\lambda^2} L(f,\cdot,y,\lambda+yz) . \tag{3.6}$$

Further, from (3.1) we get the inequalities

$$\frac{\partial}{\partial\lambda} L(f,\cdot,y,\lambda+yz)\| \le (\lambda + yz)^{-2} , \tag{3.7}$$

$$\left\|\frac{\partial^2}{\partial\lambda^2} L(f,\cdot,y,\lambda+yz)\right\| \le 2(\lambda + yz)^{-3} \tag{3.8}$$

which, by (3.6), show that both integrals $\int_0^\infty \left\|\frac{\partial^2}{\partial\lambda^2} L(f,\cdot,y,\lambda+yz)\right\|dy$ and $\int_0^\infty \left\|\tilde{A}\frac{\partial^2}{\partial\lambda^2} L(f,\cdot,y,\lambda+yz)\right\|dy$ are finite. Taking into account Corollary 2.2 we infer that

$$\int_0^\infty \frac{\partial^2}{\partial\lambda^2} L(f,\cdot,y,\lambda+yz)dy \in D(\tilde{A}) \tag{3.9}$$

and, by (3.6), (3.7) and (3.8),

$$\|\tilde{A} \int_0^\infty \frac{\partial^2}{\partial \lambda^2} L(f,\cdot,y,\lambda+yz)dy\| \le \frac{4}{\lambda z} + \frac{2\|f\|}{\lambda z^2} . \tag{3.10}$$

Let $\nu(f,x,t,\cdot)$ denote the probability distribution of the random variable $\frac{1}{t}\int_0^t f(x + X(u,\omega))du$. Put for $z > 0$

$$G(f,x,t,z) = \int_{-\infty}^{\infty} \frac{\nu(f,x,t,ds)}{z-is} .$$

It is easy to check the formula

$$G(f,x,t,z) = t \int_0^\infty \hat{\mu}(f,x,t,y)e^{-zyt}dy$$

which yields for $\lambda > 0$

$$\int_0^\infty G(f,x,t,z)te^{-\lambda t}dt = \int_0^\infty \frac{\partial^2}{\partial \lambda^2} L(f,x,y,\lambda+yz)dy .$$

Consequently, by (3.9) and (3.10) $\lambda^2 \int_0^\infty G(f,\cdot,t,z)te^{-\lambda t}dt \in D(\tilde{A})$ and $\|\tilde{A}\lambda^2 \int_0^\infty G(f,\cdot,t,z)te^{-\lambda t}dt\| \to 0$ as $\lambda \to 0$. Since $\lambda^2 \int_0^\infty G(f,\cdot,t,z)te^{-\lambda t}dt \rightharpoonup H(f,\cdot,z)$, the assertion of the Theorem is now a direct consequence of Proposition 2.4.

REFERENCES

[1] K. Ito, Random processes II, Moscov 1963 (in Russian).

[2] A.V. Skorohod, Random processes with independent increments, Moscov 1964 (in Russian).

ASYMPTOTIC ALMOST PERIODIC SOLUTIONS
FOR STOCHASTIC DIFFERENTIAL EQUATIONS

Constantin Vârsan
Department of Mathematics, INCREST
Bd.Pacii 220, 79622 Bucharest, Romania

1. INTRODUCTION

We consider stochastic differential equations, with asymptotic almost periodic coefficients and sufficient conditions for a bounded solution to be asymptotically almost periodic in distribution are given. The case of almost periodic coefficients has been discussed in many articles and the total stability property is an important concept for the existence of an asymptotically almost periodic solution for ordinary and functional differential equations, see [1], [2] for the references. A total stability concept is necessary also for stochastic differential equations but since the convergence in distribution of the initial conditions is too large for giving the convergence in $L_2(\Omega,P)$ of the corresponding solutions we need to use equivalent stochastic differential equations which are more suitable for our purpose. This presentation is based on the paper [3] and contains additionally a sufficient condition for total stability property.

2. FORMULATION OF THE PROBLEM AND MAIN RESULT

Let $f_i:I\times R^n\to R^n$, $i=0,1,\dots,m$, be continuous and asymptotic almost periodic in the variable $t\in I$, $I=[0,\infty)$, uniformly with respect to x belonging to compact sets in R^n; i.e. any sequence of real numbers $\{t_k\}\uparrow\infty$ contains a subsequence $\{t_k'\}\uparrow\infty$ such that $\{f_i(t+t_k',x)\}k\geq 1$ is convergent uniformly in $t\in I$ and x is compact sets in R^n.

Let $w(t)$, $t\geq 0$, be a m-dimensional Wiener process over the probability space $\{\Omega,F,P\}$. Let $\{F_t\}_{t\geq 0}$ be an increasing family of σ-algebras in F such that the σ-algebra generated by $w(t+s)-w(t)$, $s>0$, is independent of F_t for each $t\geq 0$. Let x_o be F_o measurable and $x_o\in L_p(\Omega,P)$; The following stochastic differential equation

$$1)\qquad dx=f_o(t,x)dt+\sum_{i=1}^{m} f_i(t,x)dw_i(t),\quad t\geq 0,\quad x(0)=x_o$$

has a unique solution $\chi(t)$, $t\in I$, if

i_1) $|f_i(t,x)|\leq K(1+|x|)$, $|f_i(t,\chi_2)-f_i(t,\chi_1)|\leq K_N|\chi_2-\chi_1|$,

for any $t\in I$, $\chi\in R^n$, $\chi_1,\chi_2\in R^n$, $|x_j|\leq N$.

In addition, for each $p\geq 2$, $E|x(t)|^p<\infty$ for any $t\geq 0$, if $E|x_o|^p<\infty$.

Denote $L^+(f_o,\ldots,f_m)$ the set of functions $g_i:I\times R^n\to R^n$ $i=0,1,\ldots m$, which are obtained as limits from f_i, i.e. $g_i(t,x)=\lim_{t_k\to\infty}(t+t_k,x)$, $i=0,1,\ldots,m$, uniformly in $t\in I$ and χ in compact sets in R^n. The solutions in (1) are studied using limiting equations

2) $$dx=g_o(t,x)dt+\sum_{i=1}^{m}g_i(t,\chi)dw_i(t),\quad t\geq 0,$$

where $(g_o,\ldots,g_m)\in L^+(f_o,\ldots,f_m)$.

REMARK 1. According to our purpose we don't need and it is suitable not to fixe in advance the probability space $\{\Omega,F,P\}$. The auxiliary probability space will be $\{\tilde\Omega,\tilde F,\tilde P\}$, where $\tilde\Omega=\bar\Omega\times\Omega$, $\tilde F=\bar F\times F$, $\tilde P=\bar P\times P$, and $\bar\Omega=[0,1)$, $\bar F$ is the σ-algebra of Borel sets in $[0,1)$, and $\bar P$ is the Lebesgue measure.

The solutions in (1) and (2) are defined considering $w(t)$, $t\geq 0$, a Wiener process on $\{\Omega,F,P\}$ with the reference family $\{F_t\}t\geq 0$, and initial condition $\chi(0)=\chi_o$ a random vector in $\{\bar\Omega,\bar F\}$.

Denote P the set of probability measures on R^n.

Let $I_1\subseteq I_2\subseteq R$ be some intervals.

DEFINITION 1. We say that $P_k:I_2\to P$, $k\geq 1$, is weakly compact uniformly with respect to $t\in I_1$, if there exists $\pi:I_2\to P$ and a subsequence $\{P_{k'}\}\subseteq\{P_k\}$ such that $\lim_{k'\to\infty}\int_{R^n}\varphi(x)P_{k'}(t,dx)=\int_{R^n}\varphi(x)\pi(t,dx)$, $t\in I_2$, uniformly with respect to $t\in I$, for any $\varphi\in C_b(R^n)$, φ Lipschitz continuous.

Let $\chi(t)$, $t\geq 0$, be a solution in (1). Denote by $P(t,dx)$, $t\geq 0$, the probability measure on R^n generated by $\chi(t)$.

DEFINITION 2. We say that a solution $\chi(t)$, $t\geq 0$, in (1) is asymptotically almost periodic in distribution if $P_k(t,dx)=P(t+t_k,dx)$, $k\geq 1$, is weakly compact uniformly in $t\varepsilon I$, for any sequence $\{t_k\}\subset I$, $\lim t_k=\infty$

Assume that (1) has a bounded solution in $L_p(\Omega,P)$, $p>2$, i.e. there exists $\tilde{\chi}(t)$, $t\varepsilon I$, solution in (1) such that

$$i_2)\ \sup_{t\geq 0} E|\tilde{x}(t)|^p\leq M<\infty \quad \text{for some} \quad p>2,$$

and we call it an L_p-bounded solution. Denote by H the set of measurable processes $h:I\times\tilde{\Omega}\to R^{n\times(m+1)}$ which are continuous in the variable $t\varepsilon I$ and for each $t\varepsilon I$, $h(t,\cdot)$ is $\tilde{F}_t$-adapted. For each $s\geq 0$, we need to consider solutions of the following equation

$$1_s)\ dx=f_o(t+s,x)dt+\sum_{i=1}^{m} f_i(t+s,\chi)dw_i(t), \quad t\geq 0,$$

$$\chi(0)=\chi_o(\bar{\omega}),$$

where $\chi_o\varepsilon L_2(\bar{\Omega},\bar{P})$ and the equation (1_s) is considered on the new space $\{\tilde{\Omega},\tilde{F},\tilde{P}\}$ (see remark 1). In a similar way, for each $(g_o,\ldots,g_m)\varepsilon$ $\varepsilon L^+(f_o,\ldots,f_m)$ and $s\geq 0$ we consider the following equation

$$2_s)\ dx=g_o(t+s,x)dt+\sum_{i=1}^{m} g_i(t+s,x)dw_i(t)\ ,$$

$$\chi(0)=\chi_o(\bar{\omega})$$

Any solution in (1_s) and (2_s) is a measurable process $\chi(t,\tilde{\omega}):I\times\tilde{\Omega}\to R^n$, where trajectories are continuous functions a.e. $(\tilde{P})$ and with $x(t,\cdot)$ $\tilde{F}_t$-adapted. Write $f=(f_o,\ldots,f_m)$, $g=(g_o,\ldots,g_m)$ and define $f_s(t,x)=$ $=(f_o(t+s,\chi),\ldots,f_m(t+s,\chi))$, $g_s(t,x)=(g_o(t+s,\chi),\ldots,g_m(t+s,\chi))$. The solutions in (1_s) and (2_s) are denoted by $x(t,f_s,\chi_o)$ and $x(t,g_s,\chi_o)$ correspondingly.

DEFINITION 3. We say that $x(t,f_s,\chi_o)$ in (1_s) is totally stable if for any $\varepsilon>0$ there exists $\delta(\varepsilon)>0$ such that

$$\sup_{t\geq 0} \tilde{E}|x(t,f_s+h,\chi_o+v_o)-x(t,f_s,\chi_o)|^2\leq\varepsilon,$$

for any $v_o\varepsilon L_2(\bar{\Omega},\bar{P})$ and $h\varepsilon H$ fulfilling $\bar{E}|v_o|^2\leq\delta(\varepsilon)$ and $\sup_{t\geq 0}\tilde{E}|h(t,\tilde{\omega})|^2\leq$

$\leq\delta(\varepsilon)$.

An L_p-bounded solution in (1_s) is defined by

$$\sup_{t\geq 0} \tilde{E}|x(t,f_s,\chi_o)|^p \leq M<\infty$$

and similarly for (2_s).

DEFINITION 4. We say that (1) (or (2)) is totally stable if any L_p-bounded solution in (1_s) (or (2_s)) is totally stable for any $s\geq 0$.

THEOREM 1. Assume (i_1) and (i_2) are fulfilled and (1) is totally stable. Then any L_p-bounded solution $x(t)$ in (1) is asymptotically almost periodic in distribution.

THEOREM 2. Assume (i_1) and (i_2) are fulfilled. Let $(g_o,\ldots,g_m)\varepsilon$ $\varepsilon L^+(f_o,\ldots,f_m)$ and the corresponding equation (2) is totally stable. Then any L_p-bounded solution $\chi(t)$ in (1) is asymptotically almost periodic in distribution.

REMARK 2. From theorem 1 or 2 follows that for any $\varphi\varepsilon C_b(R^n)$, φ Lipschitz continuous, the real function $f(t)=E\varphi(x(t,\chi_o))$ defined on $I=[0,\infty)$, is an asymptotic almost periodic function and the results in [1] apply to this case in dependence on the function φ we choose. More general, under the conditions in theorem 1 (or 2) with $p\geq 4$, the distribution function $P(t,dx)$, $t\varepsilon I$, corresponding to an L_p-bounded solution, has an almost periodic component $\Pi(t,dx):R\to(\mathcal{P},\rho)$ such that $\lim_{t\to\infty}\rho(P(t),\Pi(t))=0$ and the corresponding decomposition $P(t,dx)=$ $=\Pi(t,dx)+Q(t,dx)$, is unique, where ρ is a metric on $\mathcal{P}$ and $Q(t,dx)$ a finite radon measure.

The proof of the Theorem 1 (or 2) relies on the following

LEMMA. Assume that (1) has an L_p-bounded solution $\hat{x}(t)$, $t\geq 0$, and (1) is totally stable.

Then for any $\{t_k\}\uparrow\infty$ there exist $\{s_j\}\subset\{t_k\}$, $\{s_j\}\uparrow\infty$ and $x_j(t)$ solution in $(1s_j)$, and $\chi(t)\varepsilon L_2(\tilde{\Omega},\tilde{P})$ such that $\lim_{j\to\infty}\sup_{t\geq 0}\tilde{E}|x_j(t,\tilde{\omega})-x(t,\tilde{\omega})|^2=0$ and $P_j(t,dx)$ generated by $x_j(t)$ coincides with $\hat{P}_j(t,dx)$ generated by

$\hat{x}(t+s_j)$ for any $t\geq 0$, $j\geq 1$.

REMARK 4. As is pointed out in the Theorem 1, for any L_p-bounded solution $(p>2)$ $\hat{x}(t)$, $t\geq 0$, in (1) we get the following convergence

$$\lim_{k\to\infty} \int_{R^n} \varphi(x)\hat{P}(t+t_k,dx)=\int_{R^n} \varphi(x)P(t,dx)$$

uniformly in $t\geq 0$ for each $\varphi\varepsilon C_b(R^n)$, φ Lipschitz continuous, where $\{t_k\}\uparrow\infty$. Using the above Lemma it is obvious that this convergence and consequently the asymptotic almost periodic property can be proved for any $\varphi\varepsilon C^1(R^n)$ which fulfils $|\partial\varphi/\partial x(x)|^2\leq k(1+|x|^q)$ with $q\leq p$. It gives the possibility to obtain asymptotic almost periodic property using the definition based on the moments of the bounded solution.

3. A SUFFICIENT CONDITION FOR TOTAL STABILITY PROPERTY

The total stability property for (1) is fulfilled if we assume in addition the following.

Let f_i, $i=0,1,\ldots,m$, be of class C^1 in the variable $x\varepsilon R^n$ and write $M_i(t,x)=\partial f_i/\partial x(t,x)$

i_3) there exist $\gamma<0$, $K>0$ such that $|M_i(t,x)|\leq K$ and

$$M_o(t,x)+M_o^*(t,x)+\frac{1}{2}\sum_{i=1}^{m} M_i(t,x)M_i^*(t,x)\leq\gamma I$$

for any $(t,x)\varepsilon[0,\infty)\times R^n$, where I is the identity matrix.

THEOREM 3. Let the condition (i_3) be fulfilled. Then (1) is totally stable.

Proof

We have to prove that for each $s\geq 0$, any L_p-bounded solution in (1_s) is totally stable. It is enough to consider the case $s=0$ and write $y(t)=x(t)-\tilde{x}(t)$, where $\tilde{x}(t)=x(t,f,x_o)$, $x(t)=x(t,f+h,x_o+\bar{x}_o)$. By definition

$$dy(t)=A_o(t,\omega)y(t)dt+\sum_{i=1}^{m} A_i(t,\omega)y(t)dw_i(t)+h_o(t,\omega)dt+$$

$$+\sum_{i=1}^{m} h_i(t,\omega)dw_i(t) \ ,$$

$$y(0)=\bar{x}_o \ ,$$

where

$$A_i(t,\omega)=\int_0^1 M_i(t,\tilde{x}(t)+\theta(y(t))d\theta, \quad i=0,1,\dots,m, \text{ and } h=(h_o,\dots,h_m) .$$

Therefore $y(t)$ can be represented by

3) $$y(t)=X(t)\bar{x}_o+\int_0^t X(t)X^{-1}(s)[h_o(s,\omega)-\frac{1}{2}\sum_{i=1}^{m} A_i(s,\omega)h_i(s,\omega)]ds+$$

$$+\sum_{i=1}^{m} X(t)\int_0^t X^{-1}(s)h_i(s,\omega)dw_i(s) \ ,$$

where $X(t)$ is the solution of the matrix equation

4) $$dX=A_o(t,\omega)Xdt+\sum_{i=1}^{m} A_i(t,\omega)Xdw_i(t), \quad X(0)=I$$

We have to prove that $\sup_{t\geq 0} \tilde{E}|y(t)|^2\to 0$ if $\tilde{E}|\bar{x}_o|^2+\sup_{t\geq 0} \tilde{E}|h(t)|^2\to 0$

Using (i_3) and (4) we get that $X(t,s)=X(t)X^{-1}(s)$ fulfils

5) $$\tilde{E}|X(t,s)|^2\leq \exp \gamma(t-s), \quad t\geq s$$

where γ is the constant given in (i_3). By hypothesis $A_i(t)\triangleq A_i(t,\omega)$, $i=0,1,\dots,m$, are bounded and using (5) in (3), we get easily that the first and second term in the right hand side of (3) are approaching zero in $L_2(\tilde{\Omega},\tilde{P})$ uniformly in $t\geq 0$ if $\sup_{t\geq 0} \tilde{E}|h(t)|^2+\tilde{E}|\bar{x}_o|^2\to 0$. For the third term T_3 in (3) we have

6) $$\tilde{E}|T_3(t)|^2\leq \tilde{E}|X(t)|^2|z(t)|^2$$

where $z(t)=\sum_{i=1}^{m}\int_0^t X^{-1}(s)h_i(s)dw_i(s)$.

The following equation is fulfilled

7) $$d\tilde{E}|X(t)|^2|z(t)|^2=\tilde{E}(\text{trace } X^*(t)\tilde{A}(t)X(t))|z(t)|^2dt+$$

$$+\tilde{E}|X(t)|^2\sum_{i=1}^{m}|X^{-1}(t)|^2|h_i(t)|^2dt+$$

$$+\tilde{E}\sum_{i=1}^{m}(\text{trace}[X^*(t)(A_i^*(t)+A_i(t))X(t)]<$$

$$<z(t),\ X^{-1}(t)h_i(t)>dt$$

where $\tilde{A}(t)=A_o(t)+A_o^*(t)+\frac{1}{2}\sum_{i=1}^{m}A_i(t)A_i^*(t)$.

Denote $\varphi(t)=\tilde{E}|X(t)|^2|z(t)|^2$, $\psi(t)=\sum_{i=1}^{m}[(\tilde{E}|h_i(t)|^2)^{1/2}+\tilde{E}|h_i(t)|^2$.

Using (i_3) and the boundedness of $A_i(t)$ we get

8) $$d\varphi/dt\leq\gamma\varphi+c\psi(t),\quad t\geq 0,\quad \varphi(0)=0$$

for some constant $c>0$, provided $\sup_{t\geq 0}\psi(t)$ is sufficiently small, and finally

9) $$\varphi(t)\leq c_1\sup_{t\geq 0}\psi(t),\quad t\geq 0$$

for some constant $c_1>0$, which ensures that $\sup_{t\geq 0}\tilde{E}|T_3(t)|^2$ goes to zero when $\sup_{t\geq 0}\tilde{E}|h(t)|^2\to 0$. The proof is complete.

REFERENCES

[1] T. Yoshizawa, Stability theory and the Existence of Periodic Solutions and Almost Periodic Solutions, Springer-Verlag, Applied Mathematical Sciences 14, 1975.

[2] Y. Hino, T. Yoshizawa, Total Stability Property in Limiting Equations for a Functional Differential Equation with Infinite delay, Casopis pre pestovani matematiky, roc 111 (1986), Praha.

[3] C. Vârsan, Asymptotic almost periodic solutions for stochastic differential equations, to appear in Tohoku Mathematical Journal.

2. STOCHASTIC INFINITE DIMENSIONAL SYSTEMS

STOCHASTIC INTEGRAL WITH RESPECT TO A GENERALIZED WIENER PROCESS IN A CONUCLEAR SPACE

Tomasz Bojdecki and Jacek Jakubowski
Institute of Mathematics, University of Warsaw
00-901 Warsaw PKiN, Poland

I. Introduction

During the last years it has turned out that the duals of nuclear spaces (i.e conuclear spaces) are natural state spaces for various limit models describing some complex physical phenomena. For instance, fluctuation limits of interacting particle systems, typically, are processes in some distribution spaces. These processes are continuous Gaussian, with their martingale components being continuous Gaussian martingales, not necessarily time-homogeneous ([4],[5],[6]). In this paper we describe a construction of the stochastic integral with respect to such a martingale, which we call a generalized Wiener process. We also indicate some properties and applications of this integral. The main stress is laid upon the detailed description and characterization of the class of integrands, and on making this class as large as possible (see [7],[8],[9],[11],[12],[13] for other approaches to the stochastic integration in nuclear spaces, see also [2], where a simpler situation is considered). The detailed proofs of the results presented in this paper will be published elsewhere [3].

II. Results

We start from fixing some basic notations and definitions, see [7] or [2] for details.

Let Φ be a real vector space and let q be a separable Hilbertian seminorm (abbr. <u>H-seminorm</u>). By k_q we denote the canonical mapping $\Phi \to \Phi/_{\ker q}$, i.e. $k_q(f)=[f]_q$. We usually identify an $f \in \Phi$ and $[f]_q$. The q-completion of $\Phi/_{\ker q}$ is a separable Hilbert space denoted by Φ_q, and its dual Φ'_q is Hilbert with the norm q', the dual norm of q.

Let q and v be H-seminorms on Φ. The seminorm q is said to be <u>bounded</u> by v, written

$$q \prec v ,$$

if the identity mapping $Id:(\Phi,v)\to(\Phi,q)$ is continuous. For $q\prec v$ we have the canonical continuous mapping

$$k_{vq}: \Phi_v \to \Phi_q ,$$

being the extension of the mapping $[f]_v \to [f]_q$, for $f\in\Phi$. The conjugate mapping:

$$i_{qv}: \Phi'_q \to \Phi'_v$$

is a continuous imbedding. Analogously, if Φ is a locally convex topological vector space, Φ' denotes its strong dual, and q is a continuous seminorm on Φ, then the mapping

$$i_q: \Phi'_q \to \Phi',$$

conjugate to k_q , is a continuous imbedding.

We say that q is <u>Hilbert-Schmidt</u> <u>bounded</u> by v, written

$$q \prec_{HS} v,$$

if $q\prec v$ and k_{vq} is a Hilbert-Schmidt operator.

Let Ψ be another vector space, and u be an H-seminorm on Ψ. By $\mathcal{L}(\Phi'_q,\Psi'_u)$ (resp. $\mathcal{L}_2(\Phi'_q,\Psi'_u)$) we denote the space of all linear bounded (resp. Hilbert-Schmidt) operators from Φ'_q into Ψ'_u, and by $\|\cdot\|_{qu}$ (resp. $|\cdot|_{qu}$) we denote the norm in $\mathcal{L}(\Phi'_q,\Psi'_u)$ (resp. $\mathcal{L}_2(\Phi'_q,\Psi'_u)$). We assume that Φ is <u>nuclear</u> and <u>barrelled</u> space (typical examples are Schwartz spaces $\mathcal{S}$ and $\mathcal{D}$), and Ψ is a multi-Hilbertian space. Let $(\Omega,\mathcal{F},P)$ be a complete probability space with a filtration $(\mathcal{F}_t)_t$, which is right continuous and $\mathcal{F}_0$ contains all P-null sets.

<u>1.Definition.</u> Let q_s be a continuous H-seminorm on Φ, for $s\in R_+$, such that the function $s\to q_s(f,g)$ is Borel measurable and bounded on finite intervals for each $f,g\in\Phi$. A Φ'-valued continuous centered Gaussian process $W=(W_t)_{t\in R_+}$ is called a <u>generalized</u> <u>Wiener</u> process associated to the family $\{q_s\}_s$, if

i) W_t is $\mathcal{F}_t$- measurable for each $t\in R_+$;

ii) W_t-W_s is independent of $\mathcal{F}_s$ for $t>s$;

iii) the covariance functional $K(t',f;t'',g) = E(W_{t'}(f)W_{t''}(g))$ has the form

$$K(t',f;t'',g) = \int_0^{t'\wedge t''} q_s(f,g)ds, \qquad t',t''\in R_+,\ f,g\in\Phi. \tag{1}$$

Our aim is to define a stochastic integral with respect to a generalized Wiener process W.

In the <u>first</u> (essential) <u>step</u> we will fix $T>0$ and a continuous H-seminorm u on Ψ. We will integrate operator-valued processes X on $[0,T]$, such that $X_s(\omega)$ is a linear operator from Φ'_{q_s} into Ψ'_u. Firstly we define an integral for "step" processes. Let $0\le t_1<t_2\le T$, and let v

be a continuous H-seminorm on Φ such that

(2) $$q_s \prec v \quad \text{for } s\in]t_1,t_2] \ ,$$

and

(3) $$\left(\int_{t_1}^{t_2} q_s^2 ds \right)^{1/2} \prec v \ .$$

Since Φ is a barrelled space such v exists. We consider a process X given by the formula

(4) $$X_s(\omega) = I_{]t_1,t_2]\times F}(s,\omega) A i_{q_s v} \ ,$$

where $F\in\mathcal{F}_{t_1}$, and A has the following properties:

(5) $$A \in \mathcal{L}(\Phi'_v,\Psi'_u)$$

and

(6) $$A i_{q_s v} \in \mathcal{L}_2(\Phi'_{q_s},\Psi'_u) \quad \text{for } s\in]t_1,t_2] \ .$$

and

(7) $$\int_{t_1}^{t_2} |A i_{q_s v}|^2_{q_s u} \, ds < \infty \ .$$

2.Remark. This definition includes two important cases:

a) $A\in\mathcal{L}(\Phi'_v,\Psi'_u)$ and $i_{q_s v}\in\mathcal{L}_2(\Phi'_{q_s},\Phi'_v)$ (it is the case when $q_s \prec_{HS} v$ for $s\in]t_1,t_2]$);

b) $A\in\mathcal{L}_2(\Phi'_v,\Psi'_u)$ (of course $i_{q_s v}\in\mathcal{L}(\Phi'_{q_s},\Phi'_v)$).

The case a) permits to establish a relationship between our approach and the isometric stochastic integral in Hilbert space [10].

For X given by (4) we define the integral in the obvious form:

$$\int_0^t X_s dW_s =: I_F A(W_{t_2\wedge t}-W_{t_1\wedge t}) ,$$

where $A(W_{t_2\wedge t}-W_{t_1\wedge t})$ has the meaning explained below:

For $f\in\Phi$, by (1) and (3), we have

$$E\left[(W_{t_2}-W_{t_1})(f)\right]^2 = \int_{t_1}^{t_2} q_s^2(f) ds \le c v^2(f),$$

so $W_{t_2}-W_{t_1}$ can be uniquely extended to a continuous linear operator from Φ into $L^2(\Omega,\mathcal{F},P)$. Furthermore, by (2), for $h\in\Phi_v$

(8) $$E\left[(W_{t_2}-W_{t_1})(h)\right]^2 = \int_{t_1}^{t_2} q_s^2(k_{v q_s} h) ds$$

because both sides of (8) are continuous seminorms on Φ_v, and by (1) equality holds for a dense subset $\Phi\subset\Phi_v$. Hence $Z := (W_{t_2}-W_{t_1})A'$ is a continuous linear operator from Ψ_u into $L^2(\Omega,\mathcal{F},P)$ such that for $g\in\Psi_u$

$$E\big(Z(g)\big)^2 = \int_{t_1}^{t_2} q_s^2(k_{vq_s} A'g)\,ds .$$

Let (d_n) be an orthonormal basis in Ψ_u. Then by (7)

$$\sum_{n=1}^{\infty} E\big(Z(d_n)\big)^2 = \int_{t_1}^{t_2} |Ai_{q_s v}|^2_{q_s u}\, ds < \infty ,$$

so by the regularization theorem [7] there is a unique version of Z which is a Ψ'_u-valued random variable. Now by $A(W_{t_2}-W_{t_1})$ we mean this regular version of Z.

Our definition of the integral does not depend on v and A in this sense that if v_1 and A_1 have the same properties as v and A and $Ai_{q_s v}=A_1 i_{q_s v_1}$,then

(9) $$A(W_{t_2}-W_{t_1}) = A_1(W_{t_2}-W_{t_1}).$$

To see this, it is enough to show that both sides of (9) are equal as linear random functionals, what in turn is a consequence of the following generalization of (8): If v_1, v_2 are continuous H-seminorms on Φ satisfying (2) and (3), then for each $f\in\Phi_{v_1}$, $g\in\Phi_{v_2}$

(10) $$E\big((W_{t_2}-W_{t_1})(f)(W_{t_2}-W_{t_1})(g)\big) = \int_{t_1}^{t_2} q_s(k_{v_1 q_s} f, k_{v_2 q_s} g)\,ds$$

Of course $Y=\int XdW$ is a Ψ'_u-valued martingale, and (8) together with known estimations for moments of Gaussian variables ([7] Th.2.7.3), by the Kolmogorov theorem on continuous version, imply that Y has continuous trajectories in Ψ'_u (and hence in Ψ').

Let $\mathfrak{F}$ denote the class of linear combinations of processes defined by (4). If $X\in\mathfrak{F}$ and $X=\Sigma_{j=1}^{n}X_j$, where X_j is of the form (4) (the seminorm v may of course depend on j), we define $\int XdW = \Sigma_{j=1}^{n} \int X_j dW$.

<u>3.Proposition.</u> If $X\in\mathfrak{F}$ then $\int XdW$ is a continuous Ψ'_u-valued martingale not depending on representation of X, and

$$E\big(u'(\int_0^T XdW)\big)^2 = E\int_0^T |X_s|^2_{q_s u} ds .$$

The problem is to specify a possibly large class of processes on which our definition of stochastic integral can be extended. Looking at the conditions (5), (6), (7) and Proposition 3 it is natural to introduce the following class $\Lambda(u,T) = \Lambda(W,\Psi;u,T)$ of processes:

<u>4.Definition.</u> $\Lambda(u,T)$ is the class of processes $(X_s)_{s\le T}$ such that

i) $X_s(\omega) \in \mathcal{L}_2(\Phi'_{q_s}, \Psi'_u)$,

ii) for each $f\in\Phi$, $g\in\Psi$ the process $(q_s(X'_s g, f))_{s\le T}$ is progressively measurable,

iii) $\|X\|_\Lambda^2 := E\int_0^T |X_s|^2_{q_s u} ds$ is finite.

It can be proved that the first two conditions imply that the integral in iii) is well-defined. It is clear that $\mathcal{E} \subset \Lambda$.

5. Proposition. $\Lambda(u,T)$ is a Hilbert space with the norm $\|\cdot\|_\Lambda$.

Elements of this Hilbert space are equivalence classes, and when writing $X \in \Lambda(u,T)$ we mean that there exists a process Y satisfying conditions i), ii), iii) of Definition 4 such that $X_s(\omega) = Y_s(\omega)$ $ds \otimes dP$ - a.e. on $[0,T]\times\Omega$.

It is seen that if we treat $\mathcal{E}$ as a subspace of Λ, then $X \longrightarrow \int X dW$ is a linear isometry into $\mathcal{M}^{2,c}([0,T],\Psi'_u)$, the space of continuous square integrable martingales in Ψ'_u. $\Lambda(u,T)$ is not of the form $L^2(\nu)$ for any measure ν so it is not obvious that this isometry can be extended to the whole space $\Lambda(u,T)$. Nevertheless, it turns out that the following theorem is true:

6. Theorem. There exists exactly one linear isometry J from $\Lambda(u,T)$ into $\mathcal{M}^{2,c}([0,T],\Psi'_u)$ such that for each process X given by (4), $J(X)_t$ is given by $I_F A(W_{t\wedge t_2} - W_{t\wedge t_1})$.

For $X \in \Lambda(u,T)$ we define the integral $\int_0^t X_s dW_s := J(X)_t$.

The idea of the proof is to approximate an X in $\Lambda(u,T)$ by processes whose values are finite-dimensional operators and then to approximate the latter processes by elementary ones. We use the fact that there exists a continuous H-seminorm v such that $\left(\int_0^T q_s^2 ds\right)^{1/2} \prec_{HS} v$. Given such v, the process $(W_t)_{t\le T}$ is a continuous process in Φ'_v. But it should be stressed that our definition of the stochastic integral does not depend on the choice of the seminorm v. From the proof we get

7. Corollary. If a generalized Wiener process $W = (W_t)_{t\le T}$ lives in a Hilbert space Φ'_v, and X is a process which is integrable with respect to W in the sense of the isometric integral in Hilbert space, then the process Xi, where $i = (i_{q_s v})_s$, is integrable in our sense and both integrals coincide.

If we consider X as a process with values in Ψ' it may happen that

$X\in\Lambda(u,T)$ and at the same time $X\in\Lambda(u_1,T)$ for different continuous H-seminorms u,u_1. It can be proved that the integral as the process in Ψ' is the same for u and u_1.

Now as the <u>second step</u> we extend the set of integrands to a class which does not depend on u and T.

<u>8.Definition</u>. $\Lambda(W,\Psi)$ (abbr. Λ) is the class of processes $(X_s)_{s\in R_+}$ such that

i) $X_s(\omega) \in \mathcal{L}(\Phi'_{q_s},\Psi')$,

ii) for each $f\in\Phi$, $g\in\Psi$ the process $(q_s(X'_s g,f))_s$ is progressively measurable,

iii) there exist a sequence $(u_n)_n$ of continuous H-seminorm on Ψ and a sequence $(\vartheta_n)_n$ of bounded stopping times such that $\vartheta_n \nearrow \infty$,

$$X_s(\omega) \in \mathcal{L}_2(\Phi'_{q_s},\Psi'_{u_n}) \qquad \text{for } s\in[0,\vartheta_n(\omega)],$$

and

$$E\left(\int_0^{\vartheta_n} |X_s|^2_{q_s u_n} ds\right) < \infty \tag{11}$$

for $n=1,2,\ldots$.

<u>9.Remark.</u> The class Λ contains processes satisfying i), ii) and iii) with (11) replaced by

$$P\left(\int_0^{\vartheta_n} |X_s|^2_{q_s u_n} ds < \infty\right) = 1.$$

Of course, if $\vartheta_n \le T_n$, then $I_{[0,\vartheta_n]}X \in \Lambda(u_n,T_n)$. Therefore, by Theorem 6 we have the well defined stochastic integral $\int_0^t I_{[0,\vartheta_n]}XdW$ for $t\le T_n$. By the telescoping procedure we obtain the following theorem (here we adopt the usual notation for the process stopped at a (stopping) time ϑ: $Y^\vartheta_t = Y_{t\wedge\vartheta}$).

<u>10.Theorem.</u> Let $X\in\Lambda$. There exists a unique continuous Ψ'-valued process $\int XdW$ such that for any sequences $(u_n)_n$, $(\vartheta_n)_n$ satisfying conditions of the point iii) of Definition 8

$$\left(\int XdW\right)^{\vartheta_n} = \int I_{[0,\vartheta_n]}XdW .$$

One of applications of our definition of the stochastic integral is the possibility to represent a generalized Wiener process by means of (as the stochastic integral with respect to) another one. Namely, we have:

11.Theorem. Let W be a generalized Wiener process associated to $(r_s)_s$. Let $(q_s)_s$ be a family of continuous H-seminorms on Φ such that the function $s \to q_s(f,g)$ is Borel measurable and bounded on finite intervals for each $f,g \in \Phi$. Assume that $\dim \Phi_{r_s} = \dim \Phi_{q_s}$. Then there exist a generalized Wiener process Z associated to $(q_s)_s$ and a deterministic process $X \in \Lambda(Z,\Phi)$ such that $W = \int XdZ$.

The assumption of the equality of dimensions can be replaced by $\dim \Phi_{r_s} \le \dim \Phi_{q_s}$, but then it can be necessary to extend the underlying probability space [3]. However, if we content ourselves with representing W as a stochastic integral with respect to Z in the weak sense only, i.e. if we want $\int XdZ$ to have the same distribution as W, then no extension is needed (cf. [2]).

In the proof of Theorem 11 we use among other things two properties of the stochastic integral which are of interest by themselves.

12.Proposition. Let Ξ be a nuclear space and let X be a deterministic process such that $X \in \Lambda(W,\Xi)$. Then there exists an increasing sequence $(v_n)_n$ of H-seminorms on Ξ such that $I_{[0,n]} X \in \Lambda(W,\Xi;v_n,n)$ for n=1,2,... . The stochastic integral $\int XdW$ is a generalized Wiener process in Ξ' associated to H-seminorms $r_s(h) = q_s(X_s'h)$, $h \in \Xi$. Moreover X has the form $X_s = i_{r_s} \hat{X}_s$, where $\hat{X}_s \in \mathcal{L}(\Phi'_{q_s}, \Xi'_{r_s})$.

13.Proposition. Let Φ,Ξ be nuclear spaces, and Ψ a multi-Hilbertian space. Assume that W is a generalized Wiener process in Φ' associated to $(q_s)_s$, and X is a deterministic process, $X \in \Lambda(W,\Xi)$. Then for any $Y \in \Lambda(\int XdW,\Psi)$ the process $Y\hat{X}$ belongs to $\Lambda(W,\Psi)$ and

$$\int Yd(\int XdW) = \int (Y\hat{X})dW$$

where $\hat{X}$ is given in Proposition 12.

Theorem 11 and Propositions 12 and 13 give

14.Corollary. Let W, Z be given by Theorem 11. If $Y \in \Lambda(W,\Phi)$, then $\int YdW = \int \bar{Y}dW$ where $\bar{Y} \in \Lambda(Z,\Phi)$.

Theorem 11 makes it possible to represent a generalized Wiener process in distribution spaces such as $\mathcal{S}'$ or $\mathcal{D}'$ as a stochastic integral with respect to the <u>standard</u> <u>Wiener</u> process (i.e. process associated to the L^2-norm). For instance, the stochastic Langevin equation $dX_t = A_tX_tdt + dW_t$ discussed in [1], with X being, typically, the fluctuation limit of a particle system, can be viewed as an equation of the form $dX_t = A_tX_tdt + Y_tdZ_t$, where Z is the standard Wiener process.

References

[1] T.Bojdecki,L.G.Gorostiza, Inhomogeneous infinite dimensional Langevin equations, *Stoch.Anal.Appl.* **6** (1988), no.1, 1-9.

[2] T.Bojdecki,J.Jakubowski, Ito stochastic integral in the dual of a nuclear space, to appear in *J.Multivariate Anal.*

[3] T.Bojdecki,J.Jakubowski, Stochastic integration for inhomogeneous Wiener process in the dual of a nuclear space, to appear

[4] L.G.Gorostiza, High density limit theorems for infinite systems of unscaled branching Brownian motions, *Ann. Probab.* **11** (1983), no.2, 374-392.

[5] M.Hitsuda,I.Mitoma, Tightness problem and stochastic evolution equation arising from fluctuation phenomena for interacting diffusions, *J.Multivariate.Anal.* **19** (1986), 311-328.

[6] K.Ito, Distribution valued processes arising from independent Brownian motions, *Math.Z.* **182** (1983), 17-33.

[7] K.Ito, *Foundations of stochastic differential equations in infinite dimensional spaces*, SIAM, Philadelphia, 1984.

[8] G.Kallianpur,I.Mitoma,R.L.Wolpert, Diffusion equations in duals of nuclear spaces, *Center for Stoch. Proc. Dep. Stat., Univ. North-Car., Chapel Hill,N.C., Techn.Rep.* 234, 1988.

[9] H.Korezlioglu,C.Martias, Stochastic integration for operator valued processes on Hilbert spaces and on nuclear spaces, *Center for Stoch. Proc., Dep. Stat., Univ. North-Car., Chapel Hill,N.C., Techn.Rep.* 85, 1984.

[10] M.Metivier, *Semimaringales: A Course On Stochastic Processes*, de Gruyter, Berlin, 1982.

[11] V.Pérez Abreu, Multiple Wiener integrals and nonlinear functionals of a nuclear space valued Wiener process, *Appl. Math.Optimiz.* 16 (1987), 227-245.

[12] A.S.Ustunel, Stochastic integration on nuclear spaces and its applications, *Ann.Inst.H.Poinc.* **18** (1982), no.2, 165-200.

[13] J.B.Walsh, An introduction to stochastic partial differential equations, *Lect.Notes in Math.* **1180** (1986), Springer-Verlag, 266-439.

PERIODIC LINEAR EQUATIONS WITH GENERAL ADDITIVE NOISE IN HILBERT SPACES

Anna Chojnowska-Michalik

Institute of Mathematics, Łódź University, ul. S.Banacha 22, 90-238 Łódź, Poland

1. Introduction

Recently, in several papers ([8], [9], [6]) linear periodic stochastic differential equations have been studied. In this paper we consider the simplest linear τ-periodic system (*) below but with a general additive noise described by a process with independent periodic increments. (Such a noise seems to be as general as possible if one wants to have markovian solutions of (*).)

A τ-periodic solution to (*) will mean a solution which is a Markov process and has a τ-periodic distribution.

Our aim is to give sufficient and necessary conditions for the existence of a τ-periodic solution to (*) and to characterize τ-periodic distributions for (*).

Both the finite and infinite dimensional cases are treated of. To the author s knowledge, the results presented here are also new for the gaussian noise case. We present a complete solution of the above problem in R^d (Theorem 1; the proof can be found in the author´s paper [3]). In section 4, which is the main part of the present paper, we show that this result cannot be extended to the infinite dimension. However, we obtain some infinite dimensional counterparts of Theorem 1.

On the other hand, this paper extends the results concerning stationary measures for the case homogeneous in time, contained in [13], [5], [2].

2. Model

H,K denote real separable Hilbert spaces and, for $t \in R$, A_t is a linear operator acting in H, $a_t \in H$, $B_t \in L(K,H)$.

Let $(\Omega,\mathcal{F},P)$ be a fixed probability space. Let $(w_t)_{t\in R}$ be a K-valued Wiener process on $(\Omega,\mathcal{F},P)$ with the incremental covariance operator $\underline{W}$ and let $\underline{\eta(dt,dx)}$ be a Poisson random measure on $R \times (H \setminus \{0\})$ with intensity $\underline{M_t(dx)dt}$, independent of the process $(w_t)_{t\in R}$.

Consider

(*) $$dX_t = A_t X_t dt + a_t dt + B_t dw_t + \int_{|x|\le 1} x[\eta(dt,dx) - M_t(dx)]dt + \int_{|x|>1} x\eta(dt,dx),$$

where $A_\cdot, a_\cdot, B_\cdot$ and $M_\cdot(du)$ are τ-periodic in t; $A_\cdot, a_\cdot$, $\mathrm{tr}\, B_\cdot WB^*_\cdot$ are integrable on $[0,\tau]$ and

(1) $$\int_0^\tau \int_{H\setminus\{0\}} |x|^2 \wedge 1 \quad M_t(dx)dt < +\infty.$$

If we write (*) in the form

(*´) $$dX_t = A_t X_t dt + dZ_t,$$

then $(Z_t)_{t\in R}$ is a process on $(\Omega,\mathcal{F},P)$ with independent increments, continuous in probability, cadlag, $Z_o \equiv 0$ and dZ_t has τ-periodic Lévy characteristics $[a_t dt, B_t WB_t^* dt, M_t dt]$. The last means that the characteristic function of $Z_t - Z_s$, $t > s$, has the form:

(2) $$E(\exp i\langle y, Z_t - Z_s\rangle) = \exp \int_s^t \{i\langle y,a_u\rangle - \tfrac{1}{2}\langle y, B_u WB_u^*\rangle + \int_{H\setminus\{0\}} [\exp i\langle y,x\rangle - 1 - i\langle y,x\rangle \mathbf{1}_V(x)]M_u(dx)\}du,$$

for any $y \in H$; where $\mathbf{1}_V$ is the indicator function of the unit ball V.

Conversely, if $(Z_t)_{t\in R}$, $Z_o \equiv 0$, is a cont. in prob., cadlag process on $(\Omega,\mathcal{F},P)$ with independent τ-periodic increments characterized by (2) (where M_t is a σ-finite measure satisfying (1)), then the equation (*´) can be represented in the form (*) (see [4]) with $a_\cdot, B_\cdot, M_\cdot$ as in (2). Then the random measure $\eta(dt,dx)$ in (*) depends on the process Z_t, namely, for a Borel set $F \subset H$ separated from 0, $\eta((s,t]\times F)$ is the number of jumps (Z_t) which occurred in the time interval $(s,t]$ and belonged to F. $E\eta((s,t]\times F) = \int_s^t M_u(F)du$. Let

(3) $$\zeta_t := Z_t - \int_0^t a_s ds - \int_0^t B_s dw_s, \quad t \ge 0, \quad \text{(the jump part of } Z_t).$$

3. Final dimensional case. $H = R^d$

Let $U(t,s)$ denote the fundamental matrix for $A(t)$. Then

(**) $$X_t = U(t,0)X_o + \int_0^t U(t,s)dZ_s, \quad t \ge 0,$$

is a solution of (*) with an initial condition X_o. (X_o is assumed to be a random variable independent of $(Z_t)_{t>0}$, since we consider only markovian solutions.)

Let $\underline{C: = U(\tau,0)}$ and let $\underline{H_-},\underline{H_o},\underline{H_+}$ denote the subspace in Jordan decomposition of C associated with the eigenvalues λ such that $|\lambda| < 1$, $|\lambda| = 1$, $|\lambda| > 1$, respectively.

The following theorem describes in a complete way τ-periodic distributions for (*) in R^d and gives conditions for the existence, in terms of Lévy characteristics of (Z_t).

THEOREM 1 ([3]). I. There exists a τ-periodic solution of (*) iff the following conditions hold:

(i) Range $B_sWB_s^* \subset U(s,0)(H_-)$ for a.a.s.

(ii1) $\operatorname{supp} M_s \subset U(s,0)(H_-)$ for a.a.s.

(ii2) $\int_o^\tau \int_{R^d} \log^+ |x| M_s(dx)ds < +\infty$

(iii) there exists $h \in R^d$ s.t.

$$h - Ch = \int_o^\tau U(\tau,s)a_s ds.$$

II. If μ_t is a τ-periodic distribution for (*), then

$$(\nabla) \qquad \mu_t = \mu_t^1 * \mu_t^2 * U(t,0)\alpha * \delta_{b_t},$$

where

$$\mu_t^1 = \mathcal{L}(\int_{-\infty}^t U(t,s)B_s dw_s), \qquad \mu_t^2 = \mathcal{L}(\int_{-\infty}^t U(t,s)d\zeta_s),$$

(ζ_s) is defined by (3), α is an invariant probability measure for C (i.e. $\alpha = C\alpha$) and $b_t = U(t,0)h + \int_o^t U(t,s)a_s ds$. Moreover, $\operatorname{supp} \mu_t^i \subset U(t,0)(H_-)$, $i = 1,2$, $\operatorname{supp} \alpha \subset H_o$.

REMARKS. 1. The condition (ii2) is equivalent to

(ii2) $E \log^+ |Z_\tau| < +\infty$ (see [7]).

2. b_t is a τ-periodic solution of $\dot{x}_t = A_t x_t + a_t$.

3. The stochastic integral $\int_{-\infty}^t f(s)dZ_s$ is understood as the limit in probability (equivalently a.s., equivalently in distribution) of

the integrals $\int_u^t f(s)dZ_s$ as $u \to -\infty$.

COROLLARIES. 1. Let $H_- = H$. Then a τ-periodic solution of (*) exists iff $E \log^+ |Z_\tau| < +\infty$.

2. Let $H_+ = H$. Then a τ-periodic solution of (*) exists iff the system (*) is deterministic. In this case the τ-periodic solution is given by the formula

$$b_t = - \int_t^\infty U(t,s)a_s ds.$$

3. a) If $H_o = \{0\}$, then there exists at most one τ-periodic distribution for (*).

b) Conversely, if there exists a unique τ-periodic distribution, then $H_o = \{0\}$.

P r o o f. 1, 2 and 3.a) are obvious consequences of Theorem 1. To prove 3.b), notice that if $H_o \neq \{0\}$, then there are infinitely many invariant distributions for C. Indeed, let λ with $|\lambda| = 1$ be an eigenvalue of C. Then $\lambda = 1$ or $\lambda = -1$ or $\lambda = \exp i\delta$, $\delta \neq k\Pi$. If $\lambda = 1$ ($\lambda = -1$), C has a 1-dimensional invariant subspace G s.t. $Cx = x$ ($Cx = -x$) and every measure (every symmetric measure) concentrated on G is invariant for C. In the third case C has a 2-dimensional invariant subspace F s.t. $C_{|F}$ is a rotation and thus any measure concentrated on F invariant for rotations is invariant for C.

The following propositions are important for the proof of Theorem 1.

PROPOSITION 1. The condition (i) of Theorem 1 is equivalent to

$$\text{(ia)} \qquad \int_{-\infty}^{o} \operatorname{tr} U(0,s)B_s W B_s^* U(0,s)^* ds < +\infty.$$

REMARK 4. (ia) means that $\int_{-\infty}^{o} U(0,s)B_s dw_s$ exists.

PROPOSITION 2.

$$\text{(ii)} \qquad \int_{-\infty}^{o} U(0,s)d\zeta_s \quad \text{exists}$$

if and cnly if (ii1) and (ii2) hold.

4. Infinite dimensional case. H is a real separable Hilbert space.

We assume that the family $\{A_t\}$, $t \in R$, of linear, usually unbounded, operators generates an evolution system $\{U(t,s)\}$, $-\infty < s \leq t < +\infty$, of bounded linear operators on H. This means that the system $\{U(t,s)\}$ has the properties: $U(s,s) = Id$; $U(t,u)U(u,s) = U(t,s)$ for $t \geq u \geq s$; $U(t,\cdot)x$ and $U(\cdot,s)x$ are continuous for any $x \in H$; $x_t = U(t,s)x$, $t \geq s$, is a unique solution of the equation $\dot{x}_t = A_t x_t$ with an initial condition $x_s = x$, for x in a dense subset of H. Sufficient conditions on the family $\{A_t\}$, which guarantee the existence of an evolution system associated with $\{A_t\}$, can be found in [10], [12].

Since A is τ-periodic $U(t+\tau,s+\tau) = U(t,s)$ for any $t \geq s$.

We consider the so-called mild solution of (*) given again by

$$(**) \qquad X_t = U(t,0)X_0 + \int_0^t U(t,s)dZ_s, \qquad t \geq 0,$$

where the stochastic integral can be defined for instance as in [2].

Let, as in section 3, $C: = U(\tau,0)$. If $\dim H = +\infty$, then there are at least two counterparts of the stability subspace H_-, namely:

$$H^1 = \{x \in H : C^n x \to 0, \quad n \to +\infty\},$$

$$H^2 = \{x \in H : \sum_{n=0}^{\infty} |C^n x|^2 < +\infty\}.$$

OBSERVATIONS. 1. Notice that $C(H^i) \subset H^i$ and $C_{-1}(H^i) = H^i$, $i = 1,2$.

2. if $x \in H^1$, then $U(t,0)x \to 0$ as $t \to +\infty$.

P r o o f. Let $t_k \to +\infty$. Then $t_k = n_k\tau + s_k$, where $0 \leq s_k < \tau$ and $n_k \to +\infty$. Clearly,

$$(4) \qquad U(n\tau + s,0) = U(s + n\tau,n\tau)U(n\tau,0) = U(s,0)C^n.$$

Since $U(\cdot,0)$ is strongly continuous, $\sup_{0\leq s\leq\tau} \|U(s,0)\|$ is finite from Banach-Steinhaus theorem. Hence Observation 2 follows from (4).

PROPOSITION 3. Let H^2 be closed. Then there exist constants $\delta < 1$, L such that

$$\|C^n|_{H^2}\| \leq L\delta^n \qquad \text{for all } n.$$

Consequently, it follows from (4) that for $x \in H^2$, $U(t,0)x$ tends to

0 exponentially as $t \to +\infty$.

P r o o f. Follow the proof of Theorem 4.1 in [10], where the operator $S : H^2 \to l^2(H^2)$ is defined by $Sx = (x, Cx, C^2x, \ldots)$. ($S$ is obviously closed.)

We start the investigation of τ-periodic solutions to (*) from the following fact.

PROPOSITION 4. I. If $\int_{-\infty}^{0} U(0,s)dZ_s$ exists, then (*) has a τ-periodic solution.

II. If $H^1 = H$, then the converse implication is also true.

P r o o f. I. It is easy to verify that

$$\overline{X}_t := \int_{-\infty}^{t} U(t,s)dZ_s \text{ is a } \tau\text{-periodic solution to } (*).$$

II. Repeating the proof of Lemma 2.II [3] we obtain that $\int_{-n\tau}^{0} U(0,u)dZ_u$ is convergent in distribution as $n \to +\infty$. Hence it suffices to show that

$$(5) \qquad \int_{-n\tau+t}^{-n\tau} U(0,s)dZ_s \xrightarrow[n \to \infty]{\text{in distr.}} 0 \text{ uniformly for } t \in [-\tau,0].$$

Clearly, for $t \le 0$

$$(6) \qquad \int_{-n\tau+t}^{-n\tau} U(0,s)dZ_s \overset{\text{in distr.}}{=} \int_{t}^{0} U(0,u-n\tau)dZ_u =$$

$$= \int_{t}^{0} U(n\tau,u)dZ_u = C^n \int_{t}^{0} U(0,u)dZ_u := C^nY_t.$$

Y_t is cadlag, thus for a.a. ω the set $F(\omega) := \{Y_t(\omega),\ t \in [-\tau,0]\}$ is conditionally compact in H. Since for any $x \in H$, $C^n x \to 0$ as $n \to \infty$, there exists L such that $\|C^n\| \le L$ for all n (from Banach--Steinhaus theorem). Therefore $C^n x \to 0$ uniformly for $x \in F(\omega)$. Notice that $\overline{Y}_n := \sup_{-\tau \le t \le 0} \|C^nY_t\|$ is a random variable, since Y_t is cadlag. Consequently $\overline{Y}_n \xrightarrow{\text{a.s.}} 0$. Then (5) follows from (6) and the inequality $\sup_{-\tau \le t \le 0} P(\|C^nY_t\| \ge \varepsilon) \le P(\overline{Y}_n \ge \varepsilon)$.

COROLLARY 4 (Uniqueness). If $H^1 = H$, then there exists at most one τ-periodic distribution μ_t for $(\overset{*}{*})$. If μ_t exists, then

a) $\mu_t = \mathcal{L}(\int_{-\infty}^{t} U(t,s)dZ_s)$,

b) for any distribution ν on H and for any s, $P_{n\tau+s}(\nu) \xrightarrow[n\to\infty]{weakly} \mu_s$, where $P_t(\nu)$ denotes the distribution of X_t given by (**) with $\nu = \mathcal{L}(X_o)$.

P r o o f of b). If X_t, Y_t are solutions of (**), then $X_t - Y_t \xrightarrow{a.s.} 0$ as $t \to +\infty$. Then b) follows from Theorem I.4.1, [1].

REMARK 5. Corollary 4 gives the uniqueness in law of a τ-periodic solution to(*), if $H = H^1$. Remark that the strong uniqueness (i.e. up to modification) fails here. Let $(\tilde{Z}_t)_{t\in R}$ be an independent copy of $(Z_t)_{t\in R}$. If $\int_{-\infty}^{t} U(0,s)dZ_s$ exists, so does $\tilde{X}_o = \int_{-\infty}^{o} U(0,s)d\tilde{Z}_s$. Then $\bar{X}_t = \int_{-\infty}^{t} U(t,s)dZ_s$ and $\tilde{X}_t = U(t,0)\tilde{X}_o + \int_{o}^{t} U(t,s)dZ_s$ are τ-periodic solutions to (*), different in the strong sense.

In order to obtain <u>necessary conditions for existence of a τ-periodic solution to (*)</u>, we need the following lemma

LEMMA 1. If (*) has a τ-periodic solution, then

(ia) $$\int_{-\infty}^{o} tr\, U(0,s)B_sWB_s^*U(0,s)^*ds < +\infty$$

(v) $$\int_{-\infty}^{o} \int_{H} |U(0,s)x|^2 \wedge 1 \quad M_s(dx)ds < +\infty.$$

P r o o f. (ia). Follow the proof of Theorem 3.3 in [2] with obvious modifications.

(v). Cf. the proof of Proposition 5.1 in [2].

§4.1. The Proposition below is <u>a counterpart of Proposition 1</u> from section 3. Consider the following condition:

(i´) $$\text{Range } B_sWB_s^* \subset U(0,s)_{-1}(\text{cl } H^2) \quad \text{for a.a. } s \leq 0,$$

where cl H^2 denotes the closure of H^2. In R^d (i´) and (i) are identical, since then $U(s,0) = U(0,s)^{-1}$. Notice that, for $s \leq 0$, $U(0,s-\tau)_{-1}(H^2) = U(0,s)_{-1}[C_{-1}(H^2)] = U(0,s)_{-1}(H^2)$, cf. Observation 1. Hence the subspace $U(0,s)_{-1}(H^2)$ varies τ-periodically.

PROPOSITION 5.

I. (ia) $\Longrightarrow$ (i´).

II. If H^2 is closed, then (i´) $\Longrightarrow$ (ia).

P r o o f. Let $Q_s := B_s W B_s^*$.

I. Using (ia) and the τ-periodicity, we have for any $x \in H$:

$$(7)\qquad +\infty > \int_{-\infty}^{0} |U(0,s)Q_s^{1/2}x|^2 ds = \sum_{n=0}^{\infty} \int_{-(n+1)\tau}^{-n\tau} |U(0,s)Q_s^{1/2}x|^2 ds =$$

$$= \int_{-\tau}^{0} \sum_{n=0}^{\infty} |C^n U(0,u)Q_u^{1/2}x|^2 du.$$

Therefore $\forall\, x \in H\ \exists\, J_x \subset [-\tau,0],\ 1([-\tau,0] \setminus J_x) = 0\ \forall\, u \in J_x$, $U(0,u)Q^{1/2} x \in H^2$ (where 1 denotes Lebesgue measure). Let D be a separable dense subset of H and let $J := \bigcap_{x\in D} J_x$. Then $1([-\tau,0]\setminus J)=0$ and $\forall u \in J\ \forall x \in H$, $U(0,u)Q_u^{1/2} x \in \mathrm{cl}\, H^2$, which implies (i´).

II. Let $(e_k)_{k=1}^{\infty}$ be an orthonormal basis in H. Notice that cl (Range Q_s) = cl (Range $Q_s^{1/2}$) and hence, for a.a. $s \le 0$, $U(0,s)Q_s^{1/2} e_k \in H^2$. Then as in (7) we obtain

$$\int_{-\infty}^{0} |U(0,s)Q_s^{1/2}e_k|^2 ds \le \int_{-\infty}^{0} \sum_{n=0}^{\infty} \|C^n{}_{|H^2}\|^2 \cdot |U(0,u)Q_u^{1/2}e_k|^2 du$$

and (ia) is a consequence of Proposition 3, since $\sup_{-\tau\le u\le 0} \|U(0,u)\|$ and $\int_{-\tau}^{0} \mathrm{tr}\, Q_u du$ are finite.

REMARK 6. If H^2 is not closed, then (i´) $\not\Longrightarrow$ (ia). Consider the following, homogeneous in time, example. Let $H = L^2([0,+\infty))$ and let $(T_t)_{t\ge 0}$ be the semigroup of left translation, i.e. $(T_t x)(\theta) = x(t+\theta)$, $\theta \ge 0$; $U(t,s) = T_{t-s}$. Then $\mathrm{cl}\, H^2 = H$. Take $e_k := 1\!\!1_{[k,k+1)}$, the indicator function of $[k,k+1)$. Then $\{e_k\}$ is an orthonormal system in H and

$$(8)\qquad \int_0^{\infty} |T_t e_k|^2 dt = \int_0^{\infty}\int_t^{\infty} 1\!\!1_{[k,k+1)}(u)\,du\,dt = \int_0^{\infty} 1\!\!1_{[k,k+1)}(u) \int_0^{u} dt\,du =$$

$$= \frac{2k+1}{2}.$$

Define $Wx: = \sum_{k=1}^{\infty} \frac{1}{k^2} \langle x,e_k\rangle e_k$. Clearly, W is a self-adjoint, positive, nuclear operator. From (8) we have:

$$\int_0^{\infty} \mathrm{tr}\, T_t W T_t^* dt = \sum_{k=1}^{\infty} \int_0^{\infty} \frac{1}{k^2} |T_t e_k|^2 dt = +\infty$$

and (ia) does not hold. Consequently, the process

$$X_t = T_t X_o + \int_0^t T_{t-s} dw_s$$

has no periodic distributions. (It is easy to verify that here even Range $W^{1/2} \subset H^2$.)

§4.2. We shall now discuss whether one can <u>extend Proposition 2</u> for the infinite dimension.

PROPOSITION 6. If (*) has a τ-periodic solution, then

(ii1´) $\quad \mathrm{supp}\, M_s \subset U(0,s)_{-1}(\mathrm{cl}\, H^2)$, for a.a. $s \leq 0$.

P r o o f. From Lemma 1, the condition (v) follows. Using the τ-periodicity we obtain, similarly as in (7),

(9) $$\int_{-\infty}^{o} \int_H |U(0,s)x|^2 \wedge 1\, M_s(dx)ds =$$

$$= \int_{-\tau}^{o} \int_H \sum_{n=0}^{\infty} (|C^n U(0,u)x|^2 \wedge 1)\, M_u(dx)du.$$

Next, (v) and (9) imply that for a.a. $u \in [-\tau,0]$

(10) $$\int_H \sum_{n=0}^{\infty} (|C^n U(0,u)x|^2 \wedge 1)\, M_u(dx) < +\infty.$$

Fix $\bar{u}$ such that (10) is satisfied. Then for $M_{\bar{u}}$ a.a. x, $U(0,\bar{u})x \in H^2$, which means that $M_{\bar{u}}\{H \setminus U(0,\bar{u})_{-1}(H_2)\} = 0$ and (ii1´) holds.

REMARK 7. From (ii1´) it follows that, for any $t \leq 0$, $\int_t^o U(0,u)d\zeta_s$ takes its values in $\mathrm{cl}\, H^2$. Similarly, (i´) implies that, for any $t \leq 0$, $\int_t^o U(0,s)B_s dw_s$ takes its values in $\mathrm{cl}\, H^2$.

Proposition 7 and the remarks below show that if $\dim H = +\infty$, then Proposition 2 holds under a certain restriction and only in one direction.

PROPOSITION 7. If H^2 is closed, then the conditions: (ii1´) and

(ii2) $\quad E \log^+ |Z_\tau| < +\infty$

are sufficient for the existence of $\int_{-\infty}^{0} U(0,s)d\zeta_s$.

P r o o f. Repeat the proof of Lemma 6, b) ==> a), [3] with obvious modifications, using Remark 7. Then we obtain that

$$\sum_{n=0}^{\infty} \int_{-(n+1)\tau}^{-n\tau} U(0,s)d\zeta_s$$

is convergent. Now from the proof of Proposition 4. II the proposition follows.

REMARKS. 8. If H^2 is not closed, Proposition 7 is not still true. (See Example 8.1 in [2] for the case homogeneous in time.)

9. From Propositions 4 (with $Z_t = \zeta_t$) and 6 it follows that (ii1´) is necessary for the existence of $\int_{-\infty}^{0} U(0,s)d\zeta_s$.

10. Even if $H^2 = H$, (ii2) is not necessary for the existence of $\int_{-\infty}^{0} U(0,s)d\zeta_s$. (See Remark 6.9 in [2]).

§4.3. Sufficient conditions for existence of a τ-periodic solution and a counterpart of Theorem 1

LEMMA 2. If $\int_{-\infty}^{0} U(0,s)B_s dw_s$ and $\int_{-\infty}^{0} U(0,s)d\zeta_s$ exist and (iii) holds, then

a) there exists a τ-periodic distribution for (**),

b) any τ-periodic distribution for (**) has the form (∇)

P r o o f. a) is easy to verify.

b) Let μ_t^1, μ_t^2, b_t and h be the same as in Theorem 1. Denote

$$\nu_t^1 := \mathcal{L}\left(\int_0^t U(t,s)B_s dw_s\right), \quad \nu_t^2 := \mathcal{L}\left(\int_0^t U(t,s)d\zeta_s\right)$$

and

$$c_t := \int_0^t U(t,s)a_s ds.$$

It is easy to verify that, for arbitrary t,

(11) $$\nu^i_{n\tau+t} \xrightarrow[n\to\infty]{\text{weakly}} \mu^i_t, \qquad i = 1,2.$$

Let μ_t be a τ-periodic distribution for (**). Then for arbitrary t

(12) $$\mu_t = \mu_{n\tau+t} = U(n\tau+t,0)\mu_o * \delta_{c_{n\tau+t}} * \nu^1_{n\tau+t} * \nu^2_{n\tau+t}.$$

Denote $\kappa_s := U(s,0)\mu_o * \delta_{c_s}$. Letting $n \to \infty$ in (12), it follows from (11) and Theorem III. 2.1, [14] that $\{\kappa_{n\tau+t}\}_{n=0}^{\infty}$ is relatively weakly compact. Since the characteristic functions $\hat{\mu}^i_t$ and $\hat{\nu}^i_t$, $i = 1,2$, are everywhere different from 0,

$$\hat{\kappa}_{n\tau+t}(y) \xrightarrow[n\to\infty]{} \hat{\mu}_t(y)/[\hat{\mu}^1_t(y)\cdot\hat{\mu}^2_t(y)], \qquad \text{for } y \in H.$$

Thus from Lemma VI. 2.1, [14] we obtain that, there exists a probability measure β_t, s.t. $\kappa_{n\tau+t} \xrightarrow[n\to\infty]{w} \beta_t$. Clearly, $\beta_{t+\tau} = \beta_t$ and

(13) $$\mu_t = \beta_t * \mu^1_t * \mu^2_t.$$

From (13) it follows that $\beta_t = U(t,0)\beta_o * \delta_{c_t}$. Take $\alpha_t := \beta_t * \delta_{-b_t}$. Then it is easy to verify that $\alpha_t = U(t,0)\alpha_o$ and $\alpha_o = C\alpha_o$. Hence, from (13), the part b) follows.

REMARK 11. In R^d, the conditions of Lemma 2 are sufficient and necessary for the existence of a τ-periodic distribution for (**). In an infinite dimensional space H, the last implies the existence of $\int_{-\infty}^{0} U(0,s)B_s dw_s$ (Lemma 1) but implies neither the existence of $\int_{-\infty}^{0} U(0,s)d\zeta_s$ (cf. Examples 8.1 and 8.2 in [2]) nor (iii). (In Example 8.2, [2], there exists a stationary measure for a process

$$X_t = T_t X_o + \int_0^t T_{t-s}d\zeta_s + \int_0^t T_{t-s}a\,ds,$$

where (T_t) is a certain stable semigroup of operators but $\int_0^{\infty} T_t a\,dt$ does not exist. Let $\tau = 1$ and suppose that $\exists h \in H$ s.t. $\int_0^1 T_s a\,ds = h - T_1 h$. Then $\int_0^n T_s a\,ds = h - T_n h \xrightarrow[n\to\infty]{} h$. This, similarly as in the proof of Proposition 4. II, implies the existence of

$\int_0^\infty T_t a ds$, which gives a contradiction).

We have shown (§4.1, §4.2 and Remark 11) that one cannot extend Theorem 1 for the infinite dimension. We have only:

THEOREM 2. I. Let H^2 be closed. If the conditions (i´), (ii1´), (ii2) and (iii) are satisfied, then

a) there exists a τ-periodic distribution for (**),

b) any τ-periodic distribution for (**) has the form (∇).

II. If (**) has a τ-periodic distribution, then (i´) and (ii1´) hold.

P r o o f. Part I is an immediate consequence of Lemma 2, Propositions 5. II and 7.

Part II is the summation of Lemma 1, Propositions 5. I and 6.

§4.4. Sufficient and necessary conditions for existence of a τ-periodic solution

Theorems 3 and 4 below extend some results from [2] obtained in the case homogeneous in time. The following lemma is needed.

LEMMA 3. The conditions (v) from Lemma 1 and (vv) below are sufficient and necessary for the existence of

$$\int_{-\infty}^{0} U(0,s)d\zeta_s .$$

(vv) there exists

$$\lim_{t\to-\infty} \int_t^0 \int_H U(0,s)x[\mathbb{1}_{\{|x|\le 1\}}(U(0,s)x) - \mathbb{1}_{\{|x|\le 1\}}(x)]M_s(dx)ds$$

The proof is the same as the proof of Theorem 4.2 in [2].

THEOREM 3. Assume that the measure M_s is symmetric for a.a. s. Then:

I. there exists a τ-periodic solution to (*) iff the following conditions are satisfied: (ia), (v) and

$(\widetilde{iii})$ there exists a τ-periodic distribution for the deterministic system:

$$x_t = U(t,0)x_o + \int_0^t U(t,s)a_s ds. \tag{14}$$

II. Any τ-periodic distribution μ_t for (**) has the form

$$(\tilde{\text{v}})\qquad \mu_t = \mu_t^1 * \mu_t^2 * \beta_t,$$

where μ_t^1, μ_t^2 are such as in Theorem 1 and β_t is a τ-periodic distribution for (14).

P r o o f. If M_s is symmetric, then (v) is also sufficient for the existence of $\int_{-\infty}^{0} U(0,s)d\zeta_s$. (For any t, the integral in (vv) is equal to 0.)

Thus, from Lemma 2, it follows that (ia), (v) and $(\widetilde{\text{iii}})$ are sufficient, for the existence of a τ-periodic solution of (*).

The necessity of (ia) and (v) follows from Lemma 1. Then from the proof of Lemma 2 we obtain the necessity of $(\widetilde{\text{iii}})$ and Part II of the theorem.

REMARK 12. Clearly, (iii) implies $(\widetilde{\text{iii}})$. If Range (Id - C) is closed, then $(\widetilde{\text{iii}})$ implies (iii). (The proof of Lemma 1, [3], with slight modifications.) Obviously, $(\widetilde{\text{iii}})$ restricted to the class of distributions having expectations implies (iii).

THEOREM 4. Assume that H^2 is closed and there exists a subspace G invariant for C, such that $H = H^2 \oplus G$. Then:

I. The conditions (i´), (v), (vv) and $(\widetilde{\text{iii}})$ are sufficient and necessary for the existence of a τ-periodic solution to (*).

II. Part II of Theorem 3 holds.

P r o o f. The sufficiency of (i´), etc. follows from Proposition 5. II, Lemmas 3 and 2.

To prove that (i´), (v), (vv), $(\widetilde{\text{iii}})$ are also necessary, let Π_1, Π_2 denote the projection onto H^2, G, respectively, according to the decomposition $H = H^2 \oplus G$ and let $f := \int_0^{\tau} U(\tau,s)a_s ds$. Notice that

$$(15)\qquad \mathcal{L}\Big(\int_0^{n\tau} U(n\tau,s)dZ_s\Big) = \mathcal{L}\Big(\int_{-n\tau}^{0} U(0,s)dZ_s\Big).$$

Hence, from Remark 7, the random variables $\int_0^{n\tau} U(n\tau,s)B_s dw_s$ and $\int_0^{n\tau} U(n\tau,s)d\zeta_s$ take their values in H^2 a.s.. Then, from (**) and the assumptions on H^2 and G, we have

$$(16)\qquad \Pi_1 X_{n\tau} = C^n(\Pi_1 X_o) + \sum_{k=0}^{n-1} C^k(\Pi_1 f) + \int_o^{n\tau} U(n\tau,s)[B_s dw_s + d\zeta_s].$$

If (**) has a τ-periodic distribution, so does (16). From Proposition 3, $\sum_{k=0}^{\infty} |C^k(\Pi_1 f)| < +\infty$. Then as in the proof of Proposition 4. II we obtain that $\int_{-n\tau}^{o} U(0,s)[B_s dw_s + d\zeta_s]$ is convergent as $n \to \infty$, which implies (by the same argument as in the mentioned proof) that $\int_{-\infty}^{o} U(0,s)[B_s dw_s + d\zeta_s]$ exists. Consequently, $\int_{-\infty}^{o} U(0,s)B_s dw_s$ and $\int_{-\infty}^{o} U(0,s)d\zeta_s$ exist, which implies (i´), (v) and (vv). Finally, $(\widetilde{iii})$ and Part II follow from the proof of Lemma 2 b).

REFERENCES

[1] P. Billingsley, Convergence of Probability Measures, Wiley, New York, 1968.

[2] A. Chojnowska-Michalik, On processes of Ornstein-Uhlenbeck type in Hilbert space, Stochastic 21 (1987), 251-286.

[3] A. Chojnowska-Michalik, Periodic distributions for linear equations with general additive noise, Bull. Acad. Pol. Sci. (1989), to appear.

[4] I.I. Gikhman and A.V. Skorokhod, Theory of Random Processes, Vol. 2, Nauka, Moscow, 1973 (in Russian).

[5] J.B. Graveraux, Probabilités de Lévy sur R^d et equations differentielles stochastiques linéaires, Sem. Prob. de Rennes, 1982.

[6] A. Ichikawa, Bounded solutions and periodic solutions of a linear stochastic evolution equation, USSR - Japan Symposium on Probability Theory and Mathematical Statistics, 1986, Lect. Notes Math.

[7] Z.J. Jurek, An integral representation of operator-self-decomposable random variables, Bull. Acad. Pol. Sci. 30 (7-8) (1982), 385-393.

[8] T. Morozan, Periodic solutions of affine stochastic differential equations, Stoch. Anal. Appl. 4 (1986).

[9] E. Pardoux and M. Pignol, Etude de la stabilité de la solution d´une EDS bilinéaire à coefficients périodiques, Proc. Congrès INRIA sur l Analyse des Systèmes, Lect. Notes in Control Inf. Sci. (1984), Springer-Verlag, New York.

[10] A. Pazy, Semigroups of Linear Operators and Applications to Partial Differential Equations, Springer-Verlag, New York, 1983.

[11] K. Sato and M. Yamazato, Stationary processes of Ornstein-Uhlenbeck type, IV USSR-Japan Symposium, On Probability Theory and Mathematical Statistics, Lect. Notes Math. 1021 (1983), 541-551.

[12] H. Tanabe, Equations of Evolution, Pitman, London, 1979.

[13] J. Zabczyk, Stationary distributions for linear equations driven by general noise, Bull. Acad. Pol. Sci. 31 (3-4) (1983), 197-209.

[14] K.R. Parthasarathy, Probability Measures on Metric Spaces, Academic Press, New York and London, 1967.

Low and High Density Reaction-Diffusion Models

Peter Kotelenez

University of Utrecht

We derive existence and uniqueness of infinite dimensional Markov processes which describe nonlinear reactions with diffusion, both under low and under high density assumptions. The spatial coordinates of the processes are grid points of the "reactor" $S = [0,1]^n, n \geq 1$. If the number of particles (and their mass) are not changed and the grid size tends to 0 we obtain in the low density case a measure valued Markov process X^ϵ introduced in [2], [3] and generalized in [7], [8].

Set $S = \{r = (r_1, ..., r_n) \in \mathbb{R}^n : 0 \leq r_i \leq 1, i = 1, ..., n\}$ and $(H_0, <.,.>_0) := (L_2(S)), <.,.>_0)$, where $L_2(S)$ is the Hilbert space of real valued square integrable functions on S and $<.,.>_0$ is the standard scalar product on $L_2(S)$. Let Δ denote the Laplacian on H_0, closed with respect to homogeneous Neumann boundary conditions, $D > 0$ a diffusion constant and $R(x) = \sum_{i=0}^m c_i x^i$ a polynomial in $x \in \mathbb{R}$ such that $c_0 \geq 0$ and $c_m < 0$ if $m \geq 2$. Let us consider the following reaction-diffusion equation:

$$(1) \quad \begin{cases} \frac{\partial}{\partial t} X(t) & = D\Delta X(t) + R(X(t)) \\ X_0(r) & \geq 0. \end{cases}$$

It was shown in Kotelenez [6] that there is a unique global mild solution X of (1) if $X_0(r)$ is bounded and that X is "smooth" if X_0 is "smooth". Moreover, X is bounded if X_0 is and $m \geq 2$. Let $h^{-1} \in \mathbb{N}$ and $N := h^{-n}$. For $x = (x_1, ..., x_n) \in S$ we define

$$\pi_N x = x_N$$

such that $x_N = (x_{1,N}, ..., x_{n,N})$ and $x_{k,N} = \max\{i \in \mathbb{N} : ih < x_k\}$, $k = 1, ..., n$. Set $S_N := \pi_N S$. We can define the discrete Laplacian Δ_N satisfying the corresponding discrete homogeneous Neumann boundary conditions (Kotelenez [6]).

Now we introduce the stochastic (i.e., mezoscopic) models. Let δ_{x^i} be the Dirac measure in $x^i \in S$. Fix $\epsilon > 0$ and set for $M \in \mathbb{N}$

$$E_N^M := \{\epsilon \sum_{j=1}^M \delta_{x_N^j}, x^j \in S\}.$$

ϵ is the mass of a particle. We further set

$$E_N^\infty := \bigcup_{M=0}^\infty E_N^M,$$

where E_N^0 is an abstract point. Similarly, we define

$$E^M := \{\epsilon \sum_{j=1}^M \delta_{x^j}, x^j \in S\}$$

and

$$E^\infty := \bigcup_{M=0}^{\infty} E^M.$$

Note that both E^∞ and E_N^∞ endowed with the Prohorov metric d_p are locally compact seperable metric spaces. Following Dittrich [2], [3] we have shown in [7] the existence of an E^∞-valued Markov process X^ϵ (related to (1) as follows:
Further, let $G(t,x,y)$ be the fundamental solution (or Green's function) of $\frac{\partial}{\partial t}\xi(t) = D\Delta\xi(t)$ (cf. [6]) and $0 < \gamma_1 < \gamma_2 < \infty$ some constants.
For $k \geq 2$ $Q_k^\epsilon : S^k \to \mathbb{R}_+$ are symmetric and continuous functions such that

$$\int_{S^{k-1}} Q_k^\epsilon(x^1,...,x^k)dx^2,...,dx^k = 1 \text{ for all } x^1 \in S,$$

$$\gamma_1 Q_k^\epsilon(x^1,...,x^k) \leq \int_S G(\epsilon^{2/n},x^1,y).....G(\epsilon^{2/n},x^k,y)dy \leq$$

$$\gamma_2 Q_k^\epsilon(x^1,...,x^k) \text{ for all } (x^1,...,x^k) \in S^k.$$

If for $x = \sum_{i=1}^M \epsilon\delta_{x^i}$ $x_{i_1},...,_{i_k} := x + (sgn\ c_k)\epsilon.\sum_{j\in\{i_1,...,i_k\}} \delta_{x^j}$ and

$$(I_o f)(x) := \frac{c_0}{\epsilon}\int_S [f(x+\epsilon\delta_y) - f(x)]dy$$

then the reaction part of the generator of X^ϵ is given by

$$R^\infty f(x) := \sum_{k=1}^{m\wedge M} \frac{\epsilon^{k-1}}{k}|c_k| \sum_{i_1,...,i_k}^{M} [f(x_{i_1},...,_{i_k}) - f(x)]Q_k^\epsilon(x^{i_1},...,x^{i_k})$$

$$+(I_o f)(x) \ , \ \text{if } x \in E^M,$$

where $f \in C_0(E^\infty)$ (the continuous functions from E^∞ into $\mathbb{R}$ with support in finitely many E^M). Further, let Γ^M be the canonical mapping from S^M onto E^M, $\hat{C}(E^\infty)$ the continuous real valued functions on E^∞ vanishing at infinity with norm $|||f||| := \sup_{x\in E}|f(x)|$, and $\hat{C}^2(E^\infty)$ those f from $\hat{C}(E^\infty)$ such that $f \circ \Gamma^M$ is twice continuously differentiable and satisfies homogenous Neumann boundary conditions on S^M for any $M \in \mathbb{N}$. Denote the corresponding closed Laplacian (on $C(S^M,\mathbb{R})$) by Δ^M and the operator induced by Γ^{M*} $\Gamma^{M*}\Delta^M$. Then an infinite dimensional Laplacian on $\hat{C}^2(E^\infty)$ is given by

$$(\Delta^\infty f)(x) := (\Gamma^{M*}\Delta^M f)(x) \text{ if } x \in E^M,$$

and the generator of X^ϵ is defined (on $\hat{C}(E^\infty)$) by

$$(2) \quad A^\infty := D\Delta^\infty + R^\infty.$$

Next we construct corresponding processes on E_N^∞.

Denote by Γ_N^M the canonical mapping from S_N^M onto E_N^M and by Δ_N^M the discrete Laplacian (with homogeneous Neumann boundary conditions - defined in the same way as Δ_N) on the functions from S_N^M into $\mathbb{R}$. Then

$$(\Delta_N^\infty f)(x) = (\Gamma_N^{M*}\Delta_N^M f)(x) \text{ if } x \in E_N^M$$

defines an infinite dimensional discrete Laplacian on $\hat{C}(E_N^\infty)$.
Let π_{N*} denote the canonical mapping from $E^\infty \to E_N^\infty$ induced by π_N and for $f \in \hat{C}(E_N^\infty)$

$$\pi_N^* f = f \circ \pi_{N*} \in \hat{C}(E^\infty).$$

We will now describe three different reaction rates.

1. Discretization of R^{∞}:

(3) $(R_N^{\infty,1}f)(\pi_{N*}x) := (R^{\infty}\pi_N^* f)(x), f \in C_0(E_N^{\infty})$

2. $Q^{\epsilon}_{k,N}(x_N^1,...,x_N^k) := \begin{cases} 1 & \text{if } x_N^1 = x_N^2 = ... = x_N^k \\ 0 & \text{otherwise.} \end{cases}$

and $x_i^k := x + (sgn\ c_k)\epsilon\delta_{x^i}$ $(x \in E_N^{\infty})$ for $c_k > 0$, or for $c_k < 0$ and x has mass at the grid point x^i. If x has no mass at x^i and $c_k < 0$ we set $x_i^k := x$. Then for $x \in E_N^M$

$$(4) \quad \begin{cases} (R_N^{\infty,2}f)(x) & := \sum_{k=1}^{m\wedge M} \sum_{\substack{i_1,...,i_k \\ (\text{diff})}}^{M} |c_k|\{f(x_{i_1}^k) - f(x)\}Q^{\epsilon}_{k,N}(x_N^1,...,x_N^k) \\ & +(I_0\pi_N^* f)(x). \end{cases}$$

Note that for $f(x) =< x,\varphi >$, where $< .,. >$ is the duality extending $< .,>_0$ and φ continuous, the right handside in (4) becomes the same if we take $\frac{|c_k|}{k}\{f(x_{i_1,...,i_n}) - f(x)\}$ instead of $|c_k|\{f(x_{i_1}) - f(x)\}$. (4) means that one particle is created resp. annihilated due to the interaction with other particles, whereas in (3) we have creation resp. annihilation of k particles. If M_j is the number of particles on the j-th grid point we have $\sum_{j=1}^N M_j = M$ and obtain (using $\binom{M_i}{k} := 0$ if $k > M_i$) for $x \in E_N^M$:

$$(5) \quad (R_N^{\infty,2}f)(x) := \sum_{j=1}^{N}\sum_{k=1}^{m} |c_k|\{f(x_j^k) - f(x)\}\binom{M_j}{k} + (I_0\pi_N^* f)(x),$$

where δ_{x^j} in the definition of x_j^k is taken at the j-th grid point x^j.

3. Using the notation from (5) we set for $x \in E_N^M$

$$(6) \quad \begin{cases} (R_N^{\infty,3}f)(x) & := \sum_{j=1}^N \sum_{k=1}^m |c_k|\{f(x_j^k) - f(x)\}(M_j)^k \\ & +(I_0\pi_N^* f)(x). \end{cases}$$

Since $(M_j)^k = \sum_{i=1}^k \alpha_i^k \binom{M_j}{i}$ with $\alpha_i^k \in \mathbb{N}$ (independent of M_j) we obtain numbers c_k^+ and c_k^- such that for $x \in E_N^M$

$$(7) \quad \begin{cases} (R_N^{\infty,3}f)(x) & := \sum_{j=1}^N \sum_{k=1}^m |c_k^{\pm}|\{f(x_j^{k\pm}) - f(x)\}\binom{M_j}{k} \\ & +(I_0\pi_N^* f)(x), \end{cases}$$

where $x_j^{k\pm} := x + (sgn c_k^{\pm})\epsilon\delta_{x^j}$. Thus, (6) can be thought of as a mezoscopic reaction model belonging to another macroscopic model, i.e., with rates $\tilde{c}_k := c_k^+ + c_k^-$ instead of c_k. Define operators on $C_0(E_N^{\infty}) \subset \hat{C}(E_N^{\infty})$

(8) $A_N^{\infty,i} := D\Delta_N^{\infty} + R_N^{\infty,i}$ $i = 1,2,3.$

Theorem 1
For any $i \in \{1,2,3\}$ $A_{N_1}^{\infty,i}$ is the generator of a Feller-Markov process $X_N^{\epsilon,i}$ on $D([0,\infty); E_N^{\infty})$ the space of E_N^{∞}- valued cadlag functions).

The <u>proof</u> can be obtained from that of Th. 1 in [7]. □

Set

$$E_{N,v} := \{(\frac{M_1}{v},...,\frac{M_N}{v}) : M_j \in \mathbb{N} \cup \{0\}, j = 1,...,N\},\ v > 0.$$

Then there is a natural bijection

$$\psi_N : E_N^{\infty} \to E_{N,v}$$

$$\psi_N(x) = (\frac{M_1}{v}, ..., \frac{M_N}{v}) \text{ if } x = \sum_{j=1}^{N} M_j \epsilon \delta_{x^j},$$

where x^j is the j-th grid point. Moreover, for $\varphi \in C(S)$

$$< \varphi, x >= \epsilon \sum_{j=1}^{N} M_j \varphi(x^j).$$

On the other hand, if we consider $E_{N,v} \subset \mathrm{H}_0$ then we have

$$< \varphi, \psi_N(x) >_0 = \frac{1}{vN} \sum_{j=1}^{N} M_j \varphi(\tilde{x}^j)$$

with $\tilde{x}^j \sim x^i$. Thus we may study the reaction models 1-3 on the space of densities. For models 1-2 we will assume $v = 1$, whereas in model 3 we will assume $v > 0$ (in order to prove high density limit theorems as $v \to \infty$ and $N \to \infty$ (cf. our remarks at the end, which also explain the meaning of v and [6]). Clearly, we have

$$(9) \quad \begin{cases} \epsilon \sim \frac{1}{N} & \text{for models } 1-2 \\ \epsilon \sim \frac{1}{vN} & \text{for model } 3, \end{cases}$$

and we may interpret M_j in $(\frac{M_1}{v}, ..., \frac{M_N}{v})$ as the number of particles in the j-th cell (cf. Arnold and Theodosopulu [1] and Kotelenez [6]).

Let $C_0(E_{N,v})$ be the real valued functions on $E_{N,v}$ which depend only on finitely many points and $\hat{C}(E_{N,v})$ those real valued functions which vanish at infinity where we endow $E_{N,v} \subset \mathrm{H}_0$ with the norm associated with $< .,. >_0$. Define on $C_0(E_{N,v})$ operators

$$(10) \quad A_N^i := \psi_N^* A_N^{\infty,i} := A_N^{\infty,i} \circ \psi_{N*}, i = 1, 2, 3,$$

where $\psi_{N*} f = f \circ \psi_N$ for $f \in \hat{C}(E_{N,v})$. Further set $e_j := \frac{1}{v} 1_{[x^j)}$ (1_A is the indicator function of the set A and x^j is the j-th grid point and $[x^j) = \{(x_1, ..., x_n) : x_i^j \le x_i < x_i^j + h, i = 1, ..., n\}$) and $e_{j \pm h_p} := \frac{1}{v} 1_{[x^j \pm h_p)}$, where $h_p := (0, .., 0, h, 0...0)$ with all but the p-th coordinate zero. Moreover, denote by $\bar{M} = (\bar{M}_1, ..., \bar{M}_N)$ a typical element from $E_{N,v}$ and abbreviate

$$b_2(\bar{M}_j) = \sum_{\{k: c_k \ge 0\}} c_k \binom{\bar{M}_j}{k}, \; d_2(\bar{M}_j) = \sum_{\{k: c_k < 0\}} c_k \binom{\bar{M}_j}{k}$$

$$b_3(\bar{M}_j) = \sum_{\{k: c_k \ge 0\}} c_k v (\bar{M}_j)^k, \; d_3(\bar{M}_j) = \sum_{\{k: c_k < 0\}} c_k (\bar{M}_j)^k$$

$$D_2(\bar{M}_j) = D h^{-2} \bar{M}_j, \; D_3(\bar{M}_j) = v D h^{-2} \bar{M}_j.$$

Theorem 2
Suppose $v = 1$ for $i = 1, 2$ and $v > 0$ for $i = 3$. The closures of A_N^i on $\hat{C}(E_N)$ are generators of Feller-Markov processes X^i on $D([0, \infty); E_{N,v})$. The transition intensities for $i = 2, 3$ are given by

$$(11)\quad \begin{cases} b_i(\bar M_j) & \text{if } \bar L = \bar M + e_j =: \bar M_{+0},\\ d_i(\bar M_j) & \text{if } \bar L = \bar M - e_j =: \bar M_{-0},\\ D_i(\bar M_j) & \text{if } \bar L = \bar M + e_{j\pm h_p} - e_j =: \bar M_{\pm p}, p = 1,\dots,n \text{ with } j\pm h_p \in S^N,\\ -\sum_{p=0}^n \beta_i(\bar M, \bar M_{\pm p}) & \text{if } \bar M = \bar L,\\ 0 & \text{otherwise.}\end{cases}$$

Proof
Since $\psi_N : (E_N^\infty, d_p) \to (E_{N,v}, |.|_0)$ is a homeomorphism, (10) does define generators of Feller-Markov processes on $D([0,\infty); E_{N,v})$ (Dynkin [4] and Ethier and Kurtz [5]). We will now show (11) for $i = 3$. (the proof for $i = 2$ is the same).
Note that $C_0(E_{N,v})$ is dense in $\hat C(E_{N,v})$ and that any $f \in C_0(E_{N,v})$ can be represented by a finite sum $\sum_j < .,\varphi_j >_0$ with suitable "test functions" φ_j. Hence we easily see that for $f \in C_0(E_{N,v})$

$$(12)\quad \begin{cases} A_N^3 f(\bar M) & = \sum_{j=1}^N \sum_{p=1}^n [f(\bar M + e_{j\pm h_p} - e_j) - f(\bar M)] v N^2 D \bar M_j\\ & + \sum_{j=1}^N [f(\bar M + e_j) - f(\bar M)] b_3(\bar M_j)\\ & + \sum_{j=1}^N [f(\bar M - e_j) - f(\bar M)] d_3(\bar M_j),\end{cases}$$

where as before we define $f(\bar M - e_j)$ etc. to be equal to $f(\bar M)$ if $\bar M_j = 0$. Recalling basic definitions for Markov processes on countable state spaces finishes the proof. □

We next consider model 1. Let "$\Rightarrow$" denote weak convergence.

Theorem 3
Suppose $X_N^{1,\epsilon}(0) \Rightarrow X^\epsilon(0)$ as $N \to \infty$. Then $X_N^{1,\epsilon} \Rightarrow X^\epsilon$, as $N \to \infty$.

Proof
Following Theorem 6.1, Ch. 1, and Theorem 2.5, Ch. 4, in Ethier and Kurtz [5] we verify that for any $f \in \hat C^2(E^\infty) \cap C_0(E^\infty)$ $|||A_N^{\infty,1} \tilde\pi_N f - \tilde\pi_N A^\infty f||| \to 0$ where $\tilde\pi_N$ is the restriction of f to E_N^∞. $R_N^{\infty,1} \to R^\infty$ in this sense is evident by the continuity of $f \in C_0(E^\infty)$ (f has support in a compact domain). In this way we obtain that for $f \in C_0(E^\infty) \cap \hat C^2(E^\infty)$ there is an M such that $f(x) = 0$ if $x \notin E^M$ whence

$$|||\Delta_N^\infty \tilde\pi_N f - \tilde\pi_N \Delta^\infty f||| \le |||\Delta_N^\infty \tilde\pi_N f - \Delta^\infty f||| + |||\Delta^\infty f - \tilde\pi_N \Delta^\infty f|||$$
$$= \sup_{x\in S^M} \{|\Delta_N^M f \pi_N(x) - \Delta^M f(x)| + |\Delta^M f(x) - \Delta^M f(\pi_N x)|\}.$$

The convergence of the second term to 0 follows from the continuity of $\Delta^M f$ and that of the first term by the usual approximation of the continuous Laplacian by the discrete Laplacian. □

Remarks
Using Dynkin's formula we can write down "mezoscopic equations" for the reaction-diffusion models on $E_{N,v}$:

$$(13)\quad \begin{cases} dX_{N,v}^i(t) = & [D\Delta_N X_{N,v}^i(t) + b_i(X_{N,v}^i(t)) - d_i(X_{N,v}^i(t))]dt\\ & + \frac{1}{\sqrt{vN}} dM_{N,v}^i(t)\end{cases}$$

for $i = 2,3$, where $M_{N,v}^i(t)$ are square integrable martingales. It was proved in [6] that $X_{v,N}^3$ converges to the solution X of (1) in distribution norm, provided $N \to \infty$ and $v \to \infty$ (law of large numbers - LLN), and a central limit theorem (CLT) was proved under the condition that $N \to \infty$ and $\frac{v}{N} \to \infty$. For $m = 2$ in (1) these conditions were proved to be necessary. On the other hand in [8] it was proved that for $m = 2$ X^ϵ (on E^∞) also converges in distribution norm to X, as $\epsilon \to 0$, and a CLT was derived. The reaction in model 1 is roughly local, whereas in model 3 it is non local (if $v \to \infty$ then infinitly many particles will be in one cell and will interact since $X_{N,v}(t) \to X(t)$, i.e., the density has to be high). Obviously, model 1 is close to model 2 and by Theorem 3 and [8] we may assume that $X_N^{\epsilon,1} \sim X$ for $m = 2$ in (1).

Here only the next neighbors interact, whence the density is low in models 1 and 2. Hence we have two different approximations of (1) for $m = 2$: on the one hand by $X_N^{\epsilon,1}$ and on the other hand by $X_{N,v}^3$, where $X_{N,v}^3$ can be linked with another macroscopic equation (cf. (7)). Note, however, that (13) for $i = 3$ is the discretized macroscopic equation perturbed by a small martingale term ($vN \to \infty$). This does not hold for (13) in case $i = 2$ and it would not hold for the corresponding "mezoscopic equation" for $X_N^1(t)$. The exact relation between X_N^2 and X will be investigated in a forthcoming paper.

References

[1] Arnold, L. and Theodosopulu, M. (1980). Deterministic limit of the stochastic model of chemical reactions with diffusion. Adv. Appl. Prob. 12, 367-379.

[2] Dittrich, P. A stochastic model of a chemical reaction with diffusion (1988). Probab. Th. Rel. Fields 79, 115-128.

[3] Dittrich, P. (1988). A stochastic particle system: Fluctuations around a nonlinear reaction-diffusion equation. Stochastic Processes Appl. 30, 149-164.

[4] Dynkin, E.B. (1965). Markov processes, 1, Springer Berlin.

[5] Ethier, N.E. and Kurtz, T.G. (1986). Markov processes: Characterization and convergence. Wiley, New York.

[6] Kotelenez, P. (1988). High density limit theorems for nonlinear chemical reactions with diffusion. Probab. Th. Rel. Fields 78, 11-37.

[7] Kotelenez, P. (1988). A stochastic reaction-diffusion model. (Preprint no. 533, Department of Mathematics, University of Utrecht)

[8] Kotelenez, P. (1988). Fluctations in a Nonlinear Reaction-Diffusion Model (Preprint no. 534, Department of Mathematics, University of Utrecht)

Equations for the characteristic functional and moments of the complex stochastic evolutions - motivation and results.

Zbigniew Kotulski
Institute of Fundamental Technological Research
Polish Academy of Sciences
00-049 Warszawa, ul.Świętokrzyska 21

1. Motivation

In paper [7] the equations for the characteristic functional of the solution of the stochastic evolution equation was derived. Therefrom the complete set of the moment equations was obtained and the appropriate uniqueness and existence theorem was proved. These results were the extension of the ideas formulated in [2] and [3], where an example of some parabolic partial differential equation was considered. However, sometimes in applications it is irremissible to deal with complex space valued stochastic equations. Consider, as an example, the problem of harmonic waves, analyzed previously in [11] (see also [9]).

Let the wave be described by the following Helmholtz equation:

$$(1.1) \qquad \Delta \Phi + k_0^2\, n^2(\mathbf{r},\omega)\, \Phi = 0 \quad , \quad \mathbf{r}=(x,y,z),$$

where $n(\mathbf{r},\omega) = 1 + \varepsilon(\mathbf{r},\omega)$, $\langle\varepsilon(\mathbf{r},\omega)\rangle = 0$, $\omega \in \Omega$ and

$$\Delta = \frac{\partial^2}{\partial x^2} + \frac{\partial^2}{\partial y^2} + \frac{\partial^2}{\partial z^2} .$$

For the wave propagating along x-direction, the field Φ has the form:

$$(1.2) \qquad \Phi(\mathbf{r},\omega) = U(\mathbf{r},\omega)\, e^{ik_0x} ;$$

substituting function (1.2) into equation (1.1) and neglecting the second derivative of U with respect to x (what is justified under some conditions - see [10]), one arrives at the following parabolic differential equation (with complex coefficients):

$$(1.3) \qquad 2ik_0 \frac{\partial U(\mathbf{r},\omega)}{\partial x} + \Delta_\perp U(\mathbf{r},\omega) + k_0^2 \varepsilon(\mathbf{r},\omega) U(\mathbf{r},\omega) = 0,$$

where $\Delta_\perp = \frac{\partial^2}{\partial y^2} + \frac{\partial^2}{\partial z^2}$ is a transverse Laplacian.

To construct the required functional and moment equations for the solution of eq.(1.3) (with the appropriate initial condition), the theory developed in [7] must be extended. Such a modification is the subject of the next sections.

2. Formulation

Let $(\Omega,\mathcal{F},\mathcal{P})$ be a complete probabilistic space. Let $(\mathcal{X}, (.,.)_X)$ be a complex and $(\mathcal{Y}, (.,.)_Y)$ a real separable Hilbert space.

Consider the following stochastic evolution equation:

$$(2.1) \qquad \begin{aligned} dU &= AU\,dt + [BU]\,dW(t) \quad , \quad t\in(0,T] \quad , \\ U(0,\omega) &= U_0 \quad \mathcal{P}\ a.s. \quad , \quad U_0 \in \mathcal{X} \ , \end{aligned}$$

where:

$U = U(t,\omega)$, $t\in[0,T]$, $\omega \in \Omega$, is an $\mathcal{X}$-valued stochastic process,

$A : \mathcal{D}(A) \longrightarrow \mathcal{X}$ is a linear operator acting from the dense domain $\mathcal{D}(A) \subset \mathcal{X}$ into $\mathcal{X}$ and being the infinitesimal generator of the strongly continuous semigroup of bounded operators $K(t)$, $t\in [0,T]$,

$$K(t) : \mathcal{X} \longrightarrow \mathcal{X} \quad , \quad t\in[0,T] \ ,$$

$$(2.2) \qquad \| K(t) \| \le M\, e^{\varkappa t} \quad , \quad M \ , \ \varkappa = \text{const.}$$

$B : \mathcal{X} \longrightarrow L(\mathcal{Y},\mathcal{X})$ is a bounded bilinear operator, such that for $U\in\mathcal{X}$, $V\in\mathcal{Y}$

$$(2.3) \qquad \|[BU]V\|_X \le C\, \|U\|_X\, \|V\|_Y \quad , \quad C = \text{const.}$$

and

W is a $\mathcal{Y}$-valued Wiener process with covariance operator Q (see [4]).

The operator Q is nuclear, positive definite and self-adjoint, so it can be expanded into the series (see[4]):

$$(2.4) \qquad Q = \sum_{k=1}^{\infty} \alpha_k\, e_k \otimes_Y e_k$$

where α_k and e_k for $k=1,2,3,$ ... , are respectively, eigenvalues and orthonormal eigenvectors of Q ($\{e_k\}_{k=1,2..}$ is the Schauder basis in $\mathcal{Y}$); it is known that $\alpha_k>0$ for $k=1,2,...$ and $\Sigma\alpha_k<\infty$. Owing to formula (2.4) the Wiener process W has the representation in the form of series:

$$W(t) = \sum_{k=1}^{\infty} \sqrt{\alpha_k}\; e_k \beta_k(t) \quad , \tag{2.5}$$

where $\beta_k(t)$ for $k=1,2,...$, are real independent Wiener processes with the unit intensities.

Substitution of expression (2.5) into eq. (2.1) leads to the following stochastic differential equation:

$$dU = AU\, dt + \sum_{k=1}^{\infty} \sqrt{\alpha_k}\; [BU]e_k \; d\beta_k(t) \tag{2.6}$$

$$U(0) = U_0$$

Equation (2.6) is meant in Stratonovich interpretation; we consider its mild solution, that is the solution of the following integral equation (see [1]):

$$U(t) = K(t)U_0 + \sum_{k=1}^{\infty} \sqrt{\alpha_k} \int_0^t K(t-s)[BU]e_k d\beta_k(s) \; . \tag{2.7}$$

Characteristic functional.

Let $U(t)$, $t\in[0,T]$, be an $\mathcal{X}$-valued stochastic process. Let $\overline{U}(t)$ denotes the complex conjugate of the process $U(t)$ and λ, λ^* be two arbitrary elements of $\mathcal{X}$. The (spatial) characteristic functional of $U(t)$ is defined as:

$$F[t,\lambda,\lambda^*] = E\left\{exp[i\,(U(t),\lambda)_X + i\,(\overline{U(t)},\lambda^*)_X]\right\} \; , \quad t\in [0,T] \tag{2.8}$$

Moments.

Let us define:

$$U_1 \overset{def}{=} U \; , \qquad\qquad U_2 \overset{def}{=} \overline{U} \; , \tag{2.9}$$

where the bar, as always in this paper, denotes the complex conjugation. Let the moment of l-th order of the stochastic process $U(t)$ be defined as:

(2.10) $$\Gamma_{l-p,p}(t) = E\left\{U_{k_1}(t)\otimes U_{k_2}(t)\otimes \dots \otimes U_{k_l}(t)\right\},$$

where

$$k_i = 1 \quad \text{for} \quad i=1,2,\dots,l-p\,,$$
$$k_i = 2 \quad \text{for} \quad i=l-p+1,\dots,l\,,$$
$$\text{if} \quad p=1,2,\dots,l-1\,,$$

(2.11) or

$$k_i = 1 \quad \text{for} \quad i=1,2,\dots,l \quad \text{if} \quad p=0$$
$$k_i = 2 \quad \text{for} \quad i=1,2,\dots,l \quad \text{if} \quad p=l.$$

There are $l+1$ essentially different moments of the l-th order. Moments (2.10) can be easily obtained from the functional (2.8) by differentiation:

(2.12) $$\Gamma_{l-p,p}(t) = \left.\frac{\delta^l F[t,\lambda,\lambda^*]}{\delta\lambda^{l-p}\delta\lambda^{*p}}\right|_{\substack{\lambda=0\\ \lambda^*=0}}$$

(the derivatives are in Frechet sense).

The characteristic functional (2.8) and the moments (2.10) are calculated for the values of stochastic process $U(t)$ at fixed time t. However, on the basis of the governing equation (2.1) it is possible to derive equations which describe the evolution in time of the characteristic functional and the moments of any order.

3. Results

The derivation of the equations for the characteristic functional and the moments in the complex space case requires the analysis of two quantities: the solution process $U(t)$ itself and its conjugate $\overline{U(t)}$. Performing reasoning similiar to the considerations of paper [7] (see [6]), one obtaines the equations for the characteristic functional and for the moments of any order.

Under appropriate assumptions (see [6]) the characteristic functional $F[t,\lambda,\lambda^*]$ satisfies the following differential equation:

$$\frac{\partial}{\partial t} F[t,\lambda,\lambda^*] = \left(\frac{\delta F[t,\lambda,\lambda^*]}{\delta\lambda} , A^*\lambda \right)_X + \left(\frac{\delta F[t,\lambda,\lambda^*]}{\delta\lambda^*} , \bar{A}^*\lambda^* \right)_X +$$

$$+ \frac{I}{2}\Big\{ \sum_{k=1}^{\infty} \alpha_k \left(\left[B \frac{\delta}{\delta\lambda} \left(\left[B \frac{\delta F[t,\lambda,\lambda^*]}{\delta\lambda} \right] e_k , \lambda \right)_X \right] e_k , \lambda \right)_X +$$

(3.1)

$$+ \sum_{k=1}^{\infty} \alpha_k \left(\left[B \frac{\delta}{\delta\lambda} \left(\left[\bar{B} \frac{\delta F[t,\lambda,\lambda^*]}{\delta\lambda^*} \right] e_k , \lambda^* \right)_X \right] e_k , \lambda \right)_X +$$

$$+ \sum_{k=1}^{\infty} \alpha_k \left(\left[\bar{B} \frac{\delta}{\delta\lambda^*} \left(\left[B \frac{\delta F[t,\lambda,\lambda^*]}{\delta\lambda} \right] e_k , \lambda \right)_X \right] e_k , \lambda^* \right)_X +$$

$$+ \sum_{k=1}^{\infty} \alpha_k \left(\left[\bar{B} \frac{\delta}{\delta\lambda^*} \left(\left[\bar{B} \frac{\delta F[t,\lambda,\lambda^*]}{\delta\lambda^*} \right] e_k , \lambda^* \right)_X \right] e_k , \lambda^* \right)_X \Big\}$$

where $\lambda \in \mathcal{D}(A)$, $\lambda^* \in \mathcal{D}(\bar{A})$, and the initial and normalization conditions:

(3.2) $$F[0,\lambda,\lambda^*] = e^{i(U_o,\lambda)_X + i(\bar{U}_o,\lambda^*)_X}$$

(3.3) $$F[t,0,0] = 1$$

Remark

In paper [8] the method of constructing the characteristic functional of the solution to some particular stochastic differential equation is presented. Following the reasoning of present paper that is considering additionally the conjugate process, it is also possible to perform such a construction for complex space valued equations.

To obtain the equations for the moments of the solution of eq.(2.1), let us assume that its characteristic functional is represented in the form of series ($\lambda_1 \overset{\text{def}}{=} \lambda$, $\lambda_2 \overset{\text{def}}{=} \lambda^*$):

(3.4) $$F[t,\lambda,\lambda^*] = 1 + \sum_{l=1}^{\infty} \sum_{p=0}^{l} \frac{(i)^l}{p!(l-p)!} \Gamma_{k_1 \ldots k_l}(t) \circ \lambda_{k_1} \otimes \ldots \otimes \lambda_{k_l} ,$$

(the dot denotes the inner product of tensors), where the indexes $k_1,\ldots,k_l$ satisfy the condition given in (2.11). After substituting series (3.4) into the equation for the characteristic functional and comparing the coefficients of the like powers of λ and λ^* in the left and right hand sides, one arrives at the following moment equations:

$$\frac{\partial}{\partial t}\Gamma_{k_1..k_l}(t) = \sum_{j=1}^{l} A_j^{k_j}\,\Gamma_{k_1..k_l}(t) + \tag{3.5}$$

$$+\ \frac{1}{2}\sum_{k=1}^{\infty}\ \sum_{i,j=1}^{l} \alpha_k\ [\,B_j^{k_j}\langle[\,B_i^{k_i}\ \Gamma_{k_1..k_l}(t)\,]e_k^i\rangle\,]e_k^j\ ,$$

with the initial conditions:

$$\Gamma_{k_1..k_l}(0) = U_{k_1}(0)\otimes\ \dots\ \otimes U_{k_l}(0)\ , \tag{3.6}$$

where $l=1,2,\dots$, and the new operators are defined for the simple tensors of the form:

$$\Gamma_{k_1k_2...k_l} = \gamma_{k_1}\otimes\gamma_{k_2}\otimes\dots\otimes\gamma_{k_l} \tag{3.7}$$

as

$$A_j^{k_j}\ \Gamma_{k_1k_2...k_l} = \gamma_{k_1}\otimes\gamma_{k_2}\otimes\dots\otimes A^{k_j}\gamma_{k_j}\otimes\dots\otimes\gamma_{k_l} \tag{3.8}$$

$$[B_j^{k_j}\ \Gamma_{k_1k_2...k_l}]e_k^j = \gamma_{k_1}\otimes\gamma_{k_2}\otimes\dots\otimes[B^{k_j}\gamma_{k_j}]e_k\otimes\dots\otimes\gamma_{k_l} \tag{3.9}$$

$k_j=1,2,\quad j=1,2,\dots$, and, analogously to (2.9):

$$A^1 = A\ ,\qquad B^1 = B\ ,\qquad A^2 = \bar{A}\ ,\qquad B^2 = \bar{B}\ .$$

It is observed that the equations (3.5) are separated for each l and arbitrary fixed set of indexes $k_1, k_2, \dots, k_l$. This is the property of the linear equations with parametric or external white noise excitations.

It can be shown that the equations (3.5) for $l=1,2,\dots$, possess unique solutions. The proof is similiar to that in [7] for the real Hilbert space valued equations. This fact causes also the uniqueness of the analytical solution (that is of the form (3.4))of the functional equation (3.1). The existence needs some more restrictive assumptions on operator B; they could be precised explicitely for particular examples (see e.g. [7]).

4. The wave equation - application of the general formulae

The wave equation (1.3) written in the abstract evolutionary form (2.1) is:

$$dU = AU\,dx + [\varepsilon BU]\,dW(x)\ , \tag{4.1}$$

where now spatial variable x corresponds to time t in (2.1), and the operators are defined as:

$$A = \frac{i}{2k_0}\Delta_\perp = \frac{i}{2k_0}\left(\frac{\partial^2}{\partial y^2} + \frac{\partial^2}{\partial z^2}\right), \qquad B = \frac{ik_0}{2}\ . \tag{4.2}$$

The spaces $\mathbb{X}$ and $\mathbb{Y}$ are choosen in this model as:

$$\mathbb{X} = L^2(\mathbb{R}^2,\mathbb{C}), \qquad \mathbb{Y} = L^2(\mathbb{R}^2,\mathbb{R}). \tag{4.3}$$

We assume that random field $\varepsilon(x,y,z)$ is Gaussian with a zero mean and δ-correlated in x:

$$E\Big\{\, \varepsilon(x,y,z)\varepsilon(x',y',z') \,\Big\} = Q(y-y',z-z')\delta(x-x'). \tag{4.4}$$

The field ε is sufficiently smooth with respect to y,z , such that product εU takes values in $\mathbb{X}$. The eigenvalues and eigenvectors of operator Q are the solutions of the following integral equation:

$$\int\!\!\int Q(y-y',z-z')\ e_k(y',z')\ dy'\ dz' = \alpha_k\ e_k(y,z). \tag{4.5}$$

The covariance operator is the integral one with the kernel $Q(y-y',z-z')$ and process W in equation (4.1) is defined as

$$W(x) = \int_0^x \varepsilon(x',y,z)\ dx'. \tag{4.6}$$

The adequate conjugate operators are:

$$\overline{A} = -\frac{i}{2k_0}\Delta_\perp = -\frac{i}{2k_0}\left(\frac{\partial^2}{\partial y^2}+\frac{\partial^2}{\partial z^2}\right),$$

(4.7)

$$\overline{B} = -\frac{ik_0}{2}.$$

The equation for the characteristic functional (3.1) in this particular case of partial differential equation (1.3) is the differential equation in Volterra variational derivatives and it has the form:

$$\frac{\partial}{\partial x}F[x,\lambda,\lambda^*] = \frac{i}{2k_0}\iint dr_1\left(\Delta_\perp\frac{\delta}{\delta\lambda(r_1)}F[x,\lambda,\lambda^*]\,\overline{\lambda(r_1)}\, -\right.$$

(4.8)

$$\left. -\,\Delta_\perp\frac{\delta}{\delta\lambda^*(r_1)}F[x,\lambda,\lambda^*]\,\overline{\lambda^*(r_1)}\right) -$$

$$-\frac{1}{2}\frac{k_0^2}{4}\iint dr_1\iint dr_1'\,Q(r_1-r_1')\times$$

$$\times\left\{\overline{\lambda(r_1)}\frac{\delta}{\delta\lambda(r_1)}\left(\overline{\lambda(r_1')}\frac{\delta}{\delta\lambda(r_1')}F[x,\lambda,\lambda^*]\right)-\right.$$

$$-\,\overline{\lambda(r_1)}\frac{\delta}{\delta\lambda(r_1)}\left(\overline{\lambda^*(r_1')}\frac{\delta}{\delta\lambda^*(r_1')}F[x,\lambda,\lambda^*]\right)-$$

$$-\,\overline{\lambda^*(r_1)}\frac{\delta}{\delta\lambda^*(r_1)^*}\left(\overline{\lambda(r_1')}\frac{\delta}{\delta\lambda(r_1')}F[x,\lambda,\lambda^*]\right)+$$

$$\left.+\,\overline{\lambda^*(r_1)}\frac{\delta}{\delta\lambda^*(r_1)}\left(\overline{\lambda^*(r_1')}\frac{\delta}{\delta\lambda^*(r_1')}F[x,\lambda,\lambda^*]\right)\right\}.$$

where $r_1=(y\ ,z\ )\ ,\ r_1'=(y',z')\in\mathbb{R}^2\ ,$

$\lambda(r_1),\ \lambda^*(r_1)\in\mathcal{H} = L^2(\mathbb{R}^2,\mathbb{C})\ ,$

and the initial conditions (3.2), (3.3) are:

$$F[0,\lambda,\lambda^*] = \exp\left\{i\iint U_0(y,z)\,\overline{\lambda(y,z)}\,dy\,dz\, +\right.$$

$$+ i \int\int \overline{U_0(y,z)}\ \overline{\lambda^*(y,z)}\ dy\ dz \Big\} ,$$

$$F[x,0,0] = 1 .$$

The equations for the moments of any order are

$$\frac{\partial}{\partial x} \Gamma_{k_1 .. k_l}(x,y_1,z_1,\ldots,y_l,z_l) = \tag{4.9}$$

$$= \sum_{i=1}^{l} \frac{i}{2k_0} (-1)^{k_i+1} \left(\frac{\partial^2}{\partial y_i^2} + \frac{\partial^2}{\partial z_i^2} \right) \Gamma_{k_1 .. k_l}(x,y_1,z_1,\ldots,y_l,z_l) -$$

$$- \frac{1}{2} \frac{k_0^2}{4} \sum_{i,j=1}^{l} Q(x_i - x_j, y_i - y_j) (-1)^{k_i+k_j} \Gamma_{k_1 .. k_l}(x,y_1,z_1,\ldots,y_l,z_l)$$

for $k_1,k_2,\ldots,k_l = 1,2$, $l=1,2,\ldots$ and

$$\Gamma_{k_1 .. k_l}(0,y_1,z_1,\ldots,y_l,z_l) = U_0^{k_1}(y_1,z_1) \times \ldots \times U_0^{k_l}(y_l,z_l).$$

As it is known from the general theory for equations (3.5), equations (4.9) have unique solutions.

The equation for the characteristic functional and the moments of lover order of the form (4.8) and (4.9) coincide with the analogous equations obtained in [11] with the use of classical Furutsu- Novikov formula. In this section we have shown, how these equations can be obtained by means of our general method.

References.

[1] A. Chojnowska-Michalik, *Stochastic differential equations in Hilbert space*, in *Probability Theory*, Banach Center Publications, Vol.5, pp.53-74, Polish Scientific Publishers, Warsaw,1979.

[2] P.L.Chow, *Function-space differential equations associated with a stochastic partial differential equations*, Indiana Univ. Math. Journal, 25(1976), 609-627.

[3] P.L.Chow, *Stochastic partial differential equations in turbulence related problems*, in *Probabilistic Analysis and Related Topics*, Vol.1, A.T.Bharucha-Reid, ed., Academic Press, New York, 1978.

[4] R.Curtain, I.Prichard, *Infinite dimensional linear systems theory*, Lect.Not. in Contr. and Inf. Sc., Vol.8, Springer-Verlag, Berlin, Heidelberg, New York, 1978.

[5] Yu.Daletskii, S.N.Paramonova, *Stochastic integrals over normally distributed additive function of sets*, Dokl.Ac.Sc.USSR, **208**(1973), 512-515.

[6] Z.Kotulski, *Equations for the characteristic functional and moments of the complex stochastic evolutions*, IFTR Reports 38/1988.

[7] Z.Kotulski, *Equations for the characteristic functional and moments of the stochastic evolutions with an application*, SIAM Journal on Appl. Math., **69**(1989),No.1.

[8] Z.Kotulski, K.Sobczyk, *Characteristic functionals of randomly excited physical syctems*, Physica **123A**(1984), 261-278.

[9] L.C.Lee, *Wave propagation in a random medium. A complete set of the moment equations with different wavenumbers*, J.Math. Phys. **15**(1974), 1431-1435.

[10] K.Sobczyk, *Stochastic Wave Propagation*, Elsevier 1984.

[11] V.I.Tatarskii, *The light propagation in a medium with random inhomogenities of the refractive index and the Markov process approximation*, J.Exp.Theor.Phys., **56**(1969), 2106-2117.

ON A CLASS OF SEMILINEAR STOCHASTIC PARTIAL DIFFERENTIAL EQUATIONS

Ralf Manthey
Friedrich-Schiller-Universitat
Sektion Mathematik
DDR-6900 Jena

Consider the formal partial differential equation

$$\frac{\partial}{\partial t} u(t,x) = (\Delta u)(t,x) + f(u(t,x)) + \sigma\xi(t,x), \quad t>0, \quad x\in(0,1)$$

together with a random initial condition $u(0,x)=u_0(x), x\in[0,1]$, and (for simplicity) homogeneous boundary conditions. Here ξ denotes the space-time white noise, which precisely will be defined later. Such kind of problems arises in physics, especially in the theory of so-called critical phenomena and in stochastic quantization proposed by PARISI and WU.

Denote the above mentioned problem by (RDD). In order to deal with (RDD) from a mathematical viewpoint it is first of all necessary to give (RDD) a precise mathematical meaning.

The notion of solution

Let $(\Omega,\mathfrak{F},\mathbb{P})$ be a complete probability space. Introduce the notation $(.,.)_0$ and $\|.\|_0$ for the inner product and the norm in $\mathbb{L}_2([0,1])$ respectively. The notation $\mathcal{D}$ is used for the space of all real valued infinitely often differentiable functions with compact support in a set $\mathbb{D}\subseteq\mathbb{R}^d$. The dual of $\mathcal{D}$ is as usual denoted by $\mathcal{D}'$ (SCHWARTZ distributions).

DEFINITION. *A centered GAUSSian $\mathcal{D}'([0,\infty)\times[0,1])$-valued random variable ξ with covariance functional $C(\varphi,\psi)=(\varphi,\psi)_0$ is called GAUSSian white noise.*

DEFINITION. *A pathwise continuous GAUSSian random field $W:\Omega\times\mathbb{R}_+^2\to\mathbb{R}$ with covariance function $\mathbb{E}W(t,x)W(s,y)=(t\wedge s)\cdot(x\wedge y)$ is called two-parameter BROWNian sheet.*

Clearly, the BROWNian sheet can be represented as a $\mathcal{D}'([0,\infty)\times[0,1])$-valued random variable. In this case one observes

$$\frac{\partial^2}{\partial t \partial x} W = \xi$$

in $\mathcal{D}'$ in distribution. Having this connection in mind and making use of natural approximations of ξ in terms of usual random fields one arrives at the following stochastic integral equation (cf. [2], [3], [6]).

$$(\mathrm{S}') \qquad u(t,x) = z(t,x) + \int_0^t \int_0^1 G(t-s,x,y) f(u(s,y))\,dy\,ds,$$

where $f: \mathbb{R} \to \mathbb{R}$ and

$$z(t,x) = \int_0^t G(t,x,y) u_o(y)\,dy + \sigma \int_0^t \int_0^1 G(t-s,y,x)\,dW_{sy}.$$

The stochastic process u_o is always assumed to be continuous and independent from W. The function $G: (0,\infty)\times[0,1]\times[0,1] \to \mathbb{R}_+$ represents the fundamental solution to $\frac{\partial}{\partial t} - \Delta$ with DIRICHLET boundary conditions. One gets

$$G(t,x,y) = \sum_{n=1}^{\infty} h_n(x) h_n(y) \exp(-\lambda_n t)$$

with $h_n(x) = \sqrt{2} \cdot \sin \pi n x$, $\lambda_n = (\pi n)^2$, $n \in \mathbb{N} = \{1, 2, \ldots\}$. Note $\Delta h_n = -\lambda_n h_n$. Because of

$$\int_0^t \int_0^1 G^2(s,x,y)\,dy\,ds < \infty$$

for every $(t,x) \in \mathbb{D}_T = [0,T]\times[0,1]$ the stochastic integral in (S') is well defined. Unfortunately, this property is lost in higher dimensions of the space parameter x. Finally, let $\mathfrak{F}_t^o = \sigma\{u_o(x),\ W(s,x):\ x\in[0,1],\ s\in[0,t]\}$.

DEFINITION. *A pathwise continuous random field $u = (u(t,x))_{(t,x)\in\mathbb{D}_T}$ is called a solution to (S') if it possesses the following properties.*

(i) *$u(t,.)$ is $\mathfrak{F}_t^o$-measurable for any $t\in[0,T]$.*

(ii) *u satisfies the equation (S') $\mathbb{P}$ a.s. for every $(t,x)\in\mathbb{D}_T$.*

REMARK. In particular, the second assertion includes the existence of the r.h.s. of (S′). This is always fulfilled if f is continuous.

DEFINITION. *Every solution to* (S′) *is called solution to* (RDD).

Suitable function spaces

However, there exist other notions of solutions used in the literature. In order to discuss the relation between these notions it is useful to introduce suitable function spaces.
Let $\varphi\in\mathcal{D}([0,1])$ and define the HILBERTian norm

$$\|\varphi\|_1^2 = \sum_{n=1}^{\infty} n^4(\varphi,h_n)_0^2.$$

The corresponding inner product is given by

$$(\varphi,\psi)_1 = \sum_{n=1}^{\infty} n^4(\varphi,h_n)_0(\psi,h_n)_0, \quad \varphi,\psi\in\mathcal{D}.$$

The completion of $(\mathcal{D},\|.\|_1)$ represents a separable HILBERT space which in the following is denoted by $(\mathcal{D}_1,\|.\|_1)$. The system $\varepsilon_n:=n^{-2}h_n$, $n\in\mathbb{N}$ is an ONB in $\mathcal{D}_1$. The dual of $\mathcal{D}_1$ is denoted by $\mathcal{D}'_{-1}$. The set $\mathcal{D}'_{-1}$ equipped with the norm

$$\|f\|_{-1}^2 = \sum_{n=1}^{\infty} f^2(\varepsilon_n), \quad f\in\mathcal{D}'_{-1}$$

is a separable HILBERT space, where the corresponding inner product is given by

$$(f,g)_{-1} := \sum_{n=1}^{\infty} f(\varepsilon_n)g(\varepsilon_n), \qquad f,g\in\mathcal{D}_{-1}.$$

The dual ONB $(e_n)_{n\in\mathbb{N}}$ is defined by $e_m(\varepsilon_n)=\delta_{mn}$, $m,n\in\mathbb{N}$. If $\mathbb{L}_2^0=\{f\in\mathbb{L}_2([0,1]):\ f(0)=f(1)=0\}$ is identified with it′s dual, one has

$$\mathcal{D}_1 \subset \mathbb{L}_2^0 \subset \mathcal{D}'_{-1}.$$

The operator Δ can be extended in $\mathcal{D}'_{-1}$ from $\mathcal{D}$ to $\mathbb{L}_2^0$ by the natural ansatz

$$\Delta f = \sum_{n=1}^{\infty} (f,\Delta\varepsilon_n)_0 e_n.$$

In this case Δ is a symmetric closed operator, and the space $\mathbb{L}_2^o$ is dense in $\mathcal{D}'_{-1}$.

Brownian sheet and the cylindrical WIENER process

Denote by $\mathfrak{K}(\mathcal{D}'_{-1})$ the KOLMOGOROV σ-algebra on $\mathcal{D}'_{-1}$, i.e. the smallest σ-algebra such that the real-valued mappings $f \to f(\varphi)$ are measurable for any $\varphi \in \mathcal{D}_1$. A $\mathcal{D}'_{-1}$-valued $\mathfrak{F}/\mathfrak{K}(\mathcal{D}'_{-1})$-measurable mapping on $(\Omega, \mathfrak{F}, \mathbb{P})$ is called $\mathcal{D}'_{-1}$-valued random variable. A $\mathcal{D}'_{-1}$-valued random variable X is called centered GAUSSian $\mathcal{D}'_{-1}$-variable if the family $(X(\varphi))_{\varphi \in \mathcal{D}_1}$ is a centered GAUSSian system. Let $Y = (Y(\varphi))_{\varphi \in \mathcal{D}_1}$ be a family of real valued random variables and X a $\mathcal{D}'_{-1}$ random variable. If $X(\varphi) = Y(\varphi)$ $\mathbb{P}$ a.s. for all $\varphi \in \mathcal{D}_1$, then X is called regularization of Y. The regularization is $\mathbb{P}$ a.s. uniquely determined. A family of $\mathcal{D}'_{-1}$ random variables $X = (X_t)_{t \geq 0}$ is called $\mathcal{D}'_{-1}$ stochastic process. Such a process is called (pathwise) continuous if $X(.,\omega)$ is continuous for every $\omega \in \Omega$.

DEFINITION. *A centered continuous $\mathcal{D}'_{-1}$ GAUSSian process B is called cylindrical WIENER process if it possesses the covariance functional $C_{t,s}(\varphi,\psi) = (t \wedge s)(\varphi,\psi)_o$.*

Let $\varphi \in \mathcal{D}_1$ and W be a BROWNian sheet. Consider

$$I_t(\varphi) = \int_0^t \int_0^1 \varphi(x)\, dW_{sx}, \qquad t \geq 0.$$

Obviously, the family $I = (I_t(\varphi))_{t \geq 0, \varphi \in \mathcal{D}_1}$ is GAUSSian, centered and has the same covariance functional as in the last definition. In addition, there exists a continuous $\mathcal{D}'_{-1}$ regularization of I denoted by the same symbol. Hence I is a cylindrical WIENER process, which will be always used in the following.

Stochastic integration with respect to the cylindrical WIENER process

Let $\mathfrak{F}_t = \sigma\{W_{sx}\colon\ 0 \leq s \leq t,\ x \in [0,1]\}$ and $H\colon [0,T] \times \Omega \to \mathbb{L}_2^o$ be a $\mathfrak{F}_t$-adapted stochastic process such that

$$\mathbb{E}\int_0^t \|H_s\|_0^2 ds < \infty$$

for every $t\in[0,T]$.

DEFINITION. *The real-valued random variable*

$$\int_0^t \langle H_s, dI_s\rangle, \quad t\in[0,T],$$

given by

$$\int_0^t \langle H_s, dI_s\rangle := \sum_{n=1}^{\infty} \int_0^t (H_s, h_n)_0 dI_s(h_n)$$

is called stochastic integral of H on [0,t] *with respect to I.*

Note that $(I_t)_{t\geq 0, n\in\mathbb{N}}$ is a family of independent real-valued standard WIENER processes.
Let $\mathscr{H}(\mathbb{L}_2^o)$ denote the set of all HILBERT-SCHMIDT operators in $\mathbb{L}_2^o$ equipped with the HILBERT-SCHMIDT norm $\|.\|_{HS}$. Further, let H: $[0,T]\times\Omega\rightarrow\mathscr{H}(\mathbb{L}_2^o)$ be an $\mathfrak{F}_t$-adapted $\mathscr{H}$-valued stochastic process possesssing the property

$$\mathbb{E}\int_0^t \|H_s\|_{HS}^2 ds < \infty$$

for any $t\in[0,T]$.

DEFINITION. *The* $\mathbb{L}_2^o$*-valued random variable* $\int_0^t H_s dI_s$ *which is* ($\mathbb{P}$ *a.s.) uniquely determined by*

$$(h, \int_0^t H_s dI_s)_0 = \int_0^t \langle H_s^* h, dI_s\rangle$$

$\mathbb{P}$ *a.s. for every* $h\in\mathbb{L}_2^o$, *is called stochastic integral of H on* [0,t] *with respect to I.*
(H_s^* denotes the dual operator of H_s.)

LEMMA. *Let* $F:\mathbb{D}_T\rightarrow\mathbb{R}$ *be a non-random mapping such that* $F\in\mathbb{L}_2(\mathbb{D}_T)$ *and* $F(t,.)\in\mathbb{L}_2^o$, $t\in[0,T]$. *Then it holds*

$$\int_0^t\int_0^1 F(s,x)dW_{sx} = \int_0^t \langle F_s, dI_s\rangle \quad \mathbb{P} \text{ a.s.}$$

for every $t\in[0,T]$.

The equation (D')

Multiply the formal partial differential equation in (RDD) by $\varphi\in\mathcal{D}$ in $\mathbb{L}_2^o$. Integrate then formally from 0 to t. This leads to

$$(U_t,\varphi)_o = (U_o,\varphi)_o + \int_0^t [(U_s,\Delta\varphi)_o + (F(U_s),\varphi)_o]ds + \sigma I_t,$$

where formal expressions are replaced by corresponding terms having a precise meaning. Here U_s denotes $u(s,.)$ and $F(U_s)$ is the same as $f(u(s,.))$. Let $\mathbb{C}_o$ denote the space of all continuous functions φ defined on $[0,1]$ with $\varphi(0)=\varphi(1)=0$.

DEFINITION. *A continuous $\mathbb{C}_o$-valued stochastic process $U=(U_t)_{t\in[0,T]}$ is called solution to* (D') *if it possesses the following properties.*

(*i*) U_t *is $\mathfrak{F}_t^o$-adapted for every $t\in[0,T]$.*

(*ii*) $U=(U_s)_{s\in[0,t]}$ *satisfies* (D') $\mathbb{P}$ *a.s. for every $t\in[0,T]$, $\varphi\in\mathcal{D}$.*

Note that the continuity of U means the continuity in the $\mathbb{C}_o$-norm. As above the continuity of f is assumed.

THEOREM. *Every solution of* (S') *is also a solution to* (D') *and vice versa.*

Note that a similar theorem was proven by IWATA [1] for so-called weak solutions of (RDD) not considered here.

The equation (S)

Denote by $(T_t)_{t\geq 0}$ the family of operators in $\mathbb{L}_2^o$ given by

$$(T_t\psi)(x) = \int_0^1 G(t,x,y)\psi(y)dy, \quad t>0$$

and

$$T_o = E \quad \text{(the identical operator)}.$$

Obviously, $(T_t)_{t\geq 0}$ is a semigroup of contractions with $T_t\in\mathcal{K}$, $t>0$. Consequently, the stochastic integral

$$\int_0^t T_{t-s}dI_s$$

is well-defined and the equation

$$U_t = T_t U_o + \int_o^t T_{t-s} F(U_s)\,ds + \sigma \cdot \int_o^t T_{t-s}\,dI_s \tag{S}$$

makes sense.

DEFINITION. *A continuous $\mathbb{C}_o$-valued stochastic process $U=(U_t)_{t\in[0,T]}$ is called solution to (S) if it has the following properties.*
(i) U_t is $\mathfrak{F}^o_t$-adapted, $t\in[0,T]$.
(ii) $U=(U_s)_{s\in[0,t]}$ satisfies (S) in $\mathbb{L}^o_2$ $\mathbb{P}$ a.s. for every $t\in[0,T]$.

The equation (D)

Let $U=(U_t)_{t\in[0,T]}$ be a continuous $\mathbb{C}_o$-valued stochastic process. In this case the equation

$$U_t = U_o + \int_o^t (\Delta U_s + F(U_s))\,ds + \sigma I_t \tag{D}$$

makes sense in $\mathcal{D}'_{-1}$.

DEFINITION. *A continuous $\mathbb{C}_o$-valued stochastic process $U=(U_t)_{t\in[0,T]}$ is called solution to (D) if it has the following properties.*
(i) U_t is $\mathfrak{F}^o_t$-adapted for every $t\in[0,T]$.
(ii) $U=(U_s)_{s\in[0,t]}$ satisfies (D) in $\mathcal{D}'_{-1}$ $\mathbb{P}$ a.s. for any $t\in[0,T]$.

Equivalence

THEOREM. *If $U=(U_t)_{t\in[0,T]}$ solves one of the equations (S′), (S), (D′) or (D) then it also represents a solution of the remaining equations.*

Existence and uniqueness of solutions

It is easy to prove existence and uniqueness of a solution under standard conditions as global LIPSCHITZ continuity or local LIPSCHITZ continuity and linear growth of f.

THEOREM (c.f.[4]). *Suppose that f satisfies the following conditions.*

(F1) *f is locally LIPSCHITZ continuous.*

(F2) *There exist two nonincreasing functions g and* $h:\mathbb{R}\to\mathbb{R}$ *such that* $g\le f\le h$.

Then (RDD) *possesses a pathwise unique solution.*

Note that this theorem covers all physically interesting situations connected with (RDD).

REMARK. An analogous theorem can be proven for the CAUCHY (initial value) problem, c.f. [4].

A comparison theorem

Consider the problem (RDD) for two reaction functions f_1 and f_2.

THEOREM. *Let* f_1 *and* f_2 *be locally LIPSCHITZ continuous and such that the corresponding problems*(RDD) *in each case possess a pathwise unique solution* u_i, $i=1,2$. *In this situation the relation* $f_1\ge f_2$ *implies* $u_1\ge u_2$ $\mathbb{P}$ *a.s.*

Properties of the solution

THEOREM. *Let f be locally LIPSCHITZ continuous. In this case every solution to* (RDD) *is locally HOLDER continuous in* $t\in[0,T]$ *for any* $x\in(0,1)$ *and* $\mathbb{P}$ *a.a.* $\omega\in\Omega$.

Suppose in the following that the conditions of the above mentioned existence and uniqueness theorem are satisfied.

THEOREM. *The solution* $U=(U_t)_{t\in[0,T]}$ *is a MARKOV process in time.*

Let $(\sigma_n)_{n\in\mathbb{N}}$ be sequence with $\sigma_n\to 0$, $\sigma_n>0$, and denote by $U^{(n)}$ the solution corresponding to σ_n $U^{(0)}$ is the solution to (RDD) for $\sigma=0$.

THEOREM. *It holds*

$$\lim_{n\to\infty}\ \sup_{t\in[0,T]}\ \| U_t^{(0)}-U_t^{(n)}\| = 0, \quad \mathbb{P}\ a.s.,$$

where $\|.\|$ *denotes the* $\mathbb{C}_o$ *norm.*

Let V_o and U_o be elements of $\mathbb{C}_0$ and introduce the condition

(E) $$(u-v)\ (f(u)-f(v)) \le \pi^2(u-v)^2, \quad u\ne v.$$

THEOREM. *Under the condition* (E) *it holds*

$$\|U_t-V_t\|_0^2 \le exp(-\mu t)\cdot\|U_o-V_o\|_0^2$$

for any $t\ge 0$ *and* $\mathbb{P}$ *a.s.* $\omega\in\Omega$, *where* μ *is a positive constant.*

Note that (E) does not follow from (F2). For example, $f(u)=-u^3-\lambda u$ satisfies (F2) but it fulfils (E) only for $\lambda\le\pi^2$.

References

[1] K. IWATA: An Infinite Dimensional Stochastic Differential Equation with State Space $\mathbb{C}(\mathbb{R})$, *Prob. Th. Rel. Fields 74 (1987) 141-159*

[2] R. MANTHEY: Weak Convergence of Solutions of the Heat Equation with Gaussian Noise, *Math. Nachr. 123 (1985) 157-168*

[3] R. MANTHEY: Existence and Uniqueness of a Solution of a Reaction-Diffusion Equation with Polynomial Nonlinearity and White Noise Disturbance, *Math. Nachr.125 (1986) 121-133*

[4] R. MANTHEY: On the Cauchy Problem for Reaction-Diffusion Equations with White Noise, *Math. Nachr. 136 (1988) 209-228*

[5] R. MANTHEY: Reaktions-Diffusionsgleichungen mit weißem Rauschen, Dissertation B, Friedrich-Schiller-Universitat Jena, 1988

[6] J.B. WALSH: A Stochastic Model of Neural Response, *Adv.Appl. Prob. 13 (1981) 231-281*

STRONG FELLER PROPERTY FOR SEMILINEAR STOCHASTIC EVOLUTION EQUATIONS AND APPLICATIONS

Bohdan Maslowski
Mathematical Institute of the Czechoslovak Academy of Sciences
Žitná 25, 115 67 Praha 1, Czechoslovakia

1. Introduction

Consider a semilinear stochastic evolution equation

(1.1) $$d\xi_t = A\xi_t dt + f(\xi_t)dt + dw_t$$

on a separable Banach space $(E, \|\cdot\|)$ continuously embedded into a Hilbert space $(H, |\cdot|_H)$ $(|\cdot|_H \leq c\|\cdot\|$ for some $c > 0)$. Throughout the paper we assume the following conditions (H1) - (H5) to be fulfilled:

(H1) Linear operator A defines a C_o-semigroup $S(t)$, $t \geq 0$, on the space E, extendable to a C_o-semigroup $S_o(t)$ on H with an infinitesimal operator A_o.

(H2) Linear operator A_o is self-adjoint and negative definite. The Gaussian measure $\gamma = N(0,Q)$, where $Q = -\frac{1}{2}A_o^{-1}$, is supported by E.

(H3) Process w_t is a H-valued cylindrical Wiener process and process $Z(t) = \int_0^t S_o(t-s)dw_s$, $t \geq 0$, has a continuous E-valued modification, adapted to an increasing family of σ-fields $(\mathcal{F}_t)$ and such that for arbitrary $t \geq 0$ processes $\int_t^{t+r} S_o(t+r-s)dw_s$, $r \geq 0$, are independent of $\mathcal{F}_t$ and have the distributions identical with $Z(r)$, $r \geq 0$.

(H4) Transformation $f : E \to E$ is locally Lipschitz (i.e. Lipschitz on bounded sets) and the integral equation

$$u(t) = S(t)x + \int_0^t S(t-s)f(u(s))ds + \psi(t), \quad t \geq 0$$

has a unique global solution $u(t,x,\psi)$, $t \geq 0$, for any $x \in E$ and E-continuous function ψ such that $\psi(0) = 0$ (shortly $\psi \in C_o(0,+\infty,E)$). Moreover,
$\sup\{u(s,x,\psi);\ s \in \langle 0,t \rangle,\ x \in K,\ \psi \in C_o(0,t,E),\ |\psi| \leq R\} < \infty$

holds for any compact $K \subset E$, $t,R>0$.

(H5) There exists a real valued function $U : E \to \mathbb{R}$ bounded from above such that its restriction to an arbitrary finite dimensional subspace of E is twicely continuously differentiable and its directional derivative $U'(x;y)$ at any $x \in E$ and any direction $y \in E$ is $U'(x;y) = \langle f(x),y \rangle$, where $\langle .,. \rangle$ is the scalar product on H. We write shortly $U' = f$.

The conditions on $S(t)$ and f guaranteeing (H3) and (H4) are given in [1]. (H3) and (H4) are sufficient for existence, uniqueness and the Markov property of the solution to the equation (1.1) and (H1) - (H5) imply its symmetrizability ([17]). The corresponding invariant measure has the form

$$\mu = \alpha e^{2U(x)} \gamma(dx), \qquad \alpha^{-1} = \int e^{2U(x)} \gamma(dx). \tag{1.2}$$

Equations of the type (1.1) were studied by numerous authors especially in the important particular cases $E = H$ (see e.g. [15], [5]) and $A = d^2/dx^2$, $H = L_2(0,1)$, $E = C_o(0,1)$ (see e.g. [14], [2], [8], [13]). In the full generality they are treated in [1], [17].
Denote by C_b and $\mathbb{M}$ the sets of all bounded continuous and bounded Borel measurable, respectively, real functions on E and by $P(t,x,A)$ ($t \geq 0$, $x \in E$, $A \in \mathcal{B}(E)$ - the Borel sets on E) the transition probability function corresponding to the solution of (1.1). Set

$$T_t \varphi(x) = \int \varphi(y) P(t,x,dy) = E \varphi(\xi_t^x), \quad \varphi \in \mathbb{M}, \quad x \in E, \quad t>0.$$

Let us recall

Definition 1.1. We say the solution ξ_t of (1.1) is

(A) strongly Feller, if $T_t(\mathbb{M}) \subset C_b$ for $t>0$.

(B) strongly Feller in the restricted sense, if

$$\mathrm{Var}\,(P(t,x_n,.) - P(t,x_o,.)) \to 0, \quad x_n \to x_o \quad \text{in } E,$$

holds for all $t>0$, where Var stands for the total variation of a measure.

The above defined notions were introduced by Dynkin and Girsanov and studied e.g. by Girsanov [7]. Obviously, (B)$\Longrightarrow$(A) and (A) is stronger than the "classical" Feller property $T_t(C_b) \subset C_b$, $t>0$. Unlike the Feller property, the strong Feller property requires some nondegeneracy of the corresponding Markov process. For example, deterministic processes are not strongly Feller. A typical example of a process satisfying ((A) and) (B) is a solution of a finite-dimensional stochastic differential equation with positive definite dif-

fusion matrix.
The aim of the paper is to prove that the solution of (1.1) is strongly Feller in the restricted sense provided (H1) - (H5) is fulfilled. Some applications are also given (equivalence of transition probabilities, ergodic properties, etc.). Let us point out that if in (1.1) a covariance Wiener process (with a positive nuclear covariance operator) is considered instead of the cylindrical one, it can happen that the solution is not strongly Feller (cf. Remark 2.12 and Proposition 2.13). By (α_m, e_m) we denote the sequence of all eigenvalues and normalized eigenvectors of the operator $-A_0$ (we have $\alpha_m > 0$, $\sum 1/\alpha_m < \infty$) and $U_r(z)$ stands for an open ball with the centre z and radius r in the corresponding metric space.

2. Main results

The main result is the following

Theorem 2.1. Let (H1) - (H5) is fulfilled. Then the solution of (1.1) is strongly Feller in the restricted sense.
The proof will be given in the second part of the paper. This section contains some consequences of Theorem 2.1. We will use the following

Lemma 2.2. For every $t > 0$, $x \in E$, $\emptyset \neq U \subset E$, U open, we have

$$P(t,x,U) > 0 .$$

Proof. Given $x \in E$, $z \in E$, $t > 0$, $\varepsilon_0 > 0$, it is easy to see that there exists a function $\psi \in C_0(0,t,E)$ such that the solution $\tilde{y}$ of the integral equation

$$\tilde{y}(r) = S(r)x + \int_0^r S(r-s)f(\tilde{y}(s))ds + \psi(r) , \quad r \in \langle 0,t \rangle$$

fulfils $\tilde{y}(t) \in U_{\varepsilon_0/2}(z)$. From [17], Proposition 4 it follows that there exists a $\delta > 0$ such that for all $\hat{\psi} \in C_0(0,t,E)$ such that $\sup_{s \in \langle 0,t \rangle} \|\hat{\psi}(s) - \psi(s)\| < \delta$ and the corresponding solutions

$$\hat{y}(r) = S(r)x + \int_0^r S(r-s)f(\hat{y}(s))ds + \hat{\psi}(r) , \quad r \in \langle 0,t \rangle$$

we get $\|\hat{y}(t) - \tilde{y}(t)\| < \varepsilon_0/2$, i.e., $\hat{y}(t) \in U_{\varepsilon_0}(z)$. On the other hand, we have

$$P\left[\|\psi(s) - Z(s)\| < \delta \quad \text{for all } s \in \langle 0,t \rangle\right] > 0$$

as the (Gaussian) measure induced by the process $Z(s)$ in the space $C_0(0,t,E)$ is full.

Corollary 2.3. The transition probability measures $P(t,x,.)$, $t > 0$, $x \in E$, and the invariant measure μ are equivalent (i.e., mutually

absolutely continuous).
The proof we obtain as a simple consequence of the strong Feller property and Lemma 2.2 - see [9], pp. 197 and 198.

Corollary 2.4. The invariant measure μ is ergodic.
Proof. Assume the converse. The there exist Borel sets A_1, A_2 such that $\mu(A_i) > 0$, $P(t, x_i, A_i) = 1$ for all $t > 0$, $x_i \in A_i$, $i=1,2$. By Lemma 2.2 both A_1, A_2 are dense in E. Thus

$$T_t \chi_{A_1}(x) = P(t,x,A_1) = \begin{cases} 1, & x \in A_1 \\ 0, & x \in A_2 \end{cases}$$

is discontinuous, which contradicts the strong Feller property.

Corollary 2.5 (the strong law of large numbers). Let $\varphi : E \to \mathbb{R}$ be measurable and such that

$$(2.1) \qquad \int_E |\varphi(x)| \gamma(dx) < \infty .$$

Then

$$\frac{1}{T} \int_0^T \varphi(\xi_t^x) dt \longrightarrow \int \varphi \, d\mu \qquad \text{a.s.}$$

holds as $T \longrightarrow \infty$ for all $x \in E$.
Proof. By (H5) and (2.1) we get

$$\int_E |\varphi(x)| \mu(dx) < \infty$$

and thus, since μ is ergodic,

$$(2.2) \qquad \frac{1}{T} \int_0^T \varphi(\xi_t^y) dt \longrightarrow \int \varphi \, d\mu \qquad \text{a.s.}$$

for $y \in \mathcal{M}$, where $\mathcal{M} \subset E$ is such that $\mu(\mathcal{M}) = 1$. For any $x \in E$, $t_o > 0$, we obtain

$$P\Big[\frac{1}{T} \int_0^T \varphi(\xi_t^x) dt \longrightarrow \int \varphi \, d\mu\Big] = P\Big[\frac{1}{T} \int_{t_o}^T \varphi(\xi_t^x) dt \longrightarrow \int \varphi \, d\mu\Big] =$$
$$= \int_E P\Big[\frac{1}{T} \int_0^T \varphi(\xi_t^y) dt \longrightarrow \int \varphi \, d\mu\Big] P(t_o, x, dy) .$$

By Corollary 2.3 $P(t_o, x, E \setminus \mathcal{M}) = 0$ and hence the last term is equal to

$$\int_{\mathcal{M}} P\Big[\frac{1}{T} \int_0^T \varphi(\xi_t^y) dt \longrightarrow \int \varphi \, d\mu\Big] P(t_o, x, dy) = 1$$

by (2.2).
As a special case of Corollary 2.5 we obtain

Corollary 2.6.

$$\frac{1}{T}\int_0^T P(t,x,A)\,dt \longrightarrow \mu(A)\ , \quad T\rightarrow\infty\ ,$$

holds for every $x \in E$, $A \in \mathcal{B}(E)$. In particular, the invariant measure μ is unique.

Corollary 2.5 can be used to prove a subsequent result on a parameter identification which can be interesting from the viewpoint of applications. Consider a parameter-dependent equation

$$(2.3) \qquad d\xi_t = A\xi_t dt + U'_\Theta(\xi_t)dt + dw_t$$

whose coefficients satisfy (H1) - (H5) for every value of the parameter $\Theta \in (\Theta_1,\Theta_2)$. The unknown parameter $\Theta_o \in (\Theta_1,\Theta_2)$ can be estimated by the minimum contrast method in the following way: Let $\varphi : E \rightarrow \mathbb{R}$ be a measurable function satisfying (2.1). Set

$$\tilde{\varphi}_\infty(\Theta) = \alpha^{-1}(\Theta)\int \varphi(x)e^{2U_\Theta(x)}\gamma(dx)\ , \quad \alpha(\Theta) = \int e^{2U_\Theta(x)}\gamma(dx)$$

Define (as in [10] with $\varphi = \|\cdot\|$)

$$J_T(\Theta) = \int_0^T (\varphi(\xi_t) - \tilde{\varphi}_\infty(\Theta))^2 dt\ ,$$

where ξ_t is a solution to (2.3) with $\Theta = \Theta_o$. If $\tilde{\varphi}_\infty(\Theta) \neq \tilde{\varphi}_\infty(\Theta_o)$ is assumed for $\Theta \in (\Theta_1,\Theta_2)$, $\Theta \neq \Theta_o$, then J_T is a contrast functional in the sense of definition in [12] as

$$\int (\varphi(x)-\tilde{\varphi}_\infty(\Theta_o))^2 e^{2U_{\Theta_o}(x)}\gamma(dx) < \int (\varphi(x)-\tilde{\varphi}_\infty(\Theta))^2 e^{2U_{\Theta_o}(x)}\gamma(dx)$$

holds for any $\Theta \neq \Theta_o$. Furthermore, assuming $\tilde{\varphi}_\infty \in C^1$ we can minimize $J_T(\Theta)$: setting $J'_T(\Theta) = 0$ we get

$$(2.4) \qquad \frac{1}{T}\int_0^T \varphi(\xi_t)dt = \varphi_\infty(\hat{\Theta}_T)\ .$$

From (2.4) we obtain the desired expression for the estimator $\hat{\Theta}_T$.

Proposition 2.7. Let $\varphi : E \rightarrow \mathbb{R}$ be measurable, satisfying (2.1) and such that $\tilde{\varphi}_\infty \in C^1(\Theta_1,\Theta_2)$, $\tilde{\varphi}'_\infty(\Theta_o) \neq 0$. Denote by $\mathcal{U}$ a neighbourhood of Θ_o in which $\tilde{\varphi}'_\infty \neq 0$ and define

$$\hat{\Theta}_T = \tilde{\varphi}_\infty^{-1}\Big(\frac{1}{T}\int_0^T \varphi(\xi_t^x)dt\Big) \quad \text{if} \quad \frac{1}{T}\int_0^T \varphi(\xi_t^x)dt \in \tilde{\varphi}_\infty(\mathcal{U})\ .$$

Then $\hat{\Theta}_T \rightarrow \Theta_o$ a.s. as $T\rightarrow\infty$ for every $x \in E$ (i.e., the estimator $\hat{\Theta}_T$ is strongly consistent).

The proof follows immediately from Corollary 2.5.

Example 2.8. All the assumptions (H1) - (H5) are fulfilled for equations

$$(2.5) \qquad \frac{\partial y}{\partial t}(t,z) = \frac{\partial^2 y}{\partial z^2}(t,z) + p(y(t,z)) + \dot{w}_t \ , \quad t>0 \ ,$$

$$y(0,z) = y_o(z) \ , \ z\in(0,\pi) \ , \ y(t,0) = y(t,\pi) = 0 \ ,$$

where p is a polynomial of the form

$$p(r) = -r^{2k+1} + \sum_{j=0}^{2k} a_j r^j \ , \quad r\in\mathbb{R} \ ,$$

see e.g. [1], [2]. In this case, $H = L_2(0,\pi)$, $E = C_o(0,\pi)$, $U(x) = \int_0^{\pi} v(x(z))dz$, where $x\in E$, $v' = p$. By Theorem 2.1 the solution of (2.5) is strongly Feller in the restricted sense and Corollaries 2.3 - 2.6 can be applied. (Some of those assertions have been already proved earlier for the equation (2.5), e.g., uniqueness and ergodicity of the invariant measure follows from [14], equivalence of transition probabilities and the strong law of large numbers from [11]). Consider the important special case $p(r) = \Theta r - r^3$, where $\Theta\in\mathbb{R}$ is an unknown parameter. If $\varphi : C_o(0,\pi)\to\mathbb{R}$ is a measurable function satisfying $\int_E |\varphi(x)| \ (1+|x|_H^2)e^{2U_\Theta(x)} \gamma(dx)<\infty$ then

$$(2.6) \qquad \tilde{\varphi}'_\infty(\Theta) = \frac{\partial}{\partial\Theta}\left(\frac{1}{\alpha(\Theta)}\int_E \varphi(x)e^{2U_\Theta(x)}\gamma(dx)\right) =$$

$$= \frac{1}{\alpha(\Theta)}\int\varphi(x) \ |x|_H^2 e^{2U_\Theta(x)}\gamma(dx) -$$

$$- \frac{1}{\alpha^2(\Theta)}\left(\int\varphi(x)e^{2U_\Theta(x)}\gamma(dx)\right)\left(\int|x|_H^2 e^{2U_\Theta(x)}\gamma(dx)\right) =$$

$$= E\varphi(X_\Theta)|X_\Theta|_H^2 - E\varphi(X_\Theta)E|X_\Theta|_H^2 \ ,$$

where X_Θ is a random variable with the distribution $\alpha^{-1}(\Theta)e^{2U_\Theta(x)}\gamma(dx)$. Hence, if φ is such that the term on the right-hand side of (2.6) is nonzero on an interval (Θ_1,Θ_2) , then Proposition 2.7 can be applied to obtain $\hat{\Theta}_T\to\Theta_o$ a.s. For instance, we can set $\varphi(x) = |x|_H^2$, $(\Theta_1,\Theta_2)\subset\langle-\infty,\infty\rangle$.

Let $\emptyset\neq V\subset E$ be a closed set and denote by τ the hitting time of V by the solution ξ_t of (1.1). Another kind of applications of the strong Feller property can be following

<u>Proposition 2.9.</u> The functions $\psi(x) = E\varphi(\xi_\tau^x)$ are continuous on $E\setminus V$ for any $\varphi\in\mathbb{M}$.

Proof. For any $t>0$, $x\subset E\setminus V$ we have

(2.7) $$E\varphi(\xi_t^x) = E\varphi(\xi_\tau^x)\chi_{[\tau>t]} + E\varphi(\xi_\tau^x)\chi_{[\tau\leq t]} = \int E\varphi(\xi_\tau^{t,y})\chi_{[\tau>t]}P(t,x,dy) + E\varphi(\xi^x)\chi_{[\tau\leq t]} \,.$$

By the strong Feller property the first term on the right-hand side of (2.7) is continuous. Thus it remains to prove

Lemma 2.10.

(2.8) $$P_x[\tau\leq t]\longrightarrow 0 \quad \text{as} \quad t\searrow 0 \,,$$

locally uniformly in $x\in E\setminus V$.

Proof. Take $\varepsilon>0$, x_o and $\varepsilon_o>0$ such that $U_{2\varepsilon_o}(x_o)\subset E\setminus V$. Then a $t_o>0$ can be found such that

$$P[\,\|Z_t\|>\varepsilon_o/2 \text{ for some } t\in\langle 0,t_o\rangle] < \varepsilon$$

and $\|S(t)x_o-x_o\|<\varepsilon_o/3$, $0\leq t\leq t_o$. It follows that $\|S(t)x-x\|\leq\varepsilon_o$ for all $x\in U_{\varepsilon_o/3}(x_o)$, $t\leq t_o$. Now it is easily seen that

$$\|\xi_t^x-x\|\leq\|S(t)x-x\|+\left\|\int_0^t S(t-r)f(\xi_r^x)dr\right\|+\|Z_t\|<2\varepsilon_o$$

for all $x\in U_{\varepsilon_o/3}(x_o)$ and $t>0$, t sufficiently small, with probability at least $1-\varepsilon$.

Corollary 2.11. For $x\in E\setminus V$ denote by $\tilde\mu_x$ the distribution

$$\tilde\mu_x(\Gamma) = P[\xi_\tau^x\in\Gamma] \,, \quad \Gamma\in\mathcal{B}(V) \,.$$

The measures $\tilde\mu_x$, $x\in E\setminus V$, are absolutely continuous with respect to a certain measure $\tilde\mu$ on $\mathcal{B}(V)$.

Proof. Let $(x_n)\subset E\setminus V$ be dense in $E\setminus V$ and set

$$\tilde\mu(\Gamma) = \sum_{k=1}^{\infty} 2^{-k}\,P[\xi_\tau^{x_k}\in\Gamma] \,.$$

If $\tilde\mu(\Gamma)=0$, then $P[\xi_\tau^{x_k}\in\Gamma]=0$ for all $k\in\mathbb{N}$ and so, by Proposition 2.9, $\tilde\mu_x(\Gamma)=0$, $x\in E\setminus V$.

For other applications of the strong Feller property (and S.F.P. in the restricted sense), we refer to [7]. We close this section by a remark showing that the cylindrical Wiener process in (1.1) cannot be substituted by a covariance-type one:

Remark 2.12. Consider a linear equation

(2.9) $$d\xi_t = A\xi_t dt + d\tilde w_t$$

on a Hilbert space H, where A generates a C_o-semigroup $S(t)$ in H, $\tilde w_t$ is a H-valued Wiener process with a nuclear and positive covariance operator W. Then the solution of (2.9) need not be strongly Feller. For instance, we can take W satisfying $We_i = \lambda_i e_i$,

$i \in \mathbb{N}$, $\lambda_i > 0$, $\sum \lambda_i < \infty$ (e_i - the eigenvalues of $-A_o$). Obviously $P(t,x,.) = N(S(t)x, Q_t)$, where

$$Q_t = \int_0^t S(s)WS(s)ds \ .$$

Hence the solution of (2.9) is strongly Feller, then necessarily

$$S_t(t)(x_n - x_o) \in \text{Range } (Q_t^{1/2}) \tag{2.10}$$

for any $x_n \to x_o$ (otherwise $P(t,x_n,.)$, $P(t,x_o,.)$ are singular).

But if we take $\lambda_i = e^{-\alpha_i^3}$ and $x_n, x_o \in H$ such that $\langle x_n - x_o, e_i \rangle^2 = \frac{1}{n\alpha_i}$, $i \in \mathbb{N}$, then it is easy to check that (2.10) is false.

However, we can prove

Proposition 2.13. Consider the equation (2.9) in which $\tilde{w}_t$ is a H-valued Wiener process with a nuclear covariance operator $W > 0$. If

$$\text{Range } (S(t)) \subset \text{Range } Q_t^{1/2} \ , \quad t > 0 \ ,$$

then the solution is strongly Feller in the restricted sense.

Proof. By the closed graph theorem the operator $Q_t^{-1/2} S_t(t)$ is bounded and hence $|Q_t^{-1/2} S_t(t)(x_n - x_o)|_H \to 0$ as $x_n \to x$. By the same proof as in [16] (Theorem 4.3) we obtain

$$\text{Var } (P(t,x_n,.) - P(t,x_o,.)) =$$

$$= \int \left| \exp \left\{ \langle Q_t^{-1} S(t)(x_n - x_o), y \rangle - \tfrac{1}{2} |Q_t^{-1/2} S(t)(x_n - x_o)|^2 \right\} - 1 \right| . \\ . N(0, Q_t)(dy) \to 0 \ , \quad x_n \to x_o \ .$$

3. Proof of Theorem 2.1

The proof of Theorem 2.1 is rather technical and is based on the finite-dimensional approximation method. The fundamental step is Lemma 3.1 below.

Denote by $|\cdot|_n$ the Euclidean norm in $\mathbb{R}_n$ and by $\|\cdot\|_n$ a norm in $\mathbb{R}_n$ satisfying $|\cdot|_n \le c \|\cdot\|_n$. Let $F_n = (F_n^i)_{i=1}^n : \mathbb{R}_n \to \mathbb{R}_n$ be such that $F_n(x) = \text{grad } A_n(x)$ and

$$|A_n(x) - A_n(y)| \le M_1 \|x-y\|_n \ , \quad \|F_n(x) - F_n(y)\|_n \le M_2 \|x-y\|_n \tag{3.1}$$

$$A_n(x) \le M_3 \ , \quad \|F_n(x)\|_n \le M_4$$

holds for all $x, y \in \mathbb{R}_n$, where the constants $M_1 - M_4$ are independent of n . Consider the equation

$$d\xi_n(t) = \begin{pmatrix} -\alpha_1 & 0 & \dots & 0 \\ 0 & -\alpha_2 & & \\ 0 & \dots & \dots & -\alpha_n \end{pmatrix} \xi_n(t)dt + F_n(\xi_n(t))dt + \\ + dw_n(t) \tag{3.2}$$

in $\mathbb{R}_n$, where $w_n(t)$ is an n-dimensional standard Brownian motion. Set $\tilde{T}_t^{(n)}\varphi(x) = E\varphi(\xi_n^x(t))$ for $t>0$, $x\in\mathbb{R}_n$, $\varphi\in C_b(\mathbb{R}_n)$.

Lemma 3.1. For every fixed $t>0$ and $R_0>0$ the functions $\tilde{T}_t^{(n)}\varphi$ are continuous on $(\{\|x\|_n\le R_0\};\ \|\cdot\|_n)$ uniformly with respect to $n\ \mathbb{N}$ and $\varphi\in\{\psi\in C_b(\mathbb{R}_n);\ |\psi(x)|\le 1\ ,\ x\in\mathbb{R}_n\}$.

In the proof we need several auxiliary results (their proofs are rather standard and we omit the details)

Lemma 3.2.

(a) $E|\xi_n^x(t)|^2\le k_1$,

(b) $E\sum_{i=1}^n \beta_i|\xi_n^{i,x}(t)|\le k_2$,

(c) $E\exp\{k\sum_{i=1}^n\beta_i|\xi_n^{i,x}(t)|\}\le k_3$

for any $k>0$ and $\beta_i>0$ such that $\sum_{i=1}^\infty\beta_i<\infty$, where k_1,k_2,k_3 are independent of $n\in\mathbb{N}$ and $x=(x^i)$, $\|x\|_n\le R_0$.

Proof. Using Itô's formula on the function $V(x)=(x^i)^2$ we can get

$$E(\xi_n^i(t))^2\le (x^i)^2 e^{(-2\alpha_i+2M_4c)t} + \frac{2cM_4+1}{-2\alpha_i+2M_4c}(1-e^{(-2\alpha_i+2M_4c)t})$$

and hence we obtain (a). The part (b) is an immediate consequence of (a) and (c) we can show again by Itô's formula with $V(x) = \exp\{k\sum_{i=1}^n\beta_i|x^i|\}$ using the standard stopping time procedures and Fatou and Gronwall lemmas.

Denote by $X_n(t)$ the solution of (3.1) with $F_n=0$, i.e. $X_n^x(t) = S_n(t)x+\nu_n(t)$, where $S_n(t)x=(x^ie^{-\alpha_i t})_{i=1,\dots,n}$,

$$\nu_n(t) = (\nu^i(t))_{i=1,\dots,n} = (\int_0^t e^{-\alpha_i(t-s)}d\beta_i(s))_{i=1,\dots,n}\ ;$$

$\beta_i(t)$ - stochastically independent one-dimensional standard Wiener processes. Set $\varkappa_i(t)=E(\nu^i(t))^2=(1-e^{-2\alpha_i t})/2\alpha_i$ and denote by $\Psi_n(t,x,y)$ the transition density of $X_n(t)$, i.e.

$$\Psi_n(t,x,y) = \frac{1}{(2\pi)^{n/2}\prod_1^n\sqrt{\varkappa_i(t)}}\exp\{\sum_1^n\frac{-(y^i-x^ie^{-\alpha_i t})^2}{2\varkappa_i(t)}\}\ .$$

Further, define

$$\Phi_n(t,x,y) = \exp\{A_n(y)-A_n(x)\}\ E\exp\{\int_0^t B_n(S_n(s)x+\eta_n(s)+ H_n(s,t)(y-S_n(t)x))ds\}\ ,$$

where

$$B_n(z) = \frac{1}{2}\sum_{i=1}^{n} \alpha_i z^i F_n^i(z) - \frac{1}{2}|F_n(z)|_n^2 - \frac{1}{2}\sum_{i=1}^{n}\frac{\partial F_n^i}{\partial z^i}(z) , \quad z\in R_n ,$$

$$\eta_n(s) = (\eta^i(s))_{i=1,\dots,n} = (\nu^i(s)-H^i(s,t)\, \nu^i(t))_{i=1,\dots,n} ,$$

$$s\in\langle 0,t\rangle ,$$

$$H_n(s,t) = (H^i(s,t))_{i=1,\dots,n} = \left(\frac{E\,\nu^i(t)E\,\nu^i(s)}{E(\nu^i(t))^2}\right) = \left(\frac{e^{\alpha_i(t-s)} - e^{-\alpha_i(t+s)}}{1-e^{-2\alpha_i t}}\right) .$$

Lemma 3.3. The transition density $p_n(t,x,y)$ for the solution of (3.1) has the form

$$p_n(t,x,y) = \Psi_n(t,x,y)\,\Phi_n(t,x,y) , \quad x,y\in R_n , \quad t>0 .$$

Proof. By Girsanov formula it follows that

$$p_n(t,x,y) = \Psi_n(t,x,y)E\left[\exp\left\{\sum_{i=1}^{n}\int_0^t F_n^i(X_n^x(s))d\beta_i(s) - \right.\right.$$
$$\left.\left. - \frac{1}{2}\int_0^t |F_n(X_n^x(s))|_n^2 ds\right\} \Big| X_n^x(t) = y\right]$$

and hence we need to prove that the conditional expectation in the last term is equal to $\Phi_n(t,x,y)$. This can be done in the same way as in the similar proof in [6] (pp. 92-95).

Lemma 3.4. Let $Z_n(s)$ be R_n-valued $(\beta_i(t))$-nonanticipative processes and

$$R_n(t) = \sum_{i=1}^{n}\int_0^t F_n^i(Z_n(s))d\beta_i(s) - \frac{1}{2}\int_0^t |F_n(Z_n(s))|_n^2\, ds .$$

Then

$$E|R_n(t)|\le k_1 , \quad Ee^{kR_n(t)} \le k_2$$

for any $k>0$, where k_1, k_2 are independent of n and $Z_n(s)$.

Proof. See e.g. [4], Lemma 7.2.2.

Proof of Lemma 3.1. Take $x_1 = (x_1^i)$, $x_2 = (x_2^i)\in R_n$, $\|x_1\|,\|x_2\|\le R_o$ and $\varphi\in C_b(R_n)$, $|\varphi|\le 1$. Then

$$(3.3) \qquad |\tilde{T}_t^{(n)}\varphi(x_1) - \tilde{T}_t^{(n)}\varphi(x_2)| \le \int |p_n(t,x_1,y)-p_n(t,x_2,y)|\,dy \le$$
$$\le \int |\Psi_n(t,x_1,y)- \Psi_n(t,x_2,y)|\,\Phi_n(t,x_1,y)dy +$$
$$+\int |\Phi_n(t,x_1,y)- \Phi_n(t,x_2,y)|\,\Psi_n(t,x_2,y)dy .$$

Furthermore, if $P_n(t,x,.)$ stands for the transition probability of $\xi_n(t)$, then

(3.4)
$$\int |\Psi_n(t,x_1,y)-\Psi_n(t,x_2,y)|\, \Phi_n(t,x_1,y)dy =$$
$$= \int |\Psi_n(t,x_1,y)-\Psi_n(t,x_2,y)|\, \Psi_n^{-1}(t,x_1,y)P_n(t,x_1,dy) =$$
$$= \int \Big|1-\exp\Big\{\sum_1^n [((x_1^i)^2-(x_2^i)^2)e^{-2\alpha_i t}+2(x_2^i-x_1^i)e^{-\alpha_i t}y^i]\,.$$
$$.(2\varkappa_i(t))^{-1}\Big\}\Big| P_n(t,x_1,dy) \le |1-e^{cc_1\|x_1-x_2\|_n}| +$$
$$+ e^{cc_1\|x_1-x_2\|_n}\int \Big|1-\exp\Big\{c\|x_1-x_2\|_n \sum \frac{e^{-\alpha_i t}|y_2^i|}{\varkappa_i(t)}\Big\}\Big|\,.$$
$$.P_n(t,x_1,dy) \quad ,$$

where $c_1 = \sum_{i=1}^n R_o e^{-2\alpha_i t}(\varkappa_i(t))^{-1}$. Set $\beta_i = e^{-\alpha_i t}(\varkappa_i(t))^{-1}$.

For $R>0$ we have

(3.5)
$$\int \Big|1-\exp\{c\|x_1-x_2\|_n \sum \beta_i |y^i|\}\Big| P_n(t,x_1,dy) \le$$
$$\le E\Big|1-\exp\{c\|x_1-x_2\|_n \sum \beta_i |\xi_n^i(t)|\}\Big|\,.$$
$$.\chi_{[c\|x_1-x_2\|_n \sum \beta_i |\xi_n^i(t)|>R]} + E\Big|1-\exp\{c\|x_1-x_2\|_n\,.$$
$$.\sum \beta_i |\xi_n^i(t)|\}\Big|\chi_{[c\|x_1-x_2\|_n \sum \beta_i|\xi_n^i(t)|\le R]} \le$$
$$\le \Big[E(2+2\exp\{4R_o c\sum_1^n \beta_i|\xi_n^i(t)|\})\Big]^{1/2}.$$
$$.(P[2R_o c\sum_1^n \beta_i|\xi_n^i(t)|>R])^{1/2} + e^R c\|x_1-x_2\|_n\,.$$
$$.E\sum_1^n \beta_i|\xi_n^i(t)| \quad .$$

By Lemma 3.2 the right-hand side of (3.5) can be done arbitrarily small when $\|x_1-x_2\|_n$ is small (independently of n) and hence by (3.4) the same is true for

$$\int |\Psi_n(t,x_1,y)-\Psi_n(t,x_2,y)|\, \Phi_n(t,x_1,y)dy \,.$$

Similarly we will estimate the second term on the right-hand side of (3.3). We have

$$(3.6)\qquad \int |\Phi_n(t,x_1,y)-\Phi_n(t,x_2,y)|\,\Psi_n(t,x_2,y)dy \leq$$
$$\leq E\,\Big|\exp\Big\{A_n(X_n^{x_2}(t)) - A_n(x_1) + \int_0^t B_n(Y_n(s))ds\Big\} -$$
$$- \exp\Big\{A_n(X_n^{x_2}(t)) - A_n(x_2) + \int_0^t B_n(X_n^{x_2}(s))ds\Big\}\Big|,$$

where

$$Y_n(s) = S_n(s)x_1 + \eta_n(s) + H_n(s,t)(X_n^{x_2}(t)-S_n(t)x_1) =$$
$$= S_n(s)x_1 + \nu_n(s) + H_n(s,t)S_n(t)(x_2-x_1)\ .$$

Obviously,

$$dY_n(s) = dX_n^{x_1}(s) + \dot H_n(s,t)S_n(t)(x_2-x_1)$$

and hence by Itô's lemma

$$\int_0^t B_n(Y_n(s))ds = A_n(x_1) - A_n(Y_n(t)) + \int_0^t \sum_i F_n^i(Y_n(s))d\beta_i(s) -$$
$$- \frac{1}{2}\int_0^t |F_n(Y_n(s))|_n^2 ds + \int_0^t \sum_i F_n^i(Y_n(s)).$$
$$.\big[\alpha_i H^i(s,t)e^{-\alpha_i t}(x_2^i-x_1^i) + \dot H^i(s,t)e^{-\alpha_i t}(x_2^i-x_1^i)\big]\,ds\ .$$

Substituting it to (3.6) we get

$$(3.7)\qquad \int |\Phi_n(t,x_1,y)-\Phi_n(t,x_2,y)|\,\Psi_n(t,x_2,y)dy \leq$$
$$\leq E\,\Big|\exp\Big\{A_n(X_n^{x_2}(t))-A_n(Y_n(t)) + \int_0^t \sum F_n^i(Y_n(s))d\beta_i(s) -$$
$$- \frac{1}{2}\int_0^t |F_n(Y_n(s))|_n^2 ds + \int_0^t \sum_i F_n^i(Y_n(s)).$$
$$.\big[\alpha_i H^i(s,t)e^{-\alpha_i t}(x_2^i-x_1^i) + \dot H^i(s,t)e^{-\alpha_i t}(x_2^i-x_1^i)\big]\,ds\Big\} -$$
$$- \exp\Big\{\int_0^t \sum F_n^i(X_n^{x_2}(s))d\beta_i(s) - \frac{1}{2}\int_0^t |F_n(X_n^{x_2}(s))|_n^2 ds\Big\}\Big| =$$
$$= E\,|e^{L_1}-e^{L_2}|\ .$$

For $R>0$ we can estimate

$$(3.8)\qquad E|e^{L_1}-e^{L_2}| \leq [E(2e^{2L_1}+e^{2L_2})]^{1/2}(P[\,|L_1| + |L_2| \geq R])^{1/2} +$$
$$+ e^R E|L_1-L_2|\ .$$

We have

$$(3.9)\qquad |Y_n(s)-X_n^{x_2}(s)|_n \leq c_3\|x_1-x_2\|_n\ ,\quad s\in\langle 0,t\rangle\ ,$$

and as $(\alpha_i H^i(s,t) + \dot H^i(s,t))^2 \leq c_4\alpha_i^2$, we obtain

(3.10) $$\left|\int_0^t \sum F_n^i(Y_n(s))(x_2^i-x_1^i)(\alpha_i H^i(s,t) + \dot{H}^i(s,t)e^{-\alpha_i t})ds\right| \le$$
$$\le c_5\|x_1-x_2\|_n \quad ,$$

where c_3, c_4, c_5 are independent of n. Consequently,

(3.11) $$E|L_1-L_2| \le E|A_n(X_n^{x_2}(t))-A_n(Y_n(t))| + E\left|\sum_{i=1}^n \int_0^t (F_n^i(Y_n(s)) -\right.$$
$$\left. - F_n^i(X_n^{x_2}(s)))d\beta_i(s)\right| + \frac{1}{2} E\left|\int_0^t |F_n(Y_n(s))|_n^2 -\right.$$
$$\left.- |F_n(X_n^{x_2}(s))|_n^2 ds\right| + E\left|\int_0^t \sum_{i=1}^n F_n^i(Y_n(s))(x_2^i-x_1^i)e^{-\alpha_i t}\cdot\right.$$
$$\left.\cdot(\alpha_i H^i(s,t) + \dot{H}^i(s,t))ds\right| \le M_1 c_3\|x_1-x_2\|_n +$$
$$+ cM_2c_3t^{1/2}\|x_1-x_2\|_n + c^2M_2M_4c_3t\|x_1-x_2\|_n + c_5\|x_1-x_2\|_n \quad .$$

By Lemma 3.4 $Ee^{2L_1} \le c_6$, $Ee^{2L_2} \le c_6$, $P[|L_1| + |L_2| \ge R] \le \frac{c_7}{R}$,

where c_6, c_7 are independent of n, which together with (3.11), (3.8) and (3.7) shows that the term

$$\int |\Phi_n(t,x_1,y)-\Phi_n(t,x_2,y)| \Psi_n(t,x_2,y)dy$$

is arbitrarily small when $\|x_1-x_2\|_n$ is small, uniformly with respect to n, which remained to prove.

<u>Proof of Theorem 2.1.</u> We can use the finite-dimensional approximation procedure from [17], where $E_n = \mathrm{Lin}\{e_1,\dots,e_{m_n}\}$ and $\Pi_n : E \to E_n$ are defined in such a way that $\Pi_n x \to x$ in E as $n \to \infty$, $\|\Pi_n\| \le \tilde{c}$, $\tilde{c}$ independent of n. Furthermore the functions

$$U_n(t) = \lambda\left(\frac{|x^2|}{n^2}\right)U(\Pi_n x) \quad , \qquad x \in E \, ,$$

are defined, where λ is a nonnegative, smooth, real function satisfying $\lambda(r)=1$ for $r \in \langle -1,1\rangle$, $\lambda(r)=0$ for $|r| \ge 2$. The equations

(3.12) $$d\tilde{X}_n(t) = (A\tilde{X}_n(t) + U_n'(\tilde{X}_n(t)))dt + dW_n(t)$$

in the spaces E_n are considered. From [17], Proposition 4, it follows that $\tilde{X}_n^{\Pi_n x}(t) \to \xi_t^x$ a.s. for all $t>0$, $x \in E$. Thus we have

(3.13) $$T_t^{(n)} \varphi(\Pi_n x) \xrightarrow{n\to\infty} T_t \varphi(x) \, , \quad x \in E \, , \quad t>0 \, ,$$

for all $\varphi \in C_b(E)$, where $T_t^{(n)} \varphi(y) = E\varphi(\tilde{X}_n^y(t))$.

Let us first assume

(3.14) $|U(x)-U(y)| \leq M\|x-y\|$, $\|f(x)-f(y)\| \leq M\|x-y\|$, $\|f(x)\| \leq M$

for some $M>0$ and all $x,y\in E$. Note that (3.14) is then fulfilled also for U_n and $f_n = U_n'$ with a constant M independent of n . Thus we can use Lemma 3.1 and (3.13) to conclude that for every compact $K\subset E$ and $t>0$ the set

$$\{T_t\varphi(\cdot)|_K;\ \varphi\in C_b(E),\ |\varphi|\leq 1\}$$

is relatively compact. The strong Feller property in the restricted sense follows easily (cf. [7], p.14).

It remains to remove the additional assumptions (3.14). Take a sequence $U^n : E\to\mathbb{R}$ such that $U^n(x) = U(x)$ for $\|x\|\leq n$ and every U^n and $f^n = (U^n)'$ satisfy (3.14). By (H4) for any compact $K\subset E$ and $t>0$,

(3.15) $$\lim_{n\to\infty} P[\|\xi_s^x\| \geq n \text{ for some } s\in\langle 0,t\rangle, x\in K] = 0 .$$

Consequently, $P^n(t,x,A)\to P(t,x,A)$ as $n\to\infty$ holds uniformly with respect to $x\in K$ and $A\in \mathcal{B}(E)$. Let $x_n\to x_o$ and denote by $A_n\cup B_n = E$ the Hahn-Banach decompositions of E with respect to the signed measures $P(t,x_n,\cdot) - P(t,x_o,\cdot)$. We have

$$\begin{aligned} &\mathrm{Var}\,(P(t,x_n,\cdot)-P(t,x_o,\cdot)) \leq |P(t,x_n,A_n)-P^k(t,x_n,A_n)| + \\ &+ |P^k(t,x_n,A_n)-P^k(t,x_o,A_n)| + |P^k(t,x_o,A_n)-P(t,x_o,A_n)| + \\ &+ |P(t,x_n,B_n)-P^k(t,x_n,B_n)| + |P^k(t,x_n,B_n)-P^k(t,x_o,B_n)| + \\ &+ |P^k(t,x_o,B_n)-P(t,x_o,B_n)| \leq \varepsilon + \mathrm{Var}\,(P^k(t,x_n,\cdot)-P^k(t,x_o,\cdot)) \end{aligned}$$

for arbitrary $\varepsilon>0$ and $k\geq k_o(\varepsilon)$. As the transition probabilities P^k are already proved to be strongly Feller in the restricted sense, the proof of Theorem 2.1 is complete.

Remark. If we assume a.s. Hölder continuity of the process $Z(t) = \int_0^t S_{t-r}dw_r$ in E , we can remove the assumption

$$\sup\{|u(s,x,\psi)|; s\in\langle 0,t\rangle, x\in K, \psi\in C_o(0,t,E), |\psi|<R\}<\infty$$

from (H4). Indeed, from Proposition 4 in [17] we obtain

$$\sup\{|u(s,x,\psi)|; s\in\langle 0,t\rangle, x\in K, \psi\in C_o(0,t,E), |\psi|<R ,\ \|\psi(s_1)-\psi(s_2)\|\leq \hat{c}|s_1-s_2|^\alpha \text{ for any } s_1,s_2\in\langle 0,t\rangle\} < \infty$$

for every compact $K\subset E$, $t,R,\hat{c}>0$, $\alpha\in(0,1)$. It follows that $P^n(t,x,A)\to P(t,x,A)$ as $n\to\infty$ uniformly w.r.t. $A\in\mathcal{B}(E)$ and $x\in K$.

References

[1] G. Da Prato, J. Zabczyk, A note on semilinear stochastic equations, Differential and Integral Equations, vol.1, no.2, (1988), pp. 143-155

[2] W.G. Faris, A.V. Jona-Lasinio, Large fluctuations for a nonlinear heat equation with noise, J.Phys. A: Math.Gen., 15 (1982), pp. 3025-3055

[3] M.I. Freidlin, Random perturbations of reaction-diffusion equations: the quasi-deterministic approximation, TAMS 305 (1988), no.2, pp. 665-697

[4] A. Friedman, Stochastic Differential Equations and Applications, vol.1, AP, New York, 1975

[5] T. Funaki, Random motions of strings and related stochastic evolution equations, Nagoya Math.J., vol.89 (1983), pp. 129--193

[6] I.I. Gikhman, A.V. Skorokhod, Stokhasticheskiie differentsial'nyie uravneniia, Naukova Dumka, Kiiev, 1968

[7] I.V. Girsanov, Sil'no fellerovskiie processy, Teorija verojat. primen. 5, 1 (1960), pp. 7-28

[8] G. Jetschke, Invariant distribution of a nonlinear stochastic partial differential equations and free energy of statistical Physics, Fo.-Erg. Jena, N/87/11, N/86/20, N/86/40

[9] R.Z. Khas'minskii, Ergodicheskiie svoistva vozvratnych diffuzionnych processov i stabilizatsia reshenii zadachi Koski dlya parabolicheskikh uravnenii, Teorija verojat. primen. 5, 2 (1960), pp. 196-214

[10] T. Koski, W. Loges, On minimum contrast estimation for Hilbert space-valued stochastic differential equations, Stochastics 16 (1986), 3-4, pp. 214-225

[11] S.M. Kozlov, Some problems concerning stochastic partial differential equations, Trudy Sem. Petrovsk. 4 (1978), pp. 147--172

[12] V. Lánská, Minimum contrast estimation in diffusion processes, J.Appl.Probab. 16 (1979), pp. 65-75

[13] R. Manthey, Existence and uniqueness of a solution of a reaction-diffusion equation with polynomial nonlinearity and white noise disturbance, Math.Nachr. 125 (1986), pp. 121-136

[14] R. Marcus, Parabolic Itô equations with monotone nonlinearities, J. Functional Anal. 29 (1978), 3, pp. 275-287

[15] R. Marcus, Parabolic Itô equations, TAMS, vol.198 (1974), pp. 177-190

[16] B. Maslowski, Uniqueness and stability of invariant measures for stochastic evolution equations in Hilbert spaces, to appear in Stochastics

[17] J. Zabczyk, Symmetric solutions of semilinear stochastic equations, Preprint 416, IM PAN.

ON THE MACROSCOPIC NONEQUILIBRIUM DYNAMICS OF AN EXCLUSION PROCESS

E. Platen
Karl-Weierstraß-Institut für Mathematik, AdW der DDR
Mohrenstr. 39, Berlin, DDR-1086

1. Introduction

The exclusion process was introduced by Spitzer in /11/. Comprehensive treatments of this interacting particle system are given in /7/ and /1/.

In the exclusion model the particles attempt to move independently according to a Markov kernel on a given countable set of sites. But any jump which would take a particle to an already occupied site is suppressed. That means there is always at most one particle per site.

This paper is a continuation of /8/ where we considered the so called wide range birth and death exclusion process in random medium. We proved in /8/ a law of large numbers for this measure valued process and derived a deterministic macroscopic equation describing the evolution of the occupation rate of the sites. This macroscopic equation holds under rather general conditions. It can be specified by the choice of the Markov kernel, this means by the given jump intensity of the particles.

Within this paper we will study the case of a nonsymmetric local jump intensity which allows only jumps into the neighborhood. For instance, such jump intensity is of interest in modelling of stochastic charge transport (see /10/).

In the following we will show under suitable assumptions that the asymptotics with respect to vanishing mean jump size yields a nonlinear second order partial differential equation characterizing the evolution of the limiting particle concentration. This equation represents a continuity equation, where physical quantities as current density vector, conductivity, static potential and chemical potential can be easily identified. It seems that this continuity equation models important transport processes in physics, chemistry, biology, electronics, social sciences and other fields.

The first part of the paper generalizes the birth and death exclusion process introduced in /8/. It formulates a law of large numbers for the case that the mean number of sites per unit volume tends to infinity. The second part derives the evolution equation for the limiting particle concentration.

2. Exclusion process

2.1. Random medium

In the following we will introduce a generalization of the wide range birth and death exclusion process considered in /8/.

For unexplained notations and definitions we refer to /4/ or /5/.

Let $(\Omega, \underline{F}, P)$ denote the basic complete probability space. $\underline{F} = (\underline{F}_t)_{t\geq 0}$ is an increasing right-continuous family of complete sub-σ-fields of $\underline{F}$.

Further, $\underline{B}(E)$ represents the Borel-σ-algebra of a topological space E.

The sites are located in a closed bounded domain $\overline{Q} \subset \underline{R}^d$, $d \in \{1, 2, \ldots\}$.

We introduce a finite σ-additive measure Λ on $\underline{B}(\overline{Q})$ which is called intensity measure of sites.

$(N_n)_{n\geq 1}$ denotes a sequence of $\underline{F}_0$-measurable simple counting measures on $\underline{B}(\overline{Q})$, the so called counting measures of sites.

The parameter

$$(2.1.1) \qquad n = E\, N_n(\overline{Q}) \Big(\int_{\overline{Q}} dq \Big)^{-1}$$

can be interpreted as the mean number of sites per unit volume.

For $K \in (0, \infty)$ we denote by $\underline{C}_K$ the set of bounded Lipschitz-continuous functions $f | \overline{Q} \to [-K, K]$ with

(2.1.2) $|f(u)-f(q)| \leq K|q-u|$

for all $u,q \in \overline{Q}$, using the usual Euclidean norm. Now we assume that for each $K\in(0,\infty)$ it holds

(2.1.3) $$\lim_{n\to\infty} E \sup_{f\in \underline{\mathcal{L}}_K} \left(\int_{\overline{Q}} f(q)(\tfrac{1}{n} N_n - \Lambda)(dq)\right)^2 = 0,$$

and for all $n>1$ we have

(2.1.4) $$E(\tfrac{1}{n} N_n(\overline{Q}))^4 = K_0 < \infty$$

The counting measure N_n represents the random medium within our microscopic stochastic model . The above conditions are satisfied for a wide class of regular lattices and other point processes including the Poisson point process (see /8/).

2.2. Markovian jump mechanism

We denote by $L_{n,t}$ the counting measure of particles at time $t\geq 0$.

Further, we introduce an $\underline{F}$-adapted cadlag Poisson counting measure μ_n on $\underline{B}([0,\infty)) \otimes \underline{B}(\overline{Q}) \otimes \underline{B}(\overline{Q})$ which is characterized by its dual predictable projection which is here its intensity measure

(2.2.1) $$\nu_n(dr,du,dq)=n^{-1} w_t(u,q)N_n(du)N_n(dq)dt.$$

We assume that the jump rate $w_t(u,q)$ is nonnegative and Lipschitz-continuous with respect to u and q uniformly with respect to t.

The counting measure μ_n generates the so called possible jumps of particles. But only jumps from occupied into vacant sites will take place. Therefore, we will have at most one particle at each site.

2.3. Birth and death of particles

We include in our microscopic stochastic model also the effects of birth and death of particles.

The possible birth or death, resp., of particles is generated by the $\underline{F}$-adapted cadlag Poisson counting measures $\bar{\mu}_n$ and $\underline{\mu}_n$, resp., which are defined on $\underline{B}([0,\infty)) \otimes \underline{B}(\bar{Q})$ and characterized by their dual predictable projections

$$(2.3.1) \qquad \bar{\nu}_n(dt,du) = \bar{w}_t(u)\, N_n(du)dt$$

and

$$(2.3.2) \qquad \underline{\nu}_n(dt,du) = \underline{w}_t(u)\, N_n(du)dt$$

resp. The birth rate $\bar{w}_t(u)$ and death rate $\underline{w}_t(u)$ are assumed to be nonnegative bounded and Lipschitz-continuous with respect to $u \in \bar{Q}$ uniformely with resprect to t.

Furthermore, we suppose that μ_n, $\bar{\mu}_n$ and $\underline{\mu}_n$ are independent.

A birth (death, resp.,) at u at time t will take place only if u is vacant (occupied, resp.) at this time. So also in the case of a birth or death it is ensured that we have at most one particle at each site.

2.4. Initial condition

The function $\varphi \mid \bar{Q} \to [0,1]$ denotes the initial occupation rate of the sites. We assume that at time t=0 at most one particle is at each site, $L_{n,0}$ is $\underline{F}_0$ -measurable and for all $K \in (0,\infty)$ it holds

$$(2.4.1) \qquad \lim_{n\to\infty} E \sup_{f\in \underline{C}_K} \Big(\int_{\bar{Q}} f(q)\big(\tfrac{1}{n} L_{n,0} - \varphi(q)\lambda\big)(dq)\Big)^2 = 0.$$

2.5. Stochastic equation

Let δ_u denote the Dirac measure at $u \in \bar{Q}$ and $L_{n,t-}$ the left hand limit of the measure of particles at time t>0.

Now we can define the measure valued exclusion process $L_n = \{L_{n,t}\}_{t \geq 0}$ as unique solution of the following stochastic equation (see /8/)

$$(2.5.1) \quad L_{n,t} = L_{n,o} +$$

$$+ \int_0^t \int_{\bar{Q}} \int_{\bar{Q}} (\delta_q - \delta_u) L_{n,s-}(\{u\})(1 - L_{n,s-}(\{q\})) \mu_n(ds,du,dq)$$

$$+ \int_0^t \int_{\bar{Q}} \delta_q (1 - L_{n,s-}(\{q\})) \bar{\mu}_n(ds,dq)$$

$$- \int_0^t \int_{\bar{Q}} \delta_u L_{n,s-}(\{u\}) \underline{\mu}_n(ds,du),$$

which describes the evolution of L_n driven by $\mu_n, \bar{\mu}_n$ and $\underline{\mu}_n$. One easily notes how the $\underline{F}$-adapted, cadlag, piecewise constant and Markovian measure valued process L_n remains at any time t with at most one particle at each site. Furthermore, the interaction between the particles caused by the exclusion mechanism is reflected by the logistic nonlinearity $L_n(1-L_n)$ within the second term of the right hand side of equation (2.5.1).

2.6. Occupation rate

Let $\tilde{Q}$ denote the support of the intensity measure of sites Λ. Now, for $t \geq 0$ and $q \in \tilde{Q}$ we introduce the so called occupation rate H(t,q) which we will later interpret as the probability that a site at $q \in \tilde{Q}$ is occupied at time t. We define the occupation rate H as the unique solution (see /8/) of the following integro-differential equation

$$(2.6.1) \quad \frac{\partial}{\partial t} H(t,q) = (1 - H(t,q)) \left(\int_{\bar{Q}} w_t(u,q) H(t,u) \Lambda(du) + \bar{w}_t(q) \right)$$

$$- H(t,q) \left(\int_{\bar{Q}} w_t(q,u)(1 - H(t,u)) \Lambda(du) + \underline{w}_t(q) \right)$$

for all $t>0$, $q\in\widetilde{Q}$, with initial condition:

$$(2.6.2) \qquad H(0,q)=\varphi(q),$$

for all $q\in\widetilde{Q}$. It can be shown as in /8/ that we have for all $t\geq 0$ and $q\in\widetilde{Q}$:

$$(2.6.3) \qquad H(t,q)\in[0,1].$$

Equation (2.6.1) is a bilance equation for the macroscopic evolution of the occupation rate and allows the following interpretation: The occupation rate changes its value in dependence on the occupation rate at other points. It increases at $q\in\widetilde{Q}$ proportional to the non-occupation rate $(1-H(t,q))$ and the sum of the birth rate $\overline{w}_t(q)$ together with the occupation rates $H(t,u)$ for the sites at $u\in\widetilde{Q}\setminus\{q\}$ which are weighted by the intensity $w_t(u,q)$ for jumps from these sites into q. On the other hand $H(t,q)$ decreases proportional to its own actual value and the sum of the death rate $\underline{w}_t(q)$ together with the non-occupation rates $(1-H(t,u))$ of the other sites at $u\in\widetilde{Q}\setminus\{q\}$ which are weighted by the intensity $w_t(q,u)$ for the jumps from q into these sites.

We remark, that Groeger proved in /3/ that a unique solution exist for the bilance equation (2.6.1) together with the Poisson equation describing the self consistent static potential.

2.7. Law of large numbers

Now, we can formulate a law of large numbers for the above introduced wide range exclusion process in random medium for the case that the mean number n of sites per unit volume tends to infinity, $n\to\infty$.

<u>Theorem 2.7.1</u>

Under the above assumptions it holds for fixed $T,K\in(0,\infty)$

$$(2.7.2) \qquad \lim_{n\to\infty} E \sup_{f\in\underline{C}_K} E\Big(\sup_{0\leq t\leq T} \Big(\int_{\overline{Q}} f(q)\big(\tfrac{1}{n}L_{n,t}-H(t,q)\Lambda\big)(dq)\Big)^2 \,\Big|\, \underline{\underline{F}}_0\Big)=0.$$

The choice of the class of test functions $\underline{C}_K$ and the positioning of the expectations are crucial for the proof of the theorem which uses semimartingale methods. It can be omitted because it would be almost the same as that which is given in /8/.

The above law of large numbers shows that the random functional

$$\int_{\bar{Q}} f(q)\frac{1}{n} L_{n,t}(dq)$$

converges in the mean square sense uniformely with respect to the time $t\in[0,T]$ and all test functions $f\in\underline{C}_K$ for $n\to\infty$ to the deterministic functional

$$\int_{\bar{Q}} f(q)H(t,q)\Lambda(dq).$$

The inner conditional expectation in (2.7.2) relates to the driving Poisson counting measures μ_n, $\bar{\mu}_n$ and $\underline{\mu}_n$. The remaining outer expectation averages the random medium and the random initial occupation of the sites.

3. Asymptotics of the macroscopic nonequilibrium dynamics

3.1. Specifications for the case of a local jump intensity

For simplicity let us choose the domain

(3.1.1) $\quad Q=(0,l_1)\times\ldots\times(0,l_d)$

with

(3.1.2) $\quad l_i \in (0,\infty)$

for all $i\in\{1,\ldots,d\}$.

∂Q is the boundary of Q and $\bar{Q}$ denotes its closure. Further, $\partial Q'$ is that part of ∂Q which does not contain corners or edges.

For $i,j\in\{0,1,\ldots\}$ we denote by $\dot{\underline{C}}^i$ the set of all i-times continuous differentiable functions on $\bar{Q}$, and $\underline{C}^{ij}$ is the set of all functions on $[0,T]\times\bar{Q}$ which are i-times continuous differentiable with respect to the first and j-times with respect to the other coordinates, $T \in (0, \infty)$.

We denote by ν_q the outward unit normal of $\partial Q'$ at $q \in \partial Q'$, by div a the divergence of a vector a, by a·e the usual scalar product between two vectors a and e and by grad f the gradient of a function f on $\underline{R}^d$.

We assume that the intensity measure of sites is absolutely continuous and we write for all $q \in \overline{Q}$,

(3.1.3) $\quad \Lambda(dq) = \lambda(q)dq.$

We suppose that there exists a version of the concentration of sites λ with the properties

(3.1.4) $\quad \lambda(q) \geq K_1 > 0$

for all $q \in \overline{Q}$, where

(3.1.5) $\quad \lambda \in \underline{C}^1$.

Under this assumption we note that the support $\widetilde{Q}$ of Λ coincides with $\overline{Q}$. To formulate the local jump intensity we introduce the Lipschitz continuous probability density

(3.1.6) $\quad p(q) = \exp\{-|q|\} \left(\int_{\underline{R}^d} \exp\{-|u|\}\, du\right)^{-1}$

for all $q \in \underline{R}^d$.

We remark that the proposed approach would also work for other probability densities if they show moment properties as those listed in /9/.

For each value of a parameter $\alpha \in (0,1)$ we specify the local jump intensity for all $t \geq 0$ and $u, q \in \overline{Q}$ by the expression

(3.1.7) $\quad w_t(u,q) = \pi(t,q)^{\frac{1}{2}}\, \pi(t,u)^{-\frac{1}{2}}\, p\left(\frac{2}{\alpha}(u-q)\right) \alpha^{-(\alpha+2)}$

with

(3.1.8) $\quad \pi(t,q) = \exp\{-\Phi(t,q)\}$,

where $\Phi(t,q)$ is the so called static potential and we assume

(3.1.9) $\Phi \in \underline{C}^{1,1}$

We note that for smaller α the jump intensity w_t is more localized.

To interpret the jump intensity w_t we remark that one can show by the use of the properties of the probability density p listed in /9/ that we have for $q \in Q$ the asymptotic drift vector

$$\lim_{\alpha \to 0} \int_{\bar{Q}} (u-q) w_t(q,u) du = -b \operatorname{grad} \Phi(t,q)$$

and the asymptotic variation

$$\lim_{\alpha \to 0} \int_{\bar{Q}} (u-q)^2 \; w_t(q,u) du = bd,$$

where b is a positive constant depending on d. For instance we have for d=1 the value b=1/2 and for d=3 the value b=π.

Now, the occupation rate depends on the parameter $\alpha \in (0,1)$ and we use in the following also the notation

(3.1.6) $H_\alpha(t,q)=H(t,q)$

for all $t \in [0,T]$ and $q \in \bar{Q}$.

3.2. An integral equation

In the following we characterize the limiting dynamics of the occupation rate H_α for $\alpha \to 0$. For this purpose we assume that H_α converges pointwise to a function

$\bar{H} \mid [0,T] \times \bar{Q} \to [0,1]$ such that

(3.2.1) $\lim_{\alpha \to 0} (H_\alpha(t,q)-\bar{H}(t,q))=0$

for all $t \in [0,T]$ and $q \in \bar{Q}$.

Further, we assume that the time derivative of the

occupation rate is uniformly bounded for all $\alpha\in(0,1)$, $t\in[0,T]$ and $q\in\overline{Q}$ with

(3.2.2) $\left|\frac{\partial}{\partial t} H_\alpha(t,q)\right| \le K_2$.

Finally, let us suppose that there exists a constant

(3.2.3) $\varkappa\in(0,1/2)$

such that for all $\alpha\in(0,1)$, $q\in\overline{Q}$ and $t\in[0,T]$ we have

(3.2.4) $H_\alpha(t,q)\in[\varkappa, 1-\varkappa]$.

That means the occupation rate is never 0 or 1.

Let us denote by $\underline{\tilde{C}}^2$ the set of functions $f\in\underline{C}^2$ with

(3.2.5) $\frac{\partial}{\partial \nu_q} f(q)=0$

fo all $q\in\partial Q'$.

Now, we are able to characterize $\overline{H}$ as solution of an interesting integral equation.

<u>Theorem 3.2.6</u>

The limit $\overline{H}$ of the occupation rate H_α for $\alpha\to 0$ satisfies under the above assumptions for all $f\in\underline{\tilde{C}}^2$ and $t\in[0,T]$ the integral equation

(3.2.7) $$\int_{\overline{Q}} f(q)(\overline{H}(t,q)-\overline{H}(0,q))\lambda(q)dq$$
$$=\int_0^t\int_{\overline{Q}} \{\tfrac{b}{2}\overline{H}(s,q)[\mathrm{div}(\lambda^2(q)\mathrm{grad}\, f(q))$$
$$-\lambda^2(q)(1-\overline{H}(s,q))\mathrm{grad}\, f(q)\,\mathrm{grad}\,\phi(s,q)]$$
$$+f(q)[(1-\overline{H}(s,q))\overline{w}_s(q)-\overline{H}(s,q)\underline{w}_s(q)]\lambda(q)\}\,dq\,ds.$$

The proof of this theorem is given in /9/.

Equation (3.2.7) gives a characterization of the limit $\overline{H}$ which avoids smoothness assumptions on $\overline{H}$. Therefore one can say that (3.2.7) gives a rather weak description of $\bar{H}$. Under sufficient smoothness assumptions we will show within the next section that $\bar{H}$ is the solution of a corresponding nonlinear partial differential equation which allows a direct interpretation of the dynamics already described by the equation (3.2.7).

3.3. Asymptotic occupation rate

We assume the initial condition

(3.3.1) $\bar{H}(0,.) \in \underline{C}^2$

with

(3.3.2) $\frac{\partial}{\partial \nu_q} \bar{H}(0,q)+\bar{H}(0,q)(1-\bar{H}(0,q))\frac{\partial}{\partial \nu_q} \bar{\Phi}(0,q)=0.$

Now, we are going to introduce a function R on $[0,T]$xQ which we call asymptotic occupation rate. We suppose that there exists a function $R \in \underline{C}^{1,2}$ which is the unique solution of the nonlinear partial differential equation

(3.3.3)
$$\frac{\partial}{\partial t}R(t,q)- \lambda^{-1}(q)\frac{b}{2}\mathrm{div}(\lambda^2(q)(\mathrm{grad}\ R(t,q) + R(t,q)(1-R(t,q))\mathrm{grad}\ \bar{\Phi}(t,q))) + (1-R(t,q))\bar{w}_t(q)-R(t,q)\underline{w}_t(q)$$

for all $t>0$ and $q \in Q$ with reflecting boundary condition

(3.3.4) $\frac{\partial}{\partial \nu_q}R(t,q)+R(t,q)(1-R(t,q))\frac{\partial}{\partial \nu_q} \bar{\Phi}(t,q)=0$

for all $t \geq 0$ and $q \in \partial Q'$, and initial condition

(3.3.5) $R(0,q)=\bar{H}(0,q)$

for all $q \in \bar{Q}$.

We note that (3.3.3) represents a generalization of Burger's equation.

The following theorem shows under sufficient smoothness assumptions on the limit $\bar{H}$ of the occupation rate H_α for

$\alpha \to 0$, that $\overline{H}$ coincides with the asymptotic occupation rate R.

Theorem 3.3.6

If we assume the property

(3.3.7) $\overline{H} \in \underline{C}^{1,2}$

then we have for all $(t,q) \in [0,T] \times \overline{Q}$ the equivalence

(3.3.8) $\overline{H}(t,q)=R(t,q)$.

The proof of this assertion is given in /9/.

We remark that the smoothness assumption (3.3.7) on $\overline{H}$ could be considerably weakened by an appropriate functional analytic formulation of the nonlinear partial differential equation (3.3.3)-(3.3.5). For instance such a weaker formulation could be based on methods described in /2/ or /6/. Here we have chosen rather strong smoothness assumptions on R and $\overline{H}$ to derive the dynamics described by the equations (3.3.3) - (3.3.5) without technical difficulties in the formulation and prove of Theorem 3.3.6. It remains an interesting problem to derive these equations under much weaker assumptions.

Finally, we remark that the result could be generalized to the case of a anisotropic probability density p and a general regular domain $\overline{Q}$.

3.4. Continuity equation

For better interpretation of the asymptotic occupation rate R together with other physical quantities as particle concentration and current density vector we rewrite the equation (3.3.3) in the form of a continuity equation.

Let us introduce for all $t \geq 0$ and $q \in \overline{Q}$ the particle concentration

(3.4.1) $\rho(t,q)=R(t,q)\,\lambda(q)$

and the vector function

(3.4.2) $j(t,q)=-\frac{b}{2}\lambda^2(q)(\text{grad } R(t,q)$

$+R(t,q)(1-R(t,q)) \text{ grad } \Phi(t,q))$

which we will call current density vector.

Then it follows from (3.3.3) the continuity equation

(3.4.3) $\frac{\partial}{\partial t}\rho(t,q)= -\text{div}(j(t,q))$

$+(\lambda(q)-\rho(t,q))\overline{w}_t(q)-\rho(t,q)\underline{w}_t(q)$

for all $t>0$ and $q\in Q$, with reflecting boundary condition

(3.4.4) $j(t,q)\cdot\nu_q =0$

for all $t\geq 0$ and $q\in\partial Q$, and initial condition

(3.4.5) $\rho(0,q)=\overline{H}(0,q)\lambda(q)$

for all $q\in\overline{Q}$.

The above continuity equation relates the time derivative of the particle concentration with the current density vector.

We can also write the current density vector in the form

(3.4.6) $j(t,q)=$

$-\frac{b}{2}\lambda^2(q)R(t,q)(1-R(t,q))\text{grad }(\xi(t,q)+\Phi(t,q)),$

for all $t\geq 0$, $q\in\overline{Q}$, where

(3.4.7) $\xi(t,q)=\ln(R(t,q)(1-R(t,q))^{-1})$

is the so called chemical potential at $t\geq 0$ and $q\in\overline{Q}$.

One notes that the current density vector shows into the opposite direction of the gradient of the sum of the chemical and static potential.

The length of the current density vector is proportional to the so called conductivity

(3.4.8) $\delta(t,q)=\frac{1}{2}\, b\,\lambda^2(q)R(t,q)(1-R(t,q))$

at $t\geq 0$ and $q\in \overline{Q}$, which contains a logistic nonlinearity with respect to the asymptotic occupation rate R. The conductivity reaches its maximum at an asymptotic occupation rate with value R=1/2 which corresponds to the value of the chemical potential $\xi=0$. The logistic nonlinearity of the conductivity is caused by the interaction of the particles within the exclusion dynamics, where jumps to occupied sites are excluded. As important result of this nonlinearity it follows that the particle concentration is bounded by the concentration of sites and we have for all $t\geq 0$ and $q\in \overline{Q}$

(3.4.9) $0\leq \varrho(t,q)\leq \lambda(q)$.

Finally, we note that there is no current through the boundary $\partial Q'$ what is natural in our model. It seems that the continuity equation (3.4.3)-(3.4.5) can be used to model important transport processes with birth and death effects in physics, chemistry, biology, social sciences and other fields.

References

/1/ De Masi, A.; Ianiro, N.; Pellegrinotti, A.; Presutti, E.: A survey of the hydrodynamical behaviour of many-particle systems. Nonequilibrium phenomena II, series Stud. Statist. Mech., XI, ed. J.L. Lebowitz and E.W. Montroll, North-Holland, Amsterdam-New York, (1984).

/2/ Gajewski, H.; Groeger, K.; Zacharias, K.: Nichtlineare Operatorgleichungen und Operatordifferentialgleichungen. Akademie-Verlag, Berlin 1974.

/3/ Groeger, K.: Hopping transport of electrons influenced by a selfconsistent electrostatic potential. Preprint I-Math AdW Berlin 1988.

/4/ Ikeda, N.; Watanabe, S.: Stochastic differential equations and diffusion processes. North-Holland, Amsterdam (1981).

/5/ Jacod, J.: Calcul stochastique et problemes de martingales. Lect. Notes in Math. 714, Springer Verlag (1979).

/6/ Kačur, J.: Method of Rothe in evolution equations. Teubner, Leipzig (1985).

/7/ Liggett, T.M.: Interacting particle systems. Springer Verlag, Berlin (1985).

/8/ Platen, E.: A law of large numbers for wide range exclusion processes in random media. Stochastic processes and their Appl., to appear

/9/ Platen, E.: On a wide range exclusion process in random medium with local jump intensity. Technical Report No. 236, Center of Stoch. Proc. Univ. of North Carolina, Chapel Hill 1988.

/10/ Miller, A.; Abrahams, B. (1960) Phys. Rev. 120, 745.

/11/ Spitzer, F.: Interaction of Markov processes. Adv. in Math. 5, (1970) 256-290.

ON LARGE DEVIATIONS FOR STOCHASTIC EVOLUTION EQUATIONS

J. Zabczyk(*)

Institute of Mathematics, Polish Academy of Sciences, Sniadeckich 7, Warsaw, Poland

The Large Deviations Principle (LDP) was formulated by S.R.S. Varadhan [21] in 1966. Its validity for stochastic differential equations has been established in the sixties and in the seventies by M. Schilder [18], M. Freidlin and A. Wentzell [8] and R.G. Azencott [1].

The paper presents several large deviations theorems obtained in the eighties for stochastic equations in the infinite dimensional spaces, sketches some of their proofs and reports on new results.

1. LARGE DEVIATIONS PRINCIPLE. FORMULATION AND BASIC PROPERTIES

Let (F,ρ) be a complete, separable metric space and $(\mu_\varepsilon)_{\varepsilon>0}$ a family of probability measures on Borel subsets of F. Let $I : F \to [0, +\infty]$ be a lower semicontinuous function such that for arbitrary $r > 0$ the sets

$$K(r) = \{x \in F;\ I(x) \leq r\}$$

are compact.

The family (μ_ε) is said to satisfy the large deviation principle (LDP) or to have the large deviation property, with respect to the rate functional I if:

(i) $\overline{\lim_{\varepsilon\downarrow 0}}\ \varepsilon \ln \mu_\varepsilon(B) \leq -\mathrm{int}\{I(x);\ x \in B\}$, B an arbitrary open set

(ii) $\underline{\lim_{\varepsilon\downarrow 0}}\ \varepsilon \ln \mu_\varepsilon(B) \geq -\mathrm{int}\{I(x);\ x \in B\}$, B an arbitrary close set.

The properties (i) and (ii) can be phrased equivalently in the form of the so called exponential estimates, see [8]:

(iii) For arbitrary $x \in F$, $\delta > 0$ and $\gamma > 0$ there exists $\varepsilon_0 > 0$ such that for all $\varepsilon \in (0,\varepsilon_0)$:

$$\mu_\varepsilon(y;\rho(x,y) < \delta) \geq e^{-\frac{1}{\varepsilon}(I(x)+\gamma)}$$

(*) The final version of the paper was written while the author was at the Mathematics Institute, University of Warwick, England, Winter 1989.

(iv) For arbitrary $r > 0$, $\delta > 0$ and $\gamma > 0$ there exists $\varepsilon_0 > 0$ such that for all $\varepsilon \in (0,\varepsilon_0)$:

$$\mu_\varepsilon(y;\rho(y,K(r)) < \delta) \geq 1 - e^{-\frac{1}{\varepsilon}(r-\gamma)}.$$

The following proposition is due to S.R.S. Varadhan.

THEOREM 1: (Contraction principle). Assume that a family (μ_ε) satisfies the large deviation principle and U is a continuous mapping from F into another Polish space E. Then the family (ν_ε): $\nu_\varepsilon(B) = \mu_\varepsilon(U^{-1}(B))$, B Borel subsets of E, has the LDP with the rate function J:

(1) $J(y) = \text{int}\{I(x);\ U(x) = y\}$, $(\inf \phi = +\infty)$.

A Gaussian measure μ on a Hilbert space H with the mean vector m and the covariance operator Q will be denoted by $N(m,Q)$. Let us recall also that if L is a linear transformation from a Hilbert space H_1 into H_2 then its pseudoinverse L^{-1} maps H_0 = Image $L \subset H_2$ into H_1 according to the formula:

$L^{-1}y = x$ if and only if $Lx = y$ and $\|x\| \leq \|z\|$ for all $z \in H_1$ such that $Lz = y$.

EXAMPLE 1: Assume that $\mu_\varepsilon = N(0,\varepsilon Q)$, $\varepsilon > 0$. Then the family (μ_ε) satisfies the LDP with the rate function:

$$I(x) = \frac{1}{2}\|(Q^{\frac{1}{2}})^{-1}x\|^2, \quad x \in \text{Image } Q^{\frac{1}{2}}$$
$$= +\infty \quad , \text{otherwise.}$$

2. STOCHASTIC EVOLUTION EQUATIONS

In the present paper we are concerned with the LDP for families of probability measures naturally associated with the solutions $X^\varepsilon(t)$, $t \in [0,T]$ of stochastic semilinear equations:

(2) $dX = (AX + F(X))dt + \sqrt{\varepsilon}\, dW(t)$, $X(0) = x \in E$.

First results on the topic were obtained by W.G. Faris and G. Jona-Lasinio [7] and later by Z. Zabczyk [24], [25] and [26]. The paper [19] by W. Smolenski, R. Sztencel and J. Zabczyk gave a proof of the large deviations estimates for a large class of general semilinear equations. The estimates were extended to the white noise perturbations by G. Da Prato and J. Zabczyk [5]. Approximately, at the same time, papers by M.I. Freidlin [9] and W.M. Imajkin and A.I. Komjec [12] appeared in which the LDP was established for semilinear stochastic equations with the eliptic linear term. The LDP for the associated invariant measures was studied by M.I.

Freidlin [9], G. Jetschke [13] and J. Zabczyk [27]. Applications to the tunnelling problem were obtained by W.G. Faris and G. Jona-Lasinio [7] and M. Cassandro, E. Olivieri and P. Picco [2]. The exit problem was considered by M.I. Freidlin [9] and J. Zabczyk [26]. The above list of the papers is certainly not complete. Some new results are going to appear soon. Important problems like the LDP for stochastic evolution equations with the state dependent noise are still waiting to be resolved.

To fix ideas we will assume that the operator A is the infinitesimal generator of a C_0-semigroup $S(t)$, $t \geq 0$ on the separable, Banach space E, F is a nonlinear transformation from E into E and W is a Wiener process on a Hilbert space H containing E (or contained in E) as a dense subset. We will assume also that the semigroup $S(t)$, $t \geq 0$ can be extended (or restricted) to a C_0-semigroup $S_0(t)$, $t \geq 0$ on H with the infinitesimal generator A_0, see [5]. Without a great loss of generality one can require that the process $W(\cdot)$ has the following representation:

$$(*) \qquad W(t) = \sum_{j=1}^{+\infty} \sqrt{\lambda_j}\, \beta_j(t) e_j, \quad t \geq 0,$$

where (e_j) is an orthonormal, complete basis in H, (β_j) is a sequence of independent, real valued Wiener processes and λ_j, $j = 1,2,\dots$, are nonnegative numbers. The nonnegative bounded operator $Q : H \to H$ for which $Qe_j = \lambda_j e_j$, $j = 1,2,\dots$ is called the incremental covariance operator of $W(\cdot)$.

If Trace $Q = \sum_{j=1}^{+\infty} \lambda_j < +\infty$, then the series (*) is convergent P-a.s. uniformly on arbitrary finite interval $[0,T]$ and therefore defines a continuous H-valued process. In the sequel we will deal also with the cylindrical Wiener process corresponding to $Q = I$. The series (*) is then not convergent. However, under some additional assumptions, the meaning to the equation (2) can be given.

A solution $Z^\varepsilon(\cdot)$ to the linear stochastic equation:

$$(3) \qquad dZ^\varepsilon = A_0 Z dt + \sqrt{\varepsilon}\, dW(t), \; Z^\varepsilon(0) = 0$$

is defined by the "variation of constant formula":

$$(4) \qquad Z^\varepsilon(t) = \sqrt{\varepsilon} \int_0^t S_0(t-s) dW(s), \; t \geq 0.$$

The stochastic integral in (4) has a well defined meaning if Trace $Q < +\infty$. In that case it has even a continuous modification [4]. If W is a cylindrical Wiener process, $Q = I$, then the formula (4) defines an H-valued process if and only if, for arbitrary $T > 0$:

$$(5) \qquad \int_0^T \|S_0(t)\|_{HS}^2 dt < +\infty$$

where $\|S_0(t)\|_{HS}$, $t \geq 0$ denotes the Hilbert-Schmidt norm of the operator $S_0(t)$, $t \geq 0$. If (5) holds then the integral in (4) can be defined as the sum of a $\mathbf{P}$-a.s.convergent series:

$$(6) \qquad Z^{\varepsilon}(t) = \sum_{j=1}^{+\infty} \sqrt{\varepsilon} \int_0^t S_0(t-s)e_j d\beta_j(s), \ t \geq 0.$$

EXAMPLE 2: Let $G = \{(\alpha_k) \in \mathbb{R}^d;\ 0 \leq \alpha_k \leq L,\ k = 1,\dots,d\}$ $H = L^2(G)$ and $A_0 = (-1)^{m+1}\Delta^m$ where Δ is a Laplace operator, m a natural number and $D(A_0) = H_0^m(G) \cap H^{2m}(G)$, see [11, p. 29]. It is easy to check that the condition (5) is satisfied if and only if $m > \frac{d}{2}$. In particular if $m = d = 1$.

In several papers, see e.g. [22], [16] and [9] a solution to the linear equation (3) is given as a stochastic integral with respect to a Brownian sheet process. The fact that the definition (6) is more general can be seen from the following observation. Let $H = L^2(G)$ as in Example 2. The Brownian sheet: $B(t,a)$, $t \geq 0$, $a \in G$ can be defined as the integral of the divergent series (*):

$$B(t,a) = \sum_{j=1}^{+\infty} \beta_j(t) \int_{r \leq a} e_j(r)dr, \ t \geq 0, \ a \in G.$$

Note that by the Parceval's identity:

$$\begin{aligned} \mathbb{E}(B(t_1,a_1)B(t_2,a_2)) &= (t_1 \wedge t_2) \sum_{j=1}^{+\infty} \int_{\{r \leq a_1\}} e_j(r)dr \int_{\{r \leq a_2\}} e_j(r)dr \\ &= (t_1 \wedge t_2) \int_G I_{\{s \leq a_1\}}(u)\ I_{\{s \leq a_2\}}(u)du \\ &= (t_1 \wedge t_2) \prod_{k=1}^{d} \alpha_k^1 \wedge \alpha_k^2, \ a^1 = (\alpha_k^1),\ a^2 = (\alpha_k^2). \end{aligned}$$

Thus, formally:

$$(7) \qquad W(t,a) = \frac{\partial^d}{\partial\alpha_1 \dots \partial\alpha_d} B(t,a) \text{ and } \frac{\partial W}{\partial t}(t,a) = \frac{\partial^{d+1}}{\partial t \partial\alpha_1 \dots \partial\alpha_d} B(t,a)$$

and the Brownian sheet integration theory [22] makes the formulae (7) rigorous, see also [9, p. 668].

The nonlinear equation (2) is usually treated as an integral equation of the form:

$$(8) \qquad X(t) = S(t)x + \int_0^t S(t-s)F(X(s))ds + \sqrt{\varepsilon} \int_0^t S_0(t-s)dW(s), \ X(0) = x,$$

or equivalently as

$$(9) \qquad X(t) = S(t)x + \int_0^t S(t-s)F(X(s))ds + Z^{\varepsilon}(t), \ t \geq 0, \ X(0) = x.$$

The first step in the proof of existence of an E-valued and continuous solution to (2) consists in establishing that the process $Z^\varepsilon(\cdot)$ has E-continuous modification. This has been done by several authors under different sets of assumptions, see [22], [16] and [5]. Conditions are then formulated which imply existence and uniqueness of E-continuous solution $u(\cdot)$ to the deterministic equation:

$$(10) \quad u(t) = S(t)x + \int_0^t S(t-s)F(u(s))ds + \psi(t), \; t \geq 0$$

in which ψ is an arbitrary E-continuous function, $\psi(0) = 0$. If $u(t,x,\psi)$, $t \geq 0$, $x \in E$ is a solution to (10) then the solution $X^{x,\varepsilon}(\cdot)$ of the stochastic equation (8) can be written as

$$(\#) \quad X^{x,\varepsilon}(t) = u(t,x,Z^\varepsilon(\cdot)), \; t \geq 0, \; x \in E.$$

Sufficient conditions implying existence and uniqueness to (10) have been given in several papers, see [7], [22], [16], [5], [9] and [12].

The following three families of measures are of particular interest in the theory of evolution equations (2) with the small noise.

The laws $\mu_\varepsilon^{x,T}$ of the processes $X^{x,\varepsilon}(\cdot)$, $\varepsilon > 0$ on the space $C(0,T;E)$ of E-continuous functions.

The laws $\nu_\varepsilon^{x,T}$ of $X^{x,\varepsilon}(T)$, $\varepsilon > 0$, for fixed x and T.

The stationary distributions ν_ε^∞, of the processes $X^{x,\varepsilon}(\cdot)$, $\varepsilon > 0$.

We will describe now some representative results on the LDP for the above families.

3. THE LDP FOR ORNSTEIN-UHLENBECK PROCESSES

We start from the linear equations (3) solutions of which are often called Ornstein-Uhlenbeck processes. Without a loss of generality we will assume that the initial condition $x = 0$. Then:

$$\mu_\varepsilon^{0,T} = \mathcal{L}(Z^\varepsilon(\cdot))$$

where $\mathcal{L}(\xi)$ stands for the law of a randon variable ξ.

To show that the family $(\mu_\varepsilon^{0,T})_{\varepsilon>0}$ satisfies the LDP we will need the following result due to G. Kallianpur and H. OoDaira [14]. Assume that μ is a symmetric Gaussian measure on a separable Banach space E and let E_0 be its reproducing kernel. Define:

$$(11) \quad I(x) = \frac{1}{2}\|x\|_{E_0}, \quad x \in E_0$$
$$= +\infty \quad , \quad x \notin E_0$$

Then the following result holds [14], (a simplified proof can be found in [19]).

THEOREM 2: Let ξ be an E-valued random variable such that $\mathcal{L}(\xi) = \mu$ and let

$$\mu_\varepsilon = \mathcal{L}(\sqrt{\varepsilon}\xi),\ \varepsilon > 0.$$

Then the family $(\mu_\varepsilon)_{\varepsilon>0}$ satisfies the LDP with the rate functional (11).

Note that

$$\mu_\varepsilon^{0,T} = \mathcal{L}(\sqrt{\varepsilon}\, Z^1(\cdot)),\ \varepsilon > 0$$

so the above Theorem 2 is applicable with the measure $\mu = \mu_1^{0,T} = \mathcal{L}(Z^1(\cdot))$. The reproducing kernel for $\mu_1^{0,T}$ was calculated in [19]. Define for arbitrary $u \in L^2[0,T;H]$ a function $y = Lu$:

$$y(t) = Lu(t) = \int_0^t S_0(t-s)Q^{\frac{1}{2}}u(s)ds,\ t \in [0,T].$$

Thus $y(\cdot)$ is then a mild solution of the equation:

$$\dot{y} = A_0 y + Q^{\frac{1}{2}}u,\ y(0) = 0.$$

The following theorem holds [19]:

THEOREM 3: (i) Assume that $Z^1(\cdot)$ is an H-continuous process. Then the reproducing kernel of $\mu = \mathcal{L}(Z^1(\cdot))$ is of the form:

$$(12)\qquad I(y) = \frac{1}{2}\,\|L^{-1}(y)\|^2_{L^2[0,T;H]}.$$

(ii) If, in addition, $Z^1(\cdot)$ is an E-continuous process then the family $(\mu_\varepsilon^{0,T})$ satisfies the LDP on $C[0,T;E]$ with the rate functional (12).

Direct proofs of the part (ii) of the theorem can be found in [7], [9] and [12], usually under stronger assumptions.

EXAMPLE 3: Taking into account the second formulation of LDP expressed in (iii) and (iv), one arrives at the following corollary. For arbitrary $u \in L^2[0,T;H]$, $\delta > 0$ and $\gamma > 0$ there exists $\varepsilon_0 > 0$ such that for all $\varepsilon \in (0,\varepsilon_0)$:

$$P(\sup_{t\le T}\|\sqrt{\varepsilon}\, Z^1(t) - \int_0^t S(t-s)Q^{\frac{1}{2}}u(s)ds\|_E < \delta)$$

$$\geqq e^{-\frac{1}{2}\varepsilon(\int_s^T \|u(s)\|^2 ds + \gamma)}$$

The LDP for the second family $(\nu_\varepsilon^{0,T})$ can be obtained in a more straightforward manner because:

$$\nu_\varepsilon^{0,T} = N(0, \varepsilon R_T), \; \varepsilon > 0$$

where

$$R_T = \int_0^T S_0(r)QS_0^*(r)dr, \; T > 0.$$

By Example 1, the associated rate functional is:

$$(13) \quad I_T(x) = \frac{1}{2} \|(R_T^{\frac{1}{2}})_x^{-1}\|^2, \quad \text{if } x \in \operatorname{Im} R_T^{\frac{1}{2}}$$

$$= +\infty \qquad , \quad \text{otherwise.}$$

The rate functional does not change if the measures $\nu_\varepsilon^{0,T}$ are regarded as distributions on the Banach space E provided $\nu_\varepsilon^{0,T}(E) = 1$, $\varepsilon > 0$.

If one applies the contraction principle and Theorem 3 one gets a different expression for $I_T(\cdot)$:

$$(14) \quad I_T(x) = \frac{1}{2} \|L_T^{-1}x\|^2, \text{ if } x \in \operatorname{Im} L_T$$

$$= +\infty \qquad , \text{ otherwise.}$$

Here L_T denotes the operator acting from $L^2[0,T;H]$ into H and given by the formula:

$$L_T u = \int_0^T S_0(T-s)Q^{\frac{1}{2}}u(s)ds.$$

Consequently, as a byproduct, one obtains that:

$$\operatorname{Im} R_T^{\frac{1}{2}} = \operatorname{Im} L_T \qquad \text{and}$$

$$\|(R_T^{\frac{1}{2}})^{-1}x\| = \|L_T^{-1}x\|, \text{ for } x \in \operatorname{Im} L_T.$$

These identities are well known in control theory, see [3], [23] and [26].

More information on the rate functional is available if generator A_0 is from a special class.

EXAMPLE 4: Assume that A_0 is a self-adjoint operator with the spectrum contained in $(-\infty,-\alpha]$, $\alpha > 0$ and let $Q = I$. Then $R_T = (-2A_0)^{-1}(I-e^{2A_0T})$. Moreover

$$\|(-A_0)^{\frac{1}{2}}x\|^2 \le I_T(x) \le (1-e^{-2\alpha T})^{-1} \|(-A_0)^{\frac{1}{2}}_x\|^2, \ x \in \mathrm{Im}\, R_T^{\frac{1}{2}}.$$

So the rate functional $I_T(\cdot)$ is equivalent to the square of the "fractional-power" norm $\|\cdot\|_{\frac{1}{2}}$

$$\|x\|_{\frac{1}{2}} = \|(-A_0)^{\frac{1}{2}}x\|, \ x \in D(-A_0)^{\frac{1}{2}}.$$

The formulae from Example 4 has been recently extended by G. Da Prato [6] to analytic generators A_0, as follows.

THEOREM 4: Let A_0 be an analytic generator of a stable semigroup $S_0(t)$, $t \ge 0$ on a Hilbert space H and $\|\cdot\|_{\frac{1}{2},2}$ its interpolation norm:

$$\|x\|_{\frac{1}{2},2} = (\int_0^{+\infty} \|A_0S_0(t)x\|^2 dt)^{\frac{1}{2}}, \ x \in H.$$

Then the rate functional(13) is equivalent to $(\|\cdot\|_{\frac{1}{2},2})^2$.

The case of invariant measures (ν_ε^∞) can be treated in a direct way as well. It follows from [23] that a unique invariant distribution for the process $Z^\varepsilon(\cdot)$, $\varepsilon > 0$, exists if and only if the operator

$$R_\infty^\varepsilon = \lim_{T\uparrow+\infty} R_T^\varepsilon = \varepsilon\int_0^{+\infty} S_0(r)QS_0^*(r)dr = \varepsilon R_\infty$$

is well defined and nuclear and if the only invariant distribution for the deterministic system $\dot{Z} = A_0Z$, is $\delta_{\{0\}}$. The operator R_∞ satisfies an operator equation:

$$2\langle R_\infty A_0^* x, x\rangle + \langle Qx, x\rangle = 0, \text{ for } x \in D(A_0^*)$$

and the unique invariant measure is

$$\nu_\varepsilon^\infty = N(0, \varepsilon R_\infty), \ \varepsilon > 0.$$

The rate functional for (ν_ε^∞) is therefore

$$I_\infty(x) = \frac{1}{2} \|(R_\infty^{\frac{1}{2}})^{-1}x\|^2, \ x \in H.$$

In particular, if the operator A_0 is negative then

$$I_\infty(x) = \frac{1}{2} \|(-A_0)^{\frac{1}{2}}x\|^2, \ x \in A;$$

see [13], [9] and [27].

4. THE LDP FOR GENERAL EQUATIONS

For general evolution processes $X^{x,\varepsilon}(\cdot)$ satisfying (2) and given in the form (#) one gets the LDP via the contraction principle. The general pattern is as follows. The form of the rate functional for the families $(\mathcal{L}(Z^{\varepsilon}(\cdot)))_{\varepsilon>0}$ is known from the previous section. Assume that for $x \in E$ the mapping $U^x : \psi \to u(\cdot,x,\psi)$, see (10), transforms continuously $C_T = C[0,T;E]$ into C_T. Then the rate functional for $(\mathcal{L}(X^{x,\varepsilon}(\cdot)))_{\varepsilon>0}$ is given by (1). We have therefore the following result in the formulation of which $y^{x,\phi}(\cdot)$ stands for the mild solution of the equation

$$(15) \qquad \dot{y} = (Ay + F(y)) + Q^{\frac{1}{2}}\phi, \; y(0) = x,$$

and $K_T^x(r)$ is the set of functions $y^{x,\phi}(\cdot)$ corresponding to ϕ satisfying the inequality:

$$\frac{1}{2}\int_0^T \|\phi(s)\|_H^2 ds \le r.$$

THEOREM 5: (i) Assume that the processes $Z^{\varepsilon}(\cdot)$, $\varepsilon > 0$ are E-continuous and the transformation U^x maps continuously C_T into C_T. Then, for arbitrary $\phi \in L^2[0,T;H]$, $r > 0$, $\delta > 0$ and $\gamma > 0$ there exists $\varepsilon_0 > 0$ such that for $\varepsilon \in (0,\varepsilon_0)$

$$(16) \qquad P(\sup_{t\le T} \|X^{x,\varepsilon}(t) - y^{x,\phi}(t)\|_E < \delta) \ge e^{-\frac{1}{\varepsilon}(\frac{1}{2}\int_0^T \|\phi(s)\|_H^2 ds+\gamma)}$$

and

$$(17) \qquad P(\text{distance}_E\,(X^{x,\varepsilon},K_T^x(r)) < \delta) \ge 1-e^{-\frac{1}{\varepsilon}(r-\gamma)}.$$

(ii) Assume, in addition, that the transformations U^x are uniformly locally Lipschitz: for arbitrary $\alpha > 0$ there exists $\beta > 0$ such that $\|U^x(\psi_1) - U^x(\psi_2)\|_{C_T} \le \beta\,\|\psi_1-\psi_2\|_{C_T}$ for all ψ_1,ψ_2 and x satisfying $\|\psi_1\|_{C_T} \le \alpha$, $\|\psi_2\| \le \alpha$, $\|x\|_E \le \alpha$. Then the estimates (16) and (17) hold uniformly with respect to x in bounded sets.

Thus the validity of the exponential estimates (16) and (17) depends on the properties of the transformations U^x. Sufficient conditions for (local) Lipschitzianity of U^x have been given in a number of papers. The case of the operator $A_0 = \frac{d^2}{dz^2}$ and $H = L^2[0,L;R^d]$ is covered in papers [16] and [9]. The nonlinearity F was that of the "coordinate-wise" type: $F(x)(r) = f(x(r))$ $r \in [0,L]$, $f : R^d \to R^d$ a locally Lipschitz function satisfying some growth or dissipativity conditions to assure non-explosion of the local solutions to (2). The case of A_0 being the Laplace operator Δ was considered in [12]. General infinitesimal generators A_0 (and A) were treated in [5].

The contraction principle implies also the LDP for the family $(\mathcal{L}(X^{x,\varepsilon}(T)))_{\varepsilon>0}$.

THEOREM 6: Assume that the conditions of part (i) of Theorem 5 hold. Then the family $(\mathcal{L}(X^{x,\varepsilon}(T)))_{\varepsilon>0}$ satisfies the LDP with the rate functional I_T^x:

$$I_T^x(z) = \frac{1}{2} \inf \{\int_0^T \|\phi(s)\|^2 ds;\ \ y^{x,\phi}(T) = z\}.$$

Thus $I_T^x(z)$ is half of the energy needed to transfer x to z via the control system (15). This is in a complete agreement with the results obtained for the linear systems in the previous sections.

There are no general results on the LDP for invariant measures (ν_ε^∞) associated with $(X^{x,\varepsilon}(\cdot))$. Some are available however for special classes of systems discussed in the next sections.

5. THE LDP FOR GRADIENT SYSTEMS

For gradient systems the nonlinear transformation F is of the form:

$$(18)\qquad F(x) = - U'(x),\ x \in E,$$

and the operator A_0 is negative definite, see [ ].

Here U is a real valued function on E and U' is its derivative. To be more precise denote by D(x;h) the directional derivative of the function U at a point x and a direction h:

$$D(x;h) = \lim_{s\downarrow 0} s^{-1}(U(x + sh) - U(x)).$$

We will assume that D(x;h) exists for all $x \in E$, $h \in H$ and that there exists a vector $U'(x) \in H$ such that:

$$\langle U'(x),h\rangle_H = D(x;h),\ x \in E,\ h \in H.$$

EXAMPLE: Assume that $H = H^2(G)$ and $E = C(G)$, compare Example . Define

$$U(x) = \int_G p(x(r))dr,\ x \in E,$$

where $p : R^1 \to R^1$ is a continuously differentiable function. Then $U'(x) \in E \subset H$ for $x \in E$ and

$$U'(x)(r) = \frac{dp}{dz}(x(r)),\ r \in G.$$

All the results formulated in the previous sections apply to the gradient systems. In particular for $(\mathcal{L}(X^{x,\varepsilon}(\cdot)))_{\varepsilon>0}$ the Theorem 5 is applicable. More explicit formulae are available for the families $(\mathcal{L}(X^{x,\varepsilon}(T)))_{\varepsilon>0}$ and for the invariant measures $(\nu_\varepsilon^\infty)_{\varepsilon>0}$.

Let us assume that $E \subset H$ and that A_0 is a negative definite operator. Denote by $z^x(t)$, $t \geq 0$ the mild solution of the equation:

$$\dot{z} = A_0 z - U'(z), \quad z(0) = x$$

we have the following result:

THEOREM 7: Assume that $D(-A_0)^{\frac{1}{2}} \subset E$, $Q = I$, and that the functional U is of class C^2 and satisfies $U(0) = 0$, $U(x) \geq 0$ for $x \in E$ and $U'(0) = 0$. Define $I_\infty(x) = \inf_{T>0} I_T^0(x)$, $x \in E$.

(i) If $x \notin D(-A_0)^{1/2}$ then $I_\infty(x) = +\infty$

(ii) If $x \in D(-A_0)^{1/2}$ and $z^x(t) \to 0$ as $t \uparrow +\infty$ then

$$I_\infty(x) = \|(-A_0)^{1/2} x\|^2 + 2U(x).$$

The theorem was obtained by M. Freidlin [9] for the case $A_0 = \frac{d^2}{dz^2}$, $H = L^2(0,L;R^d)$ $E = C(0,L;R^d)$. The general case stated in the theorem seems to be new. Its proof, see [28], uses similar ideas as the one from [9] but requires more general tools. It can be divided into several steps. One proves first that trajectories $y^{0,\phi}(\cdot)$ of the controlled system:

$$(19) \quad \dot{y} = A_0 y - U'(y) + \phi, \quad y(0) = 0$$

are contained in $D(-A_0)^{1/2}$. This follows from the linear theory, see Example 4, and implies (i). Then one shows that the state 0 can be transferred to an arbitrary point from a $D(-A_0)^{1/2}$ - neighbourhood of 0 with an energy smaller than any positive number fixed in advance. This can be done in a similar way as in [15]. Differentiability of U' at 0 is crucial here. Another important tool is the following identity

$$(20) \quad \frac{1}{2}\int_0^T \|\phi(s)\|^2 ds = \frac{1}{2}\int_0^T \|\dot{y}(s) + A_0 y(s) - U'(y(s))\|^2 ds$$
$$+ \left(\|(-A_0)^{1/2} y(T)\|^2 - \|(-A_0)^{1/2} y(0)\|^2 \right.$$
$$\left. + 2U(y(T)) - 2U(y(0)) \right),$$

where $y(\cdot)$ is a solution to (19) with the initial and final conditions $y(0)$ and $y(T)$ respectively. This implies that

$$(21) \quad I_T^a(x) \geq \|(-A_0)^{1/2}x\|^2 - \|(-A_0)^{1/2}a\|^2 + 2(U(x) - U(a)), \; x, a \in D(-A_0)^{1/2}.$$

Define $y(t) = z^x(T-t)$, $t \in [0,T]$ and let $b = z^x(T)$ then (20) gives

$$(22) \quad I_T^b(x) = \|(-A_0)^{1/2}x\|^2 - \|(-A_0)^{1/2}b\|^2 + 2(U(x) - U(b)).$$

Taking into account the controllability result mentioned above and the formula (20) one obtains

$$\lim_{T\uparrow+\infty} I_T^0(x) = \|(-A_0)^{1/2}x\|^2 + 2U(x).$$

REMARK: In the proof an important estimate (21) was established. The part (ii) shows that the estimate is sharp.

As far as the invariant measures are concerned we have the following theorem [27].

THEOREM 8: Assume that the assumptions of Theorem 7 hold and that $z^b(t) \to 0$, as $t \uparrow +\infty$ for all $b \in E$. Then, for arbitrary $\varepsilon > 0$, there exists a unique invariant measure ν_ε^∞ for the equation (2). The family (ν_ε^∞) satisfies the LDP with the rate functional I_∞ (see Theorem 7).

The uniqueness result follows from B. Maslowsky [17]. The case corresponding to $A_0 = \frac{d^2}{dz^2}$ was obtained by G. Jetschke [13] and by M.I. Freidlin [9].

6. RATE FUNCTIONALS FOR SOME SECOND ORDER SYSTEMS

Assume, as before, that A_0 is a negative definite operator on a Hilbert space H and consider the following equation:

$$\dot{y} = v$$

$$\dot{v} = A_0 y - U'(y) - \beta v + \phi,$$

in which β is a positive constant. One can show, see [28], that the minimal energy $I\binom{\bar{y}}{\bar{v}}$ needed to steer the state $\binom{0}{0}$ to $\binom{\bar{y}}{\bar{v}} \in \begin{matrix} D(-A_0)^{1/2} \\ \times \\ H \end{matrix}$ is exactly

$$I\binom{\bar{y}}{\bar{v}} = \beta(\|\bar{v}\|^2 + \|(-A_0)^{1/2}\bar{y}\|^2 + 2U(\bar{y})).$$

It turns out also that the rate functional for invariant measures associated with the solutions to the stochastic equations:

$$dX = Y\,dt$$

$$dY = (A_0X - U'(X) - \beta Y)dt + \sqrt{\varepsilon}\,dW$$

is precisely the functional I, see [28].

REFERENCES

[1] R.G. Azencott, Sur les grand deviations, Ecole d'Eté de Probabilité Saint-Flour, Lecture Notes in Mathematics 774 (1978).

[2] M. Cassandro, E. Olivieri and P. Picco, Small random perturbations of infinite dimensional dynamical systems in nucleation theory, Ann. Inst. Henri Poincaré, Physique Théorique, 44 (1986), 343-396.

[3] R.F. Curtain, Linear-quadratic control problem with fixed endpoints in infinite dimensions, JOTA, 44 (1984), 55-74.

[4] G. Da Prato, S. Kwapien and J. Zabczyk, Regularity of solutions of linear stochastic equations in Hilbert spaces. Stochastics 23 (1987), 1-23.

[5] G. Da Prato and J. Zabczyk, A note on semilinear stochastic equations, Diff. and Int. Equs, 1 (1988), 143-155.

[6] G. Da Prato, Controllability for parabolic equations, preprint.

[7] W.G. Faris and G. Jona-Lasinio, Large fluctuations for a nonlinear heat equation with noise, J. Phys., A: Math. Gen., 19 (1982), 3025-3055.

[8] M. Freidlin and A. Wentzell, Random perturbations of dynamical systems. Springer-Verlag, 1987.

[9] M.I. Freidlin, Random perturbations of reaction diffusion equations, TAMS, 305 (1988), 665-697.

[10] T. Funaki, Random motion of strings and related stochastic evolution equations, Nagoya Math. J. 89 (1983), 129-196.

[11] D. Henry, Geometric theory of semilinear parabolic equations, Lecture Notes in Mathematics, 840 (1981).

[12] W.M. Imajkin and A.I. Komječ, On large deviations for solutions of stochastic nonlinear equations, Proceedings of the Petrovski's Seminar, 13 (1988), 177-195 (in Russian).

[13] G. Jetschke, Invariant distributions of a nonlinear stochastic partial differential equs. Fo.-Evg. Jena, 86/11,20,40.

[14] G. Kallianpur and H. OoDaira, Freidlin-Wentzell type estimates for abstract Wiener spaces, Sankpyá, 40 (1978) Series A, Pt.2, 116-137.

[15] K. Magnusson, A.J. Pritchard and M.D. Quinn, The application of fixed point theorems to global nonlinear controllability problems. Banach Center Publications, 14 (1985), 319-344.

[16] R. Manthey, Existence and uniqueness of a solution of a reaction-diffusion equation with polynomial nonlinearity and white noise disturbances, Math. Nachr., 125 (1986), 121-133.

[17] B. Maslowsky, Strong Feller property for semilinear stochastic evolution equations and applications, This Proceedings.

[18] M. Schilder, Some asymptotic formulas for Wiener integrals. TAMS, 125 (1966), pp. 63-85.

[19] W. Smolenski, R. Sztencel and J. Zabczyk, Large deviations estimates for semi-linear equations, Proc. 5th IFIP Conference on Stochast. Diff. Syst. Eisenach, 1986.

[20] W. Smolenski and R. Sztencel, Large deviations for non-linear radonifications of white noise Proc. Conf. Stoch. PDEs, Trento 1988 (to appear).

[21] S.R.S. Varadhan, Asymptotic probabilities and differential equations, Comm. Pure. Appl. Math., 19 (1966), pp. 261-286.

[22] J.B. Walsh, An introduction to stochastic partial differential equations, Ecole d'Eté de Probabilités de Saint Flour, Lecture Notes in Mathematics, 1180 (1984), 263-439.

[23] J. Zabczyk, Structural properties and limit behaviour of linear stochastic systems in Hilbert spaces, Banach Center Publications, Vol. 14 (1985), 591-609.

[24] J. Zabczyk, Exit problem and control theory, systems and control letters, 6 (1985), 165-172.

[25] J. Zabczyk, Stability under small perturbations, Proceedings of the 3rd Bad Honet Conference on Stochastic Systems, Bonn 1985, LNCIN 78 (1986), 362-367.

[26] J. Zabczyk, Exit problem for infinite dimensional systems, LNiM 1236, 1987, 239-257.

[27] J. Zabczyk, Symmetric solutions of semilinear stochastic equations, Proc. Conf. Stoch. PDEs and Appl. Trento 1988 (to appear).

[28] J. Zabczyk, Minimum energy problems and small noise evolutions, in preparation.

3. STOCHASTIC CONTROL AND ESTIMATION

APPROXIMATION OF ZAKAI EQUATION

BY THE SPLITTING UP METHOD

A. Bensoussan[(1)] - *R. Glowinski*[(2)]

Abstract

The objective of this article is to apply an operator splitting method to the time integration of Zakaï equation. Using this approach one can decompose the numerical integration into a stochastic step and a deterministic one, both of them much simpler to handle than the original problem. A strong convergence theorem is given, in the spirit of existing results for deterministic problems.

Keywords : Nonlinear filtering - Fractional step methods - Zakaï equation.

Introduction

We consider in this article the Zakaï equation of non linear filtering written as follows

(1) $$dy + A^*(t)y\,dt = B(t)y \cdot dw$$

where

$$A(t)\varphi = -\sum_{i,j} a_{ij} \frac{\partial^2 \varphi}{\partial x_i \partial x_j} - \sum_i g_i \frac{\partial \varphi}{\partial x_i}$$

is the diffusion operator and A^* its formal adjoint ;

$$B(t)\varphi = \varphi\, h$$

where $h(x,t)$ is a given function which belongs to $L^\infty(0,T;R^m)$, and corresponds to the observation process. In (1) w is a standard m dimensional Wiener process.

(1) *University of Paris Dauphine and INRIA*

(2) *University of Houston and INRIA*

We apply the idea of splitting up, considering $A(t)y\,dt - B(t)y \cdot dw$ as the sum of two operators.

Hence one writes a sequence of problems of the form

$$d\varphi + A(t)\varphi\,dt = 0$$
$$d\psi = B(t)\psi \cdot dw$$

which are considerably simpler than (1). Indeed the φ equation is deterministic and the ψ equation has a closed form solution.

The technique of splitting up for deterministic partial differential equations, has been used extensively by many authors. We refer here to the work of R. TEMAM [8], which is used as the background for our developments and also to R. GLOWINSKI [3] and G.I. MARCHUK [6] for applications in Mathematical Physics.

We refer to F. LEGLAND [4] and R.J. ELLIOTT, R. GLOWINSKI [2] for semi-discretization schemes of the Zakaï equation, which, although not related to the splitting up method, bear some analogies with those developed in this article.

We limit the presentation to a brief sketch up. The full developments will be published elsewhere. At the conference, similarities with the work of A. Rascanu have been noticed. The final paper will be of the three authors.

1. Setting of the problem

1.1. Notation - Assumptions.

Consider functions

(1.1) $g(x,t), \sigma(x,t)$ from $R^n \times (0,\infty)$ to R^n and $\mathcal{L}(R^n;R^n)$ respectively, satisfying : g, σ are measurable bounded, and Lipschitz with respect to x, uniformly in t.

(1.2) $$h(x,t) \in L^\infty(R^n \times (0,\infty); R^m)$$

Let $\Omega, \mathcal{A}, P$ be a probability space on which exists an m dimensional standard Wiener process $w(t)$, and let

$$F^t = \sigma(w(s), s \leq t).$$

Define the 2nd order differential operator

(1.3) $$A(t)\varphi = -\sum_{i,j} a_{ij}\frac{\partial^2\varphi}{\partial x_i \partial x_j} - \sum_i g_i \frac{\partial \varphi}{\partial x_i}$$

where we have set

(1.4) $$a = \frac{1}{2}\sigma\sigma^* \quad (a = \text{ matrix } a_{ij}).$$

We assume that

$$a_{ij}(x,t)\xi_i\xi_j \geq \alpha|\xi|^2, \quad \forall \xi \in R^m, \quad \alpha > 0. \tag{1.5}$$

We shall also define the operator

$$B(t)\varphi \equiv \varphi(x)h(x,t). \tag{1.6}$$

Formally Zakaï equation is written as

$$\begin{aligned} & dy + A^*(t)y\,dt = B(t)y \cdot dw \\ & y(0) = y_0. \end{aligned} \tag{1.7}$$

To write (1.7) in a convenient functional framework, we use the formalism of E. PARDOUX [7] (see also A. BENSOUSSAN [1]).

1.2. Functional set up

Following the variational formulation of P.D.E. due to J.L. LIONS [5], we introduce the Hilbert spaces

$$H = L^2(R^n), \quad V = H^1(R^n)$$

and identify H with its dual. We denote by V' the dual of V.

We denote by

$$(\varphi, \psi) = \int_{R^n} \varphi\psi\, dx$$

the scalar product in H, and by

$$((\varphi, \psi)) = \int_{R^n} (\varphi\psi + D\varphi \cdot D\psi) dx$$

the scalar product in V. The duality between V and V' is referred as $<\,,\,>$.

We now write $A(t)$ in divergence form as

$$A(t) = -\frac{\partial}{\partial x_i}(a_{ij}\frac{\partial}{\partial x_j}) + a_i\frac{\partial}{\partial x_i}$$

where we have set

$$a_i = \frac{\partial a_{ij}}{\partial x_j} - g_i$$

and

$$A^*(t) = -\frac{\partial}{\partial x_i}(a_{ij}\frac{\partial}{\partial x_j}) - \frac{\partial}{\partial x_i}(a_i).$$

The operator $A(t)$ belongs to $L^\infty(0,T;\mathcal{L}(V;V'))$ and satisfies the coercivity condition

$$< A(t)\varphi, \varphi > + \lambda|\varphi|^2 \geq \beta||\varphi||^2 \qquad \forall \varphi \in V, \quad \beta > 0, \quad \lambda \geq 0. \tag{1.8}$$

Now the operator $B(t) \in L^\infty(0,T,\mathcal{L}(H;H^m))$, and we may write more clearly

$$B(t)y \cdot dw = \sum_{j=1}^{m} B^j(t)y\, dw_j.$$

where $B^j(t) \in L^\infty(0,T;\mathcal{L}(H;H))$ corresponds to

$$B^j(t)\varphi = \varphi h_j(x,t).$$

We use the notation $L^2_F(0,T;V)$ to denote the Hilbert space of processes $z(t)$ with values in V such that $E\int_0^T \|z(t)\|^2 dt < \infty$, and such that a.e.t., $z(t)$ is F^t measurable. Naturally, we can replace V by H or any Hilbert space.

We can state the classical result of existence and uniqueness for (1.7) (cf. E. PARDOUX [7], cf. also A. BENSOUSSAN [1]),

Theorem 1.1. *Assume (1.1), (1.2), (1.5) then for $y_0 \in H$, there exists a unique solution of (1.7) in the functional space*

$$y(\cdot) \in L^2_F(0,T;V) \cap L^2(\Omega,\mathcal{A},P;C(O,T;H))$$

□

The equation (1.7) can be interpreted as an Ito differential in V', since

$$y(t) = y_0 - \int_0^t A^*(s)y(s)ds + \sum_j \int_0^t B^j(s)y(s)dw_j.$$

In addition the following Ito's calculus rule holds (equivalent of an energy equality)

$$d|y(t)|^2 + 2 < A(t)y(t), y(t) > dt = 2\sum_j (y, B^j y)dw_j + \sum_j |B^j y|^2 dt \tag{1.9}$$

Note that the integrand in the stochastic integral at the right hand side of (1.9) $(y, B^j y)$ is a.s. in $L^\infty(0,T)$ but does not belong to $L^2_F(0,T)$. This is the source of technical (although not fundamental) difficulties. To avoid them, one can rely on the following additional result.

Proposition 1.1. *The process $y(\cdot)$ satisfies*

$$y(\cdot) \in L^\infty(0,T;L^4(\Omega,\mathcal{A},P;H)).$$

In the sequel we shall replace (1.7) by

$$\begin{aligned} &dy + (A^*(t)y + \mu y)dt = B(t)y \cdot dw \\ &y(0) = y_0 \end{aligned} \tag{1.10}$$

where μ is a convenient positive constant. Since, one derives (1.10) from (1.7) by the transformation

$$y \to ye^{-\mu t}$$

it suffices to consider (1.10).

2. The splitting up approximation scheme

2.1. The algorithm

Let N be an integer, which will tend to $+\infty$, and set

$$k = {}^{T}\!/_{N+1}$$

We shall define two processes y_{1k}, y_{2k} depending on k. We split $[0,T]$ in steps $0, k, \ldots, (N+1)k$. Consider an interval $[rk, (r+1)k[\quad, r = 0 \ldots N$, then y_{1k}, y_{2k} are defined on this interval by the relations

$$\begin{aligned} &dy_{1k} + (A^*(t)y_{1k} + \frac{\mu}{2}y_{1k})dt = 0 \\ &dy_{2k} + \frac{\mu}{2}y_{2k}dt = B(t)y_{2k} \cdot dw \\ &y_{1k}(rk) = y_k^r \\ &y_{2k}(rk) = y_k^{r+\frac{1}{2}} \end{aligned} \tag{2.1}$$

and the sequences $y_k^r, y_k^{r+\frac{1}{2}}$ are defined as follows

$$\begin{aligned} y_k^{r+\frac{1}{2}} &= y_{1k}((r+1)k - 0) \\ y_k^{r+1} &= y_{2k}((r+1)k - 0). \end{aligned} \tag{2.2}$$

Clearly (2.1), (2.2) define completely y_{1k}, y_{2k} in $[rk, (r+1)k[$ once y_k^r is given. As a starting point we set

$$y_k^o = y_0 \tag{2.3}$$

and (2.1), (2.2) define completely y_{1k}, y_{2k} in $[0, T[$. In (2.2) μ is a parameter which will be fixed later on. The processes y_{1k}, y_{2k} are right continuous and their discontinuity points are $k, \ldots Nk$ (on $[0, T[$). Since the equation for y_{1k} is deterministic we have

$$\begin{aligned} &y_k^r, y_k^{r+\frac{1}{2}} \text{ are } F^{kr} \text{ measurable (with values in } H) \\ &y_{1k}(t) \text{ is } F^{kr} \text{ measurable } \forall t \in [kr, (k+1)r[\\ &y_{2k}(t) \text{ is } F^t \text{ measurable } \forall t \end{aligned} \tag{2.4}$$

We can state the following existence result for (2.1)

Proposition 2.1. *The system (2.1), (2.2) define in a unique way y_{1k}, y_{2k} in $L^2_F(0, T, V)$, $L^2_F(0, T; H)$ respectively*

2.2 A priori estimates

We begin by establishing a priori estimates

Proposition 2.2. *The processes* y_{1k}, y_{2k} *satisfy*

$$E\int_0^T \|y_{1k}\|^2 dt \leq C \tag{2.5}$$

$$E|y_{1k}(t)|^4, E|y_{2k}(t)|^4 \leq C, \quad \forall t \in [0,T[, \tag{2.6}$$

where C *does not depend on* T *(for a convenient choice of* k*).*

3. Convergence

3.1 Statement of the main result

Our main result is the following

Theorem 3.1. *Assume* (1.1), (1.2), (1.5). *Then one has :*

$$y_{1k}, y_{2k} \to y \text{ in } L^2_F(0,T;V) \text{ and } L^2_F(0,T;H) \text{ respectively} \tag{3.1}$$

$$\begin{aligned} &y_{1k}(t), y_{2k}(t) \to y(t) \text{ in } L^2(\Omega,\mathcal{A},P;H) \quad \forall t \in [0,T[. \\ &y_{1k}(T-0), y_{2k}(T-0) \to y(T) \text{ in } L^2(\Omega,\mathcal{A},P;H) \end{aligned} \tag{3.2}$$

3.2. Weak convergence.

We can extract subsequences, still denoted y_{1k}, y_{2k} such that :

$$y_{1k} \to y_1 \text{ in } L^2_F(0,T;V) \text{ weakly}$$

$$y_{2k} \to y_2 \text{ in } L^2_F(0,T;H) \text{ weakly}$$

and

$$y_{1k}, y_{2k} \to y_1, y_2 \text{ in } L^\infty(0,T;L^4(\Omega,\mathcal{A},P;H)) \text{ weak star.}$$

We first have :

Lemma 3.1. *The functions* y_1 *and* y_2 *are equal to a common function* η.

Naturally $\eta \in L^2_F(0,T;V) \cap L^\infty(0,T;L^4(\Omega,\mathcal{A},P;H))$.

Our objective now is to check that η satisfies the equation (1.7) and thus $\eta = y$, by the uniqueness.

Lemma 3.2. $\eta = y$

From the uniqueness of the limit, we can assert that :

$$y_{1k} \to y \text{ in } L^2_F(0,T;V)) \text{ weakly}$$
$$y_{2k} \to y \text{ in } L^2_F(0,T;H) \text{ weakly}$$

and in $L^\infty(0,T;L^4(\Omega,\mathcal{A},P;H))$ weak star.

3.3. Strong convergence.
Consider the expression

$$\begin{aligned} X_k(t) =& E|y(t)-y_{1k}(t)|^2 + 2E\int_0^t < A(y-y_{1k}), y-y_{1k} > ds \\ &+ E\int_0^t \mu|y-y_{1k}|^2 ds + E\int_0^{[\frac{t}{k}]k} (\mu|y-y_{2k}|^2 - \sum_j |B^j(y-y_{2k})|^2) ds \\ =& X_k^1(t) + X_k^2(t) + X_k^3(t) \end{aligned}$$

with

$$\begin{aligned} X_k^1(t) =& E|y(t)|^2 + 2E\int_0^t < Ay, y > \quad ds \\ &+ E\int_0^t \mu|y|^2 ds + E\int_0^{[\frac{t}{k}]k} (\mu|y|^2 - \sum_j |B^j y|^2) ds \\ &\to E|y(t)|^2 + 2E\int_0^t < Ay, y > ds + 2\mu E\int_0^t |y|^2 ds - E\int_0^t \sum_j |B^j y|^2 ds \\ =& |y_0|^2 \end{aligned}$$

$$\begin{aligned} X_k^2(t) =& -2 \quad E(y(t), y_{1k}(t)) - 2 \quad E\int_0^t (< Ay, y_{1k} > + \\ & + < Ay_{1k}, y >) ds - 2\mu E\int_0^t (y, y_{1k}) ds \\ & - 2\mu E\int_0^{[\frac{t}{k}]k} (y, y_{2k}) ds + 2E\int_0^{[\frac{t}{k}]k} \sum_j (B^j y, B^j y_{2k}) ds \\ & \to -2E|y(t)|^2 - 4E\int_0^t < Ay, y > ds \\ & - 4\mu E\int_0^t |y|^2 ds + 2E\int_0^t \sum_j |B^j y|^2 ds \\ =& -2|y|^2 \end{aligned}$$

Moreover

$$X_k^3(t) = |y_0|^2$$

Therefore $X_k(t) \to 0, \forall t$.

This implies

(3.3)
$$\begin{aligned}
&y_{1k} \to y \text{ in } L^2_F(0,T;V) \text{ strongly}\\
&y_{2k} \to y \text{ in } L^2_F(0,T,H) \text{ strongly}\\
&y_{1k}(t) \to y(t) \text{ in } L^2(\Omega,\mathcal{A},P;H) \quad \forall t \in [0,T[,\text{ strongly}\\
&y_{1k}(T-0) \to y(T-0) \text{ in } L^2(\Omega,\mathcal{A},P;H),\text{ strongly}
\end{aligned}$$

Constructing an expression similar to $X_k(t)$, we can prove the remaining of the results.

4. Remarks.

4.1 Explicit solution.

The equation for y_{2k} can be explicitly solved, namely

$$y_{2k}(x,t) = y_k^{r+\frac{1}{2}}(x)e^{-\frac{\mu}{2}k}\exp\int_{rk}^{t}\sum_j h_j(x,s)dw_j(s) - \frac{1}{2}\int_{rk}^{t}|h(x,s)|^2 ds$$

whereas y_{1k} is solution of the classical Fokker Plank equation. Therefore we can write for y_{1k} the following equation :

$$\begin{aligned}
&\frac{\partial y_{1k}}{\partial t} + (A*(t)y_{1k} + \frac{\mu}{2}y_{1k}) = \sum_{r=1\ldots N}\delta_{rk}(t)y_{1k}(x,t-0)\\
&\{\exp[-\frac{\mu k}{2} + \int_{t-k}^{t}\sum_j h_j dw_j - \frac{1}{2}\int_{t-k}^{t}|h|^2 ds] - 1\}\\
&y_{1k}(0) = y_0.
\end{aligned}$$

which can be considered as the approximation of the original Zakaï equation.

One can understand the right hand side as follows. For k small and assuming h continuous in time, the term within brackets is equivalent to $-\frac{\mu k}{2} + \sum_j h_j(x,t)(w_j(t) - w_j(t-k))$, up to 2nd order terms. Comparing with the original Zakaï equation, it means that we have replaced the term :

$$y(x,t)(-\frac{\mu}{2}dt + \sum_j h_j(x,t)dw_j)$$

with the sum of impulses.

$$\sum_{r=1\ldots N}\delta_{rk}(t)y(x,t-0)(-\frac{\mu k}{2} + \sum_j h_j(x,t)(w_j(t) - w_j(t-k))$$

4.2. Fully numerical scheme.

It remains of course to discretize completely the equation (4.1) both in time and space. This can be done using classical tools of numerical analysis. Numerical results will be reported elsewhere.

4.3. Extension.

Our variational techniques directly inspired from the deterministic case bear two serious limitations. Firstly g, h must be bounded, which leaves out of the framework of the linear case. More importantly, the case when there is correlation between the system noise and the observation noise, which leads to an operator B involving the gradient of y, seems to be out of the scope of our theory.

The first limitation is purely technical, and can be overcome using Sobolev spaces with weights.

References

[1] A. BENSOUSSAN, On a General Class of Stochastic Partial Differential Equations, *Journal of Hydrology and Hydraulics*, ed. T.E. Unny, vol. 1, n^0 4, 1987, pp. 297-303.

[2] R.J. ELLIOTT, R. GLOWINSKI, Approximations to solutions of the Zakaï filtering equation, Technical Report, Dpt. of Statistics and Applied Probabilities, University of Alberta, Edmonton, 1988 (to appear in *Stochastics*).

[3] R. GLOWINSKI, *Numerical Methods for Nonlinear Variational Problems*, Springer, New-York, 1984.

[4] F. LEGLAND, Estimation de paramètres dans les processus stochastiques en observation incomplète, Thèse, Université Paris Dauphine, 1981.

[5] J.L. LIONS, *Contrôle Optimal des Systèmes gouvernés par des équations aux dérivées partielles*, Dunod, Paris, 1968.

[6] G.I. MARCHUK, *Methods of Numerical Mathematics*, Springer, New-York, 1975.

[7] E. PARDOUX, Stochastic partial differential equations and filtering of diffusion processes, *Stochastics*, 3, 1979, pp. 127-168.

[8] R. TEMAM, Sur la stabilité et la convergence de la méthode des pas fractionnaires, Thèse, Paris, 1967.

An ergodic control problem on whole Euclidean space

A. Bensoussan[(1)] - H. Nagai[(2)]

0. Introduction.

In this paper we study an ergodic control problem on whole Euclidean space arising from the eigenvalue problem of an elliptic operator. Ergodic control problems for quasi-linear equations on bounded regions with periodic or Neuman boundary conditions have been studied by several authors (cf. [2],[5] and [6]). But their methods cannot apply for the equation on unbounded regions. We specialize the Bellman equation to the case where the Hamiltonian is $|p|^2/2 + V(x)$, but treat it on whole Euclidean space. Noticing the relationship between the equation and the eigenvalue problem of the Schrödinger operator $-\Delta/2 + V(x)$, we study asymptotic behaviour of the solution of the equation.

1. Setting of the problem.

Let $V(x)$ be a function on N- dimensional Euclidean space R^N such that

(1.1) $V(x) \geq 0$, smooth, $V(x) \to \infty$ as $|x| \to \infty$.

We then consider the following eigenvalue problem of the Schrödinger operator $-\Delta/2 + V(x)$ in $L^2(R^N)$:

(1.2) $$-\frac{1}{2}\Delta\Phi + V\Phi = \lambda\Phi.$$

We first note that the operator $-\Delta/2 + V$ on $C_0^\infty(R^N)$ is essentially self-adjoint in $L^2(R^N)$ (cf. [7]). Let H be the unique self-adjoint extention in $L^2(R^N)$ of the operator. In other words H is the self-adjoint operator corresponding to the bilinear form $\mathcal{E}^V$ defined by

$$\mathcal{E}^V(u,v) = \frac{1}{2}\int \nabla u \cdot \nabla v dx + \int Vuvdx,$$

(1) University of Paris Dauphine and INRIA, France

(2) College of General Education, Nagaya University, Japan

where the domain $\mathcal{D}[\mathcal{E}^V]$ of the form is described as

$$\mathcal{D}[\mathcal{E}^V] = H_V^1 \equiv \{ u \in L^2(R^N) \mid \int|\nabla u|^2dx + \int Vu^2dx < +\infty \}.$$

On the operator H the following facts are known (cf. [7],[9]). The resolvent operator $G_\gamma = (\gamma + H)^{-1}$ is compact for each $\gamma > 0$ and as a result H has a purely discrete spectrum:

$$0 \leq \lambda_1 < \lambda_2 \leq \lambda_3 \leq \cdots\cdots \to \infty.$$

Moreover the principal eigenvalue λ_1 is simple. Let Φ be the eigenfunction corresponding to λ_1 normalized as $\int\Phi(x)^2dx = 1$, then Φ satisfies the estimates

$$(1.3) \qquad 0 < \Phi(x) \leq A\exp(-B|x|)$$

for some positive constants A and B.

Thus we can suppose that we are given a function $\Phi \in \mathcal{D}[\mathcal{E}^V]$ satisfying (1.2) with $\lambda = \lambda_1$ and (1.3). Now let us set

$$(1.4) \qquad w(x) = -\log\Phi(x) + \int\Phi(x)^2\log\Phi(x)dx,$$

then we have

$$(1.5) \qquad -\frac{1}{2}\Delta w + \frac{1}{2}|\nabla w|^2 + \lambda_1 = V(x),$$

$$\int w\Phi^2dx = 0 \quad \text{and} \quad \int|\nabla w|^2\Phi^2dx < +\infty.$$

The equation (1.5) looks like the Bellman equation of ergodic control. Indeed (1.5) is rewritten as

$$(1.6) \qquad -\frac{1}{2}\Delta w + \lambda_1 = \inf_{z\in R^N}\{z\cdot\nabla w + \frac{1}{2}|z|^2 + V(x)\}.$$

Therefore as the corresponding Bellman equation of discounted type we have

$$(1.7) \qquad -\frac{1}{2}\Delta v_\alpha + \alpha v_\alpha = \inf_{z\in R^N}\{z\cdot\nabla v_\alpha + \frac{1}{2}|z|^2 + V(x)\},$$

which is nothing but

$$(1.8) \qquad -\frac{1}{2}\Delta v_\alpha + \frac{1}{2}|\nabla v_\alpha|^2 + \alpha v_\alpha = V.$$

We shall study first the equation (1.8) and then asymptotic behaviour as $\alpha \to 0$ of the positive solution v_α of (1.8).

Remark. In the equation (1.6) the infimum is attained by $z = -\nabla w$, therefore the diffusion process on R^N with the generator :

$$L \equiv -\frac{1}{2}\Delta + \nabla w \cdot \nabla = -\frac{1}{2}\Delta - \frac{\nabla w}{\Phi} \cdot \nabla$$

is considered "optimal" and it is a symmetric ergodic one with the invariant measure $\Phi^2(x)dx$.

2. Transformation.

We study (1.8) through the transformation

(2.1) $\quad v_\alpha = -\log u_\alpha, \qquad 0 < u_\alpha \leq 1$

and thus obtain

(2.2) $\quad -\frac{1}{2}\Delta u_\alpha + Vu_\alpha = -\alpha u_\alpha \log u_\alpha, \qquad 0 < u_\alpha \leq 1.$

If we have a positive solution v_α of (1.7), then the infimum in (1.7) is attained by $z = -\nabla v_\alpha$. Therefore the diffusion process with the generator :

$$L_\alpha \equiv -\frac{1}{2}\Delta + \nabla v_\alpha \cdot \nabla = -\frac{1}{2}\Delta - \frac{\nabla u_\alpha}{u_\alpha} \cdot \nabla$$

is considered "optimal" and it is a symmetric ergodic one with the invariant measure $u_\alpha(x)^2/\|u_\alpha\|^2\, dx$ if $u_\alpha \in L^2(R^N)$. We thus concerned with L^2-solution of (2.2).

3. Sub- and Supersolutions of (2.2).

Let us consider the equation (2.2). We take a constant c such that $\sup_x c\Phi(x) = 1$ and set

(3.1) $\quad \Phi_\alpha(x) = c\exp(-\lambda_1/\alpha)\Phi(x),$

then $\Phi_\alpha(x)$ is a subsolution:

(3.2) $\quad -\frac{1}{2}\Delta\Phi_\alpha + V\Phi_\alpha \leq -\alpha\Phi_\alpha \log\Phi_\alpha$

since it follows that $\lambda_1\Phi_\alpha \leq -\alpha\Phi_\alpha\log\Phi_\alpha$ from $\Phi_\alpha \leq \exp(-\lambda_1/\alpha)$.

To find a supersolution of (2.2) we consider the following equation:

$$(3.3) \qquad -\frac{1}{2}\Delta X_\alpha + (V-\lambda_1)X_\alpha + \alpha X_\alpha = \alpha\exp(-\lambda_1/\alpha).$$

If we have a positive solution X_α of (3.3), then it satisfies

$$\begin{aligned} -\frac{1}{2}\Delta X_\alpha + VX_\alpha &= \alpha\exp(-\lambda_1/\alpha) + (\lambda_1-\alpha)X_\alpha \\ &\geq \inf_{\xi\in R^1}\{\alpha\exp(-1-\xi/\alpha) + \xi X_\alpha\} \\ &= -\alpha X_\alpha\log X_\alpha. \end{aligned}$$

Thus X_α turns out to be a supersolution of (2.2). To solve the equation (3.3) let us set

$$L^2_\mu = \{\, z \mid \beta_\mu z \in L^2(R^N) \,\}, \qquad \beta_\mu(x) = \exp(-\mu(1+|x|^2)^{1/2})$$

$$L^2_{V,\mu} = \{\, z \in L^2_\mu \mid \int V\beta_\mu{}^2 z^2 dx < +\infty \,\},$$

$$H^1_{V,\mu} = \{\, z \in L^2_{V,\mu} \mid \int|\nabla(\beta_\mu z)|^2 dx < +\infty \,\},$$

and define the norm

$$\|z\|^2_{H^1_{V,\mu}} = \int z^2\beta_\mu^2 V dx + \int z^2\beta_\mu^2 dx + \int|\nabla(z\beta_\mu)|^2 dx.$$

Let us consider the bilinear form on $H^1_{V,\mu}$

$$\begin{aligned} (3.4) \qquad a(z_1,z_2) = \frac{1}{2}\int\nabla z_1\cdot(\nabla z_2\beta_\mu^2 - 2\mu z_2\beta_\mu^2 x/(1+|x|^2)^{1/2})dx \\ + \int(V-\lambda_1+\alpha)z_1 z_2\beta_\mu^2 dx. \end{aligned}$$

Then we have

<u>Lemma 1</u>. For $0 < \mu^2/2 < \alpha$ the bilinear form a is continuous and coercive on $H^1_{V,\mu}$.

For the proof see [4]. Because of Lemma 1 we have a unique solution X_α of the equation (3.3).

In the follwings we need an L^∞ estimate on X_α. To have it we take up a solution $u(x,t)$ of

$$\frac{\partial u}{\partial t} - \frac{1}{2}\Delta u + (V-\lambda_1)u = 0, \quad u(x,0) = 1,$$

then we have the formula

$$X_\alpha(x) = \alpha\exp(-\lambda_1/\alpha)\int_0^\infty e^{-\alpha t}u(x,t)dt. \tag{3.5}$$

And we have the following estimate due to B. Simon [8]:

$$u(x,t) \leq \begin{cases} \exp(\lambda_1 t), & t \leq 1, \\ c_N t^{N/2}, & t \geq 1, \end{cases} \tag{3.6}$$

where c_N is a positive constant. Thus by (3.5) we obtain the estimate

$$X_\alpha(x) \leq \exp(\lambda_1+\alpha-\lambda_1/\alpha) + c_N\exp(-\lambda_1/\alpha)\alpha^{-N/\alpha}\Gamma(N/2+1). \tag{3.7}$$

4. Existence and Uniqueness.

Let us consider the following equation:

$$-\frac{1}{2}\Delta\zeta + V\zeta + \gamma\zeta = \gamma z - \alpha z\log z, \quad \zeta \in H^1_{V,\mu}. \tag{4.1}$$

We can easily check that the following bilinear form :

$$b(z_1,z_2) = \frac{1}{2}\int\nabla z_1\cdot(\nabla z_2\beta_\mu^2 - 2\mu z_2\beta_\mu^2 x/(1+|x|^2)^{1/2})dx \tag{4.2}$$

$$+ \int(V+\gamma)z_1z_2\beta_\mu^2\, dx.$$

is continuous and coercive on $H^1_{V,\mu}$ for $\gamma > \mu^2/2$. Therefore (4.1) has a unique solution for a given function $z \in L^2_{V,\mu}$ because $z \in L^2_{V,\mu}$, with $z > 0$ implies $\gamma z - \alpha z\log z \in L^2_\mu$.

For $z \in L^2_{V,\mu}$ let $\zeta = T_{\gamma,\alpha}z$ be the solution of (4.1). We note that the map $y \rightarrow \gamma y - \alpha y\log y$ is monotone on $0 < y \leq X_\alpha(x)$ for sufficintly large γ because of the estimate (3.7). Therefore the operator $T_{\gamma,\alpha}$ is monotone on $K^0_\alpha = \{ z \in L^2_{V,\mu} \mid 0 < z \leq X_\alpha \}$. We set

$$K_\alpha = \{ z \in L^2_{V,\mu} \mid \Phi_\alpha \leq z \leq X_\alpha \wedge 1 \},$$

then we have

<u>Lemma 2</u>. The operator $T_{\gamma,\alpha}$ maps K_α into itself for sufficiently large γ (cf. [4]).

To prove existence of the solution of (2.2) we can now employ Tartar's methods which were useful for studying quasi-variational inequalities (cf. [3]) and we obtain

<u>Theorem 1</u>. Assume (1.1), then the set of solutions of (2.2) in K_α is not empty and has a minimum and a maximum solutions.

Moreover we have

<u>Theorem 2</u>. Assume (1.1) and

(4.3) $\exp(-\delta\sqrt{V}) \in L^2(R^N)$ for some $\delta > 0$,

then the positive solution in $H^1_{V,\mu}$ is unique and belongs to H^1_V for $0 < \alpha < 1/\delta$.

(cf. [4] for detailed proofs).

5. <u>Asymptotic behaviour</u>.

We first have the following theorem on asymptotic behaviour of the solution u_α of (2.2).

<u>Theorem 3</u>. Assume (1.1), then for any solution u_α of (2.2) in $K_\alpha \cap H^1_{V,\mu}$ one has

(5.1) $$\lim_{\alpha \to 0} (-\alpha \log u_\alpha(x)) = \lambda_1.$$

The proof of Theorem 3 is easy. Indeed we have

(5.2) $$-\alpha \log \Phi_\alpha(x) \geq -\alpha \log u_\alpha(x) \geq -\alpha \log X_\alpha(x).$$

Therefore (5.1) follows from (3.1),(3.7) and (5.2).

To state our theorem on asymptotic behaviour of the solution (1.8) let us introduce a function space:

$$H^1_\Phi = \{ v \mid \int v^2\Phi^2\, dx + \int |\nabla v|^2\Phi^2\, dx < +\infty \}.$$

H^1_Φ is a Hilbert space with inner product

$$(f,g) = \frac{1}{2}\int \nabla f\cdot\nabla g\Phi^2 dx + \int fg\Phi^2 dx$$

Let u_α be the solution of (2.2) in H^1_V, then it is locally smooth by regularity properties of elliptic equations. Therefore $v_\alpha = -\log u_\alpha$ satisfies the equation (1.8). Moreover, if we assume (4.3), then $u_\alpha \in H^1_V$ for $0 < \alpha < 1/\delta$ and we have $v_\alpha \in H^1_\Phi$ because

$$-\Phi_\alpha \log u_\alpha \leq -u_\alpha \log u_\alpha \leq \alpha\exp(-1-\sqrt{V}/\alpha) + \sqrt{V}u_\alpha$$

and

$$|\nabla v_\alpha|\Phi = |\nabla u_\alpha|\Phi / u_\alpha \leq |\nabla u_\alpha|\Phi / \Phi_\alpha .$$

We then have

Theorem 4. Assume (1.1) and (4.3). Let u_α be the positive solution of (2.2) in H^1_V and $v_\alpha = -\log u_\alpha$, then $v_\alpha - \int v_\alpha\Phi^2 dx$ converges to $w = -\log\Phi + \int\Phi^2\log\Phi dx$ in H^1_Φ.

In the proof of Theorem 4 the following formula plays a key role:

$$\frac{1}{2}\int|\nabla f|^2\Phi^2 dx = \frac{1}{2}\int|\nabla(\Phi f)|^2 dx + \int V(\Phi f)^2 dx - \lambda_1\int(\Phi f)^2 dx, \quad f \in H^1_\Phi. \tag{5.3}$$

(5.3) is due to Albeverio - Hoeph-Krohn - Streit (cf. [1], [4]).

References

[1] S. Albeverio, R. Hoeph-Krohn & L. Streit, Energy forms, Hamiltonians and distorted Brownian paths, J. Math. Phys, 18 (1977) 907-917.

[2] A. Bensoussan & J. Frehse, On Bellman equations of ergodic type with quadratic growth Hamiltonian, Contributions to Modern Calculus of Variations, Pitman Res. Notes in Math. Seris, ed. L. Cesari, Longman, vol. 148, (1987) 13-26.

[3] A. Bensoussan & J.L. Lions, Impulsive control and quasi-variational inequalities, Gauthier-Villars, Paris (1984).

[4] A. Bensoussan & H. Nagai, An ergodic control problem arising from the principal eigenfunction of an elliptic operator, to appear.

[5] F. Gimbert, Problems de Neuman quasi-linéaires, J. Func. Anal., 62 (1985) 65-72.
[6] P.L. Lions, Quelques remarques sur les problèmes elliptiques quasi-linéares du 2e ordre, J. Anal. Math., 45 (1985) 234-254.
[7] M. Reed & B. Simon, Methods of modern mathematical physics I,II,III ,IV, Academic press, New York (1978).
[8] B. Simon, Brownian motion, L^p properties of Schrödinger operators and the localization of binding, J. Func. Anal., 35 (1980) 215-229.
[9] B. Simon, Schrodinger semi-groups, Bull. A.M.S., 7 (1982) 447-526.

On Limit Control Principle for Singularly Perturbed Markov Processes

T. Bielecki*, Ł. Stettner**

* Institute of Econometrics, Main College of Planning and Statistics, Al. Niepodległości 162, 00-554 Warsaw, Poland

** Institute of Mathematics Polish Academy of Sciences, Śniadeckich 8, 00-950 Warsaw, Poland

1. Construction of singularly perturbed Markov processes. Convergence of semigroups. Limit control principle.

Let E_i be a compact separable space endowed with a Borel σ-field $\mathcal{E}_i$, $i=1,2$, $E=E_1\times E_2$, $\mathcal{E}=\mathcal{E}_1\times\mathcal{E}_2$.

Assume

(A1) A is a generator of a strongly continuous semigroup of contractions on the space C(E) of continuous bounded functions on E, $A1=0$ and $Af(x,y)=A^y f(x,y)$ for $f\in D(A)$ the domain of A, where for each $y\in E_2$, A^y is a generator of a strongly continuous semigroup of contractions on $C(E_1)$,

(A2) B is a generator of a strongly continuous semigroup on C(E), $B1=0$ and for $f\in D(B)$, $Bf(x,y)=\bar{B}f(x,y)$, where $\bar{B}$ stands for a generator of a strongly continuous semigroup $S(t)$ of contractions on $C(E_2)$.

Moreover let $S(t)$ be uniformly ergodic on $C(E_2)$ i.e.

(E1) there exist a probability measure μ and constants $\beta,\gamma>0$ such that for $t>0$, $f\in C(E_2)$

$$\|S(t)f-\mu(f)\| \le \beta e^{-\gamma t}\|f\| \tag{1}$$

Assume

(A3) closure $(D(A)\cap D(B))=C(E)$

(A4) $\exists_{\varepsilon>0}\ \forall_{0\le\varepsilon\le\varepsilon_0}\ \exists_{\lambda>0}$ Range $(\lambda-A^\varepsilon) = C(E)$, where $A^\varepsilon=\varepsilon A+B$

Then, under (A2) there exists a right continuous Markov process $Y=(\Omega_2,y_t,F_t^2,F^2,P_y)$ on E_2 with semigroup $(S(t))$. Define A_ε=closure $\varepsilon^{-1}A^\varepsilon$, $\bar{f}(x) \overset{def}{=} Pf(x) \overset{def}{=} \int_{E_2} f(x,y)\mu(dy)$ for $f\in C(E)$.

By virtue of Proposition 2 of [2], for $\varepsilon<\varepsilon_0$, A_ε is a generator of a strongly continuous semigroup of contractions $T_\varepsilon(t)$ on C(E), to which there corresponds a right continuous Feller process $(XY)^\varepsilon$ =

$(\Omega=\Omega_1\times\Omega_2,(x_t,y_{t/\varepsilon}),F_t=F^1_t\times F^2_{t/\varepsilon}, F=F^1\times F^2, P^\varepsilon_{xy})$ on E. Let us notice that the x-coordinate (X^ε) of $(XY)^\varepsilon$ is then singularly perturbed with the process $\varepsilon^{-1}Y=(y_{t/\varepsilon})_{t\geq 0}$. Under the assumptions

(A5) $D_1(A)=\bigcap_{y\in E_2} D(A^y)$ is dense in $C(E_1)$

(A6) for $f\in D_1(A)$, $A^y f(x)\in C(E)$

(A7) closure Range $(\lambda-C)=C(E_1)$ for some $\lambda>0$, where C=closure PA,

$X^\varepsilon\Rightarrow X$, weakly in $D([0,\infty),E_1)$ as $\varepsilon\to 0$, where $X=(\Omega_1,x_t,F^1_t,F^1,P_x)$ is a Feller Markov process on E_1, with semigroup $T(t)$ and infinitesimal generator C (see Theorem 3 of [2]).

Moreover, recalling Corollary 2.6 of [6] we have

Proposition 1. Under (A1)-(A7), (E1) for $0<a<b$, $f\in C(E_1)$

$$\lim_{\varepsilon\to 0}\ \sup_{s\in[a,b]}\ \sup_{(x,y)\in E} |T_\varepsilon(s)f(x,y)-T(s)\bar{f}(x)|=0 \qquad (2)$$

Proof.Let $T^\varepsilon(s)=T_\varepsilon(s\varepsilon)$. Obviously A^ε is the infinitesimal generator of $(T^\varepsilon(s))$, and for $f\in C(E)$, $T^\varepsilon(s)f \longrightarrow S(s)f$ in $C(E)$, uniformly for s from bounded intervals, as $\varepsilon\longrightarrow 0$. For $\delta>0$, by virtue of (1) one can choose $t>0$ such that $\|S(t)(f-\bar{f})\|\leq\delta$. Then for $\varepsilon<st^{-1}$ we have

$$\sup_{(x,y)\in E} |T_\varepsilon(s)f(x,y) - T(s)\bar{f}(x)| \leq \|T_\varepsilon(s-\varepsilon t)T_\varepsilon(\varepsilon t)(f-\bar{f}) - T_\varepsilon(s-\varepsilon t)S(t)(f-\bar{f})\| + \|T_\varepsilon(s-\varepsilon t)S(t)(f-\bar{f})\| + \|T_\varepsilon(s)\bar{f} - T(s)\bar{f}\|$$
$$\leq\|T_\varepsilon(\varepsilon t)(f-\bar{f})-S(t)(f-\bar{f})\|+\|S(t)(f-\bar{f})\|+\|T_\varepsilon(s)\bar{f}-T(s)\bar{f}\|$$

Letting $\varepsilon\longrightarrow 0$ and taking into account Theorem 2.1 of [5] and the fact that δ could be chosen arbitrarily small we obtain (2).

Let us extend $(T(s))$ to the semigroup $\bar{T}(s)$ on $C(E)$ by the formula $\bar{T}(s)f(x,y)=T(s)\bar{f}(x)$ for $s>0$, $T(0)f(x,y)=f(x,y)$. Clearly $(\bar{T}(s))$ is a semigroup of contractions, discontinuous at $s=0$, and by virtue of (2) is a limit semigroup for $T_\varepsilon(s)$, $\varepsilon>0$, as $\varepsilon\longrightarrow 0$.

In the paper we study control problems for singularly perturbed processes $(XY)^\varepsilon$ with ε small, maybe unknown.

If for any $\delta>0$ there is a δ-optimal control of X, which applied to $(XY)^\varepsilon$ is still nearly optimal for a sufficiently small ε, then we say that limit control principle is satisfied.

Our purpose is to present control problems for which limit control principle is holds. We consider first optimal stopping and impulsive control since these problems involve the properties of the semigroups $T_\varepsilon(s)$, $T(s)$, $\bar{T}(s)$ only. Namely, in section 2 optimal stopping and impulsive control with discounted cost functional is studied. The same control problems with ergodic functionals are investigated in section 3. Finally in section 4, we consider a particular diffusion control model, independently of the assumptions and results of section 1, and then show limit control principle. It should be emphasized that the independence of y-coordinate of $(XY)^{\varepsilon}$ of x^{ε} is crucial for the constructions and proofs in the paper.

The results are sketched or announced without proofs. The detailed proofs will appear in [2] and [3].

The idea of limit control principle appeared first in [8]. Singular control problems with continuously acting control and discounted cost functional were studied in [1]. Some results concerning singular stopping and impulsive control with discounted cost functional can be found in [11].

2. Discounted Stopping and Impulsive Control.

Consider first optimal stopping problem. For a given $f,c\in C(E)$, $\alpha>0$, $\varepsilon<\varepsilon_0$ we minimize

$$I_{xy}^{\varepsilon,h}(\tau)=E_{xy}^{\varepsilon}\left(\int_0^{\tau+h} f(x_s,y_{s/\varepsilon})e^{-\alpha s}ds+e^{-\alpha(\tau+h)}c(x_{\tau+h},y_{(\tau+h)/\varepsilon})\right) \tag{3}$$

and characterize the value function $w_{\varepsilon}(x,y)=\inf J_{xy}^{\varepsilon,h}(\tau)$, where infimum is over all (F_t) Markov times. The constants $\alpha,h>0$ have the interpretations as a discount rate and a delay of stopping respectively. Our "limit" stopping problem consists in minimizing

$$I_x^h(\tau)=E_x\left(\int_0^{\tau+h} \bar{f}(x_s)e^{-\alpha s}ds+e^{-\alpha(\tau+h)}\bar{c}(x_{\tau+h})\right) \tag{4}$$

over all (F_t^1) Markov times, where according to the definition of the limit process X, expected value E_x is over measure P_x on Ω_1.

For a Borel set $B\in\mathcal{E}$ let D_B denote first entry time to the set B.

Define

$$v_{\varepsilon}(x,y)=\inf_{\tau} E_{xy}^{\varepsilon}\left(\int_0^{\tau} T_{\varepsilon}(h)f(x_s,y_{s/\varepsilon})e^{-\alpha s}ds+e^{-\alpha\tau}T_{\varepsilon}(h)c(x_{\tau},y_{\tau/\varepsilon})\right) \tag{5}$$

and

$$v(x)=\inf_{\tau} E_x\left(\int_0^{\tau} T(h)\bar{f}(x_s)e^{-\alpha s}ds+e^{-\alpha\tau}T(h)\bar{c}(x_{\tau})\right) \tag{6}$$

with infimum over (F_t), (F_t^1) Markov times respectively.

Theorem 1. We have

$$w_{\varepsilon}(x,y)\longrightarrow w(x)\overset{\text{def}}{=\!=}\inf_{\tau} I_x^h(\tau),\quad v_{\varepsilon}(x,y)\longrightarrow v(x),\ \text{as}\ \varepsilon\longrightarrow 0 \tag{7}$$

uniformly in $(x,y)\in E$. Moreover, if $\|v_{\varepsilon}-v\|\le 3^{-1}\delta$, $\|T(h)\bar{c}-T_{\varepsilon}(h)c\|\le 3^{-1}\delta$, then $D_{C_{\delta}\times E_2}$, where

$$C_{\delta}=\{x:\ v(x)\ge T(h)\bar{c}(x)-\delta\} \tag{8}$$

is 2δ optimal stopping time for $(XY)^{\varepsilon}$ with functional (3).

Proof. The convergence (7) follows from [11, theorem 3.1] To prove the remaining part of the theorem it is sufficient to show that $C^{\varepsilon}_{\delta/3} \subset C_{\delta} \times E_2$, where

$$C^{\varepsilon}_{\delta/3} = \{(x,y):\ v_{\varepsilon}(x,y) \ge T_{\varepsilon}(h)c(x,y) - \delta/3\} \tag{9}$$

For details see Theorem 8 of [2].

The Markov time $D_{C_{\delta}}$ is δ optimal for the optimal stopping of the limit process X with the cost functional (4). Thus according to Theorem 1, the limit control principle is satisfied. The delay of stopping h was important to apply Proposition 1 in the proof of Theorem 1. One can expect the limit control principle to be satisfied for nondelayed case (h=0). Nevertheless this problem is still open.

Consider now the case when $(XY)^{\varepsilon}$ can be controlled with the use of impulses, i.e. at a chosen Markov time τ one can shift the x-coordinate (X^{ε}) to a new state $\xi \in U$, where ξ is F_{τ} measurable U-valued state random variable and U is a given compact subset of E_1. Since, because of the same reasons as above we assume a delay h of stopping, the impulsive strategy $V=(\tau_i,\xi_i)$ consists of Markov times $\tau_{i+1} \ge \tau_i + h$, and F_{τ_i}-measurable state variables ξ_i. For a shift from x to ξ a strictly positive cost $c(x,\xi) > a > 0$ is incurred and our purpose is to minimize

$$J^{\varepsilon}_{xy}(V) \overset{\text{def}}{=\!=} E^{\varepsilon,V}_{xy}\Big(\int_0^{\infty} e^{-\alpha s} f(x_s,y_s)ds + \sum_{i=1}^{\infty} \exp(-\alpha(\tau_i+h))c(x_{\tau_i+h},\xi_i)\Big) \tag{10}$$

where $P^{\varepsilon,V}_{xy}$ denotes the measure corresponding to control V (for the construction see [9]).

The limit control problem consists in shifting of the process X, at (F^1_t) Markov times τ_i+h to $F^1_{\tau_i}$-measurable state variables ξ_i, with the purpose to minimize

$$J_x(V) \overset{\text{def}}{=\!=} E^{V}_{xy}\Big(\int_0^{\infty} e^{-\alpha s} \bar{f}(x_s)ds + \sum_{i=1}^{\infty} \exp(-\alpha(\tau_i+h))c(x_{\tau_i+h},\xi_i)\Big) \tag{11}$$

over all impulsive strategies $V=(\tau_i,\xi_i)$.

Theorem 2. For any $\delta>0$, there exist a set $I_{\delta} \in \mathcal{E}_1$ and a Borel measurable function $\xi: E_1 \to U$ such that the strategy $V=(\hat{\tau}_i,\hat{\xi}_i)$, where $\hat{\tau}_1 = D_{I_{\delta}}$, $\hat{\xi}_1 = \xi(x_{\hat{\tau}_1+h})$, $\hat{\tau}_{i+1} = \hat{\tau}_i + h + D_{I_{\delta}} \circ \theta_{\hat{\tau}_i+h}$, $\hat{\xi}_{i+1} = \xi(x_{\hat{\tau}_{i+1}+h})$ is δ-optimal for the controlled limit process X with the functional (11).

Moreover if ε is sufficiently small, then the strategy $(\hat{\tau}'_i,\hat{\xi}'_i)$, where $\hat{\tau}'_i = D_{I_{\delta}\times E_2}$, $\hat{\xi}'_1 = \xi(x_{\hat{\tau}'_1+h})$, $\hat{\tau}'_{i+1} = \hat{\tau}'_i + h + D_{I_{\delta}\times E_2} \circ \theta_{\hat{\tau}'_i+h}$, $\hat{\xi}'_{i+1} = \xi(x_{\hat{\tau}'_{i+1}+h})$ and (x_s) stands now for the x-coordinate of the controlled process $(XY)^{\varepsilon}$, is 2δ-optimal for the singularly perturbed controlled process $(XY)^{\varepsilon}$ with the functional (10).

In the other words the limit control principle is again satisfied.

For the proof as well as the form of the set I_{δ} and the function ξ we refer to [2, theorem 9].

3. Convergence of invariant measures. Ergodic stopping and impulsive control.

In this section in addition to the previous assumptions we shall require the uniform ergodicity of $T_{\varepsilon}(t)$ on $C(E)$ i.e.

(E2) there exist constants $\bar{\beta},\bar{\gamma}>0$ and measures π_{ε} such that for $f\in C(E)$

$$\|T_{\varepsilon}(t)f-\pi_{\varepsilon}(f)\| \le \bar{\beta}e^{-\bar{\gamma}t}\|f\| \tag{12}$$

Applying some ideas of [4] one can show (see [3] example following Lemma 2) that if the process Y has a bounded away from 0 density of the transition probability with respect to some Borel measure m_Y on $(E_2,\mathcal{E}_2)$, and x-coordinate of $(XY)^{\varepsilon}$ is a diffusion with reflection in a bounded domain with bounded drift $b(y_{s/\varepsilon})$ at time s then T_{ε} satisfies (E2).

Proposition 2. Under (E1), (E2), there exists a unique invariant measure $\pi\otimes\mu$ for the extended semigroup $(\bar{T}(s))$. Moreover $\pi_{\varepsilon}\Rightarrow\pi\otimes\mu$ weakly as $\varepsilon\to 0$.

Proof. Under (E2) we obtain the uniform ergodicity of T(s) on $C(E_1)$ (see Lemma 1 [3]). Then the assumptions of Proposition 3 of [3] are satisfied, from which the assertion of Proposition 2 follows.

We are in position now to study ergodic stopping. The problem consists in minimizing the following cost functional

$$I^{\varepsilon}_{xy}(\tau)=\liminf_{t\to\infty} E^{\varepsilon}_{xy}\left(\int_0^{\tau\wedge t} f(x_s,y_{s/\varepsilon})ds + c(x_{\tau\wedge t})\right) \tag{13}$$

over all (F_t) Markov times. The corresponding limit control problem is to minimize

$$I_x(\tau)=\liminf_{t\to\infty} E_x\left(\int_0^{\tau\wedge t} \bar{f}(x_s)ds + c(x_{\tau\wedge t})\right) \tag{14}$$

over all (F^1_t) Markov times.

Let $v_{\varepsilon}(x)$, $v(x)$ be the value functions corresponding to the functionals (13), (14) respectively.

Theorem 3. Suppose $\pi(f)>0$. Then $v_{\varepsilon}\to v$ uniformly as $\varepsilon\to 0$. Moreover for $\varepsilon<\varepsilon_0$ such that $\|v_{\varepsilon}-v\|\le\delta/2$, the first entry time $D_{C_{\delta}\times E_2}$ to the set

$$C_{\delta}=\{x\in E_1:\ v(x)\ge c(x)-\delta\}$$

is 2δ optimal stopping time for $(XY)^\varepsilon$ with the cost functional (13)

Proof. Under (E2) minimizing (13) and (14) it is sufficient to restrict ourselves to Markov times with expected values uniformly in ε, x, y bounded with respect to measures P^ε_{xy}, P_x respectively. Therefore the problems reduce to optimal stopping over finite time interval. The latter problems can be uniformly approximated by discrete finite valued optimal stopping for which the convergence of the value functions as $\varepsilon\to 0$ can be easily shown (for details see Theorem 1 of [3]). The 2δ-optimality of $D_{C_\delta\wedge E_2}$ can be proved in a similar way as in Theorem 2.

The impulsive control of $(XY)^\varepsilon$, we consider in this section consists of F_t-Markov times τ_i and E_j-valued random variables ξ_i^j. We assume that

$$\tau_{i+1}=\tau_i+\sigma_{i+1}\circ\Theta_{\tau_i} \tag{15}$$

where σ_i are F_t-Markov times and ξ^j_{i+1}, j=1,2, are adapted to the information we obtain between τ_i and τ_{i+1}. At Markov time τ_i the x,y coordinates of the controlled process $(XY)^\varepsilon$ are shifted to ξ_i^1,ξ_i^2 respectively. For such shift a cost $c(x^{i-1}_{\tau_i})+d(\xi_i^1,\xi_i^2)$, where $x^{i-1}_{\tau_i}$ stands for the position of x-coordinate before an immediate shift to ξ_i^1, is incurred. The cost functional we minimize is of the form

$$J^\varepsilon_{xy}(V)=\liminf_{t\to\infty} t^{-1}\ E^{\varepsilon,V}_{xy}\Big(\int_0^t f(x_s,y_{s/\varepsilon})ds+\sum_{i=1}^{\infty}\chi_{\tau_i\le t}[c(x^{i-1}_{\tau_i})+d(\xi_i^1,\xi_i^2)]\Big) \tag{16}$$

where $f,d\in C(E)$, $c\in C(E_1)$, $c(x)>a>0$ for $x\in E_1$.

The impulsive control problem is to shift the limit process X at F_t^1 Markov times τ_i (of the form (15) with σ_i-(F_t^1) Markov times) to E_1-valued random variables ξ_i^1, adapted to the observation of X available after (i-1)-st anf before i-th shift. The cost functional corresponding to our control problem is of the form (16) with $P^{\varepsilon,V}_{xy}$ replaced by P^V_x and functions f,d replaced by $\bar f$, $\hat d(x)=\inf_{y\in E_2} d(x,y)$.

Define

$$\lambda_\varepsilon=\inf_{(x,y)\in E}\ \inf_\tau\ [E^\varepsilon_{xy}(\int_0^\tau f(x_s,y_{s/\varepsilon})ds+c(x_\tau))+d(y)](E_x\tau)^{-1} \tag{17}$$

$$\lambda=\inf_{x\in E_1}\inf_\tau [E^\varepsilon_x(\int_0^\tau \bar f(x_s)ds+c(x_\tau))+d(y)](E_x\tau)^{-1} \tag{18}$$

and

$$w_\varepsilon(x,y)=\inf_\tau E^\varepsilon_{xy}\{\int_0^\tau (f(x_s,y_{s/\varepsilon})-\lambda_\varepsilon)ds+c(x_\tau)\} \tag{19}$$

$$w(x)=\inf_\tau E_x\{\int_0^\tau (\bar{f}(x_s)-\lambda)ds+c(x_\tau)\} \tag{20}$$

where in formulas (17)-(19), (18)-(20) the infima are over (F_t), (F^1_t) Markov times τ respectively.

From a general theory, see [13] as well as [10] it is known that the optimal values of cost functionals for $(XY)^\varepsilon$ and X are equal to $\lambda_\varepsilon,\lambda$ respectively, and w_ε, w play a role of Bellman functions with the use of which we construct optimal strategies.

Theorem 4. The limit control principle for ergodic impulsive control introduced above is satisfied. Moreover $\lambda_\varepsilon\to\lambda$ and if $\pi(\bar{f})=\lambda$, then it is optimal "to do nothing" in the limit problem as well as δ-optimal for $(XY)^\varepsilon$ with the functional (16) corresponding to ε sufficiently small. If $\pi(\bar{f})>\lambda$, then $\|w_\varepsilon-w\|\to 0$, $w_\varepsilon,w\in C(E)$ and denoting ε_0 such that $\|w_\varepsilon-w\|\le\delta/2$ for $\varepsilon<\varepsilon_0$, $(\bar{x},\bar{y})$ such that $w(\bar{x})+d(\bar{x}.\bar{y})=0$, we have that the strategy to shift $(XY)^\varepsilon$ to $(\bar{x},\bar{y})$ at the first entry time to $\bar{C}_\delta=\{(x,y)\in E,\ w(x)\ge c(x)-\delta\}$ is $5\delta\lambda_\varepsilon(2a-4\delta)^{-1}$ optimal for $\delta<a/2$, $\varepsilon<\varepsilon_0$.

Proof. Combine theorem 2 [4] together with Remarks 5,6,7 of [4].

4. Continuously acting control with long run average cost.

We consider now a particular model of a singularly perturbed diffusion. We start from the following system of SDE's:

$$\begin{aligned} dx^1_t&=f_1(x_t)dt\\ dx^2_t&=f_2(x_t)dt+\sigma(x^2_t)dw_t \end{aligned} \tag{21}$$

$x^1_0=x^1$, $x^2_0=x^2$, where $x^1_t\in R^{d_1}$, $x^2_t\in R^{d_2}$, (w_t) is a standard Wiener process on a probability space (Ω,F,P),

(T1) σ is a bounded, continuous, uniformly positive definite on R^{d_2}, $d_2\ge 1$

(T2) $\|f_1(x)-f_1(y)\|+\|f_2(x)-f_2(y)\|+\|\sigma(x)-\sigma(y)\|\le\|x-y\|$

$\|f_1(x)\|+\|f_2(x)\|\le B(1+\|x\|)$, for some constant B independent of x with $\|\ \|$ standing for an Euclidean norm.

Clearly there exists a pathwise unique solution to (21).

Let $U\subset R^n$ be a compact set of the values of the family of all admissible controls $\mathcal{A}$, which are measurable functions $u:R^d\to U$. Assume

(T3) $b:R^d\times U\to R^{d_2}$, $k:R^d\times U\to R$ are bounded measurable and continuous in u for fixed $x \in R^{d_1}$, where $d = d_1 + d_2$

(T4) for each $x \in R^d$ the set

$$\left\{ \begin{bmatrix} b(x,u) \\ k(x,u) \end{bmatrix}, \ u \in U \right\} \text{ is convex in } R^{d+1}$$

(T5) $f_3 : R^d \times E_2 \to R^{d_2}$ is bounded measurable

Let $Y = (\Omega_2, F_t^2, y_t, P_y)$ be an independent of (x_t) (under P) right continuous Feller Markov Process on E_2. Assume moreover Y is uniformly ergodic on bounded Borel functions $b\mathcal{E}$, i.e. there exists an invariant measure μ and constants $\beta, \gamma > 0$ such that for $f \in b\mathcal{E}$

$$\sup_{y \in E_2} |E_y(f(x_t)) - \mu(f)| \leq \beta e^{-\gamma t} \|f\|$$

Denote by * the transposition of vectors and matrices and define

$$\zeta_\varepsilon^t(u) = \int_0^t (b^*(x_s, u(x_s, y_{s/\varepsilon})) + f_3^*(x_s, y_{s/\varepsilon}))(\sigma^*(x_s^2))^{-1} dw_s$$

$$-1/2 \int_0^t [b^*(x_s, u(x_s, y_{s/\varepsilon})) + f_3^*(x_s, y_{s/\varepsilon})](\sigma^*(x_s^2))^{-1}(\sigma(x_s^2))^{-1} \tag{22}$$

$$[b(x_s, u(x_s, y_{s/\varepsilon})) + f_3(x_s, y_{s/\varepsilon})]ds$$

Let $P_{xy}^{u,\varepsilon}$ be a new measure on (Ω, F) such that the restrictions to F_t satisfy

$$dP_{xy}^{u,\varepsilon}|F_t = \exp(\zeta_0^t(u)) dP_{xy}|F_t \ , \ t \geq 0, \ (x,y) \in R^d \times E_2$$

Under $P_{xy}^{u,\varepsilon}$ the pair $(x_t, y_{t/\varepsilon})$ corresponds to singularly perturbed process $(XY)^\varepsilon$ studied in sections 1-3.

Define the cost functional

$$I_\varepsilon(u,x,y) = \limsup_{t \to \infty} t^{-1} E_{xy}^{\varepsilon,u}\{ \int_0^t k(x_s, u(x_s, y_{s/\varepsilon}))ds\} \tag{23}$$

to be minimize over all $u \in \mathcal{A}$.

To study the limit control problem we restrict the family of controls. Let $\mathcal{A}^r$ be the set of all measurable functions $u : R^{d_1} \to U$. For $u \in \mathcal{A}^r$ define $\zeta^t(u)$ substituting $\bar{f}_3$ for f_3 in (22) and P_x^u as a measure on (Ω, F) such that the restrictions of P_x^u satisfy

$$dP_x^u|F_t = \exp(\zeta^t(u)) dP_x|F_t, \text{ for } t \geq 0, \ x \in R^{d_2}$$

Then under P_x^u, (x_t) corresponds to limit process X from sections 1-3, and the limit cost functional is

$$I(u,x) = \limsup_{t \to \infty} t^{-1} E_x^u\{ \int_0^t k(x_s, u(x_s))ds\} \tag{24}$$

Before we formulate the results we need several assumptions, for which the sufficient conditions can be found in [3], [7], [14]:

(T6) for $u\in\mathcal{A}$, $\bar{u}\in\mathcal{A}^r$ under $P^{u,\varepsilon}_{xy}$, $P^{\bar{u}}_{x}$ respectively, (x_t) is a regular diffusion process,

(F1) $\sup_{u\in\mathcal{A}} \sup_{x\in\Gamma_1, y\in E_2, \varepsilon>0} E^{u,\varepsilon}_{xy}\{(T_\Gamma)^2\}<\infty \qquad \sup_{u\in\mathcal{A}^r} \sup_{x\in\Gamma_1} E^{u}_{x}\{(T_\Gamma)^2\}<\infty$

(F2) $\sup_{u\in\mathcal{A}} \sup_{x\in\Gamma_1, y\in E_2, \varepsilon>0} P^{u,\varepsilon}_{xy}(T_{S^c_n}>T_\Gamma)\to 1 \qquad \sup_{u\in\mathcal{A}^r} \sup_{x\in\Gamma_1} P^{u}_{x}(T_{S^c_n}>T_\Gamma)\to 1 \quad n\to\infty$

where $S_n=\{x\in R^d: \|x\|\le n\}$, $\Gamma=\partial S_n$, $\Gamma_1=\partial S_{n+k}$, $k>0$,

(F3) for any $\varepsilon>0$, $u\in\mathcal{A}$ there exists a unique invariant probability measure π^ε_u for $(x_t,y_{t/\varepsilon})$ under $P^{u,\varepsilon}_{xy}$; for $\bar{u}\in\mathcal{A}^r$, there exists a unique invariant measure $\pi^{\bar{u}}$ for (x_t) under $P^{\bar{u}}_{x}$,

(F4) $P_x(x_t\in A)\to 0$, uniformly on compact sets as $\lambda(A)\to 0$, $t\ge 0$, $P_y(y_{t/\varepsilon}\in B)\to 0$ uniformly in $y\in E_2$ for $B\in\mathcal{E}_2$, as $\mu(B)\to 0$.

Theorem 5. Under the above assumptions the optimal values λ, λ^ε of the functionals I_ε, I over $u\in\mathcal{A}$ or $u\in\mathcal{A}^r$ respectively, do not depend on x,y. Moreover there exist optimal controls $u^\varepsilon\in\mathcal{A}$, $u\in\mathcal{A}^r$, for which the values λ^ε, λ of the functionals (23), (24) are achieved. In addition $\lambda^\varepsilon\to\lambda$, as $\varepsilon\to 0$, and the limit control principle is satisfied.

Proof. For the existence of optimal strategies see [7], [14], where under (T4) the Roxin construction [12] of Borel selectors was applied. The remaining part follows from theorem 5 of [3].

References

[1] Bensoussan A., Methodes de Perturbation en Contrôle Optimal, Dunod 1988,

[2] Bielecki T., Stettner Ł., On Some Problems Arising in Asymptotic Analysis of Markov Processes with Singularly Perturbed Generators, Stoch. Anal. Appl. 6 (2) (1988), 129-168,

[3] Bielecki T., Stettner Ł., On Ergodic Control Problems for Singularly Perturbed Markov Processes, to appear in JAMO,

[4] Kogan Yu. A., On Optimal Control of Nonterminating Diffusion with Reflection, Theor. Prob. Appl. 14 (1969), 469-502,

[5] Kurtz T., A Limit Theorem for Perturbed Operator Semigroups with Applications to Random Evolutions, J. Funct. Anal. 12 (1973), 55-67,

[6] Kurtz T., Applications of an Abstract Perturbation Theorem to Ordinary Differential Equations, Houston J. Math. 3 (1977), 67-82,

[7] Kushner H.J., Optimality Conditions for the Average Cost per Unit Time Problem with a Diffusion Model, SIAM J. Control Optimiz. 16

(1978), 330-346,

[8] Kushner H.J., Runggaldier W., Nearly Optimal Feedback Controls for Stochastic Systems with Wideband Noise Disturbances, SIAM J. Control Optimiz. 25 (1987), 298-315,

[9] Robin M., Controle Impulsionnel de Processus de Markov, Thesis, University of Paris IX 1978,

[10] Robin M., On Some Impulse Control Problems with Long Run Average Cost, SIAM J. Control Optimiz. 19 (1981), 333-358,

[11] Robin M., On Some Perturbation Problems in Optimal Stopping and Impulse Control, in IMA Volume 10, Stochastics Differential Systems, Stochastic Control Theory and Applications, Ed. W. Fleming, P.L. Lions, 473-500,

[12] Roxin E., The Existence of Optimal Control, Michigan Math. J. 9 (1962), 109-119,

[13] Stettner L., On some Stopping and Impulsive Control Problems with a General Discount Rate Criteria, to appear in Prob. Math. Statistics,

[14] Stettner L., On the Existence of an Optimal per Unit Time Control for a Degenerate Diffusion Model, Bull. Polish Acad. Sci. 34, (1986), 746-769.

SOME SOLVABLE STOCHASTIC CONTROL PROBLEMS IN SYMMETRIC SPACES OF TYPE IV*

T. E. DUNCAN
Department of Mathematics
University of Kansas
Lawrence, KS 66045

1. Introduction.

A number of quite general results are available on the existence or the existence and the uniqueness of optimal stochastic controls for the control of a wide range of stochastic systems. However, only a relatively few examples of controlled diffusions are available where the optimal control is expressed explicitly as a function of the state of the system. This lack of examples is especially apparent for controlled diffusions that are described by nonlinear stochastic differential equations. Some examples of explicitly solvable stochastic optimal control problems in Euclidean spaces besides the well known linear regulator problem are given in [1, 2, 13, 15].

For the control of nonlinear stochastic systems it is natural to investigate systems that possess some inherent geometry. This consideration motivated the study of controlled diffusions in manifolds which was initiated in [4-6]. A natural family of manifolds to consider for explicitly solvable control problems arise from operations on Lie groups. An important family of such manifolds are symmetric spaces. Apparently the first example of a solvable stochastic control problem for a controlled diffusion in a noncompact manifold with nonzero curvature is given in [7] where the manifold is real hyperbolic three space, a noncompact symmetric space of rank one. This example was extended to other cost functionals and to all noncompact symmetric spaces of rank one and dimension > 2 in [10]. Since compact manifolds are also important models for some physical phenomena stochastic control problems in compact symmetric spaces were investigated. An example of a solvable stochastic control problem in spheres of dimension > 2 is given in [8]. These spheres are compact symmetric spaces of rank one. This example in the spheres was extended to other cost functionals and to all compact

* Research partially supported by NSF Grant ECS-8718026.

symmetric spaces of rank one and dimension > 2 in [9]. The techniques that are used for the compact symmetric spaces are inherently somewhat different from those for the noncompact symmetric spaces.

While the aforementioned symmetric spaces, which provided solvable stochastic control problems, represent an interesting family of manifolds, all of these symmetric spaces are of rank one. In this paper a countable family of nontrivially distinct stochastic control problems are formulated and explicitly solved in each irreducible symmetric space of type IV in É. Cartan's classification. These noncompact (Riemannian globally) symmetric spaces are the spaces G/K where G is a connected Lie group whose Lie algebra is $\mathfrak{g}^{\mathbb{R}}$ where $\mathfrak{g}$ is a simple Lie algebra over $\mathbb{C}$ and K is a maximal compact subgroup of G.

In following the approach of the examples in [7, 10] one important difference for symmetric spaces of type IV is that useful spherical functions are not as readily available. For rank one symmetric spaces the eigenvalue problem that defines the spherical functions is a second order linear ordinary differential equation whose solutions can be identified with the hypergeometric functions. Among these hypergeometric functions are polynomials that have suitable monotonicity properties and can be used as cost functionals for the control problem. For symmetric spaces of rank greater than one, the eigenvalue problem is given by a system of differential operators so it is more difficult to find spherical functions that can be used for the control problem. However, for symmetric spaces of type IV the restricted roots each have multiplicity two so that some root space properties can be used to find spherical functions that are polynomials and that possess a suitable monotonicity property in a positive Weyl chamber. A countable family of such spherical functions are exhibited.

The stochastic control problem is the control of Brownian motion by a drift term so that this controlled diffusion remains close to a fixed point in the manifold. This model is the same one that occurs in [7-10].

In addition to the difference of finding useful spherical functions for symmetric spaces of rank > 1 there are other differences that arise in the stochastic control problem studied here as compared with the corresponding control problem in rank one symmetric spaces. By some symmetry properties of the control problem it can be reduced to the analysis of a stochastic differential equation in a positive Weyl chamber. For noncompact symmetric spaces of rank one a Weyl chamber is a half line and the investigation of the boundary behavior of the controlled diffusion is only the analysis of two points. Since the dimension of a Weyl chamber is the rank, the investigation of the boundary behavior of the controlled diffusion in this paper requires analysis on submanifolds that are of all dimensions less than the rank. Furthermore the dimension of the control vector is the rank of the symmetric space.

2. Preliminaries

Initially a few results from semisimple Lie theory are reviewed for their use in the analysis of the noncompact symmetric spaces that are used here. A good reference for this material is Helgason [14]. Let $\mathfrak{g}$ be a semisimple Lie algebra over R and let $\mathfrak{g}_c$ be its complexification. A direct sum decomposition $\mathfrak{g} = \mathfrak{k} + \mathfrak{p}$ into a subalgebra $\mathfrak{k}$ and a vector space $\mathfrak{p}$ is called a Cartan decomposition if there exists a compact real form $\mathfrak{u}$ such that

$$\sigma(\mathfrak{u}) \subset \mathfrak{u} \quad , \quad \mathfrak{u} \cap \mathfrak{g} = \mathfrak{k} \quad , \quad \sqrt{-1}\mathfrak{u} \cap \mathfrak{g} = \mathfrak{p}$$

where σ is the conjugation of $\mathfrak{g}_c$ with respect to $\mathfrak{g}$. Every semisimple Lie algebra $\mathfrak{g}$ over R admits a Cartan decomposition. Let W_Σ be the group generated by the reflections of Σ where Σ is the set of roots of the pair $(\mathfrak{g},\mathfrak{a}_p)$ where $\mathfrak{a}_p$ is a maximal abelian subspace in $\mathfrak{p}$, $\mathfrak{g} = \mathfrak{k} + \mathfrak{p}$ is a Cartan decomposition and the rank of the associated symmetric space is the dimension of $\mathfrak{a}_p$. Let Δ be the roots of $(\mathfrak{g}_c,\mathfrak{a}_c)$ where $\mathfrak{a} = \mathfrak{a}_{\mathfrak{k}} + \mathfrak{a}_p$ is the extension of $\mathfrak{a}_p$ to a Cartan subalgebra $\mathfrak{a}$ of $\mathfrak{g}$ and $\mathfrak{a}_c$ is the complexification of $\mathfrak{a}$ and is a Cartan subalgebra of $\mathfrak{g}_c$. Let Δ_p be the elements of Δ that do not vanish identically on $\mathfrak{a}_p$. The set of restrictions of Δ_p to $\mathfrak{a}_p$ is Σ as defined above and thus the elements of Σ are often called the restricted roots.

An element $H \in \mathfrak{a}$ is called regular if $\lambda(H) \neq 0$ for all $\lambda \in \Sigma$, otherwise it is called singular. The subset $\mathfrak{a}' \subset \mathfrak{a}$ of regular elements consists of the complement of finitely many hyperplanes, and its components are called Weyl chambers. Fix a Weyl chamber $\mathfrak{a}^+$ and call a root λ positive if λ has positive values in $\mathfrak{a}^+$. Let $A^+ = \exp \mathfrak{a}^+$.

Let $\mathfrak{g} = \mathfrak{k} + \mathfrak{p}$ be the direct sum decomposition (that is Cartan decomposition) of the Lie algebra $\mathfrak{g}$ of G into the Lie algebra $\mathfrak{k}$ of K and its orthogonal complement $\mathfrak{p}$ with respect to the Killing form of $\mathfrak{g}$. The dimension of a maximal abelian subspace, $\mathfrak{a}$, in $\mathfrak{p}$ is the rank of the symmetric space and there is a root space (direct sum) decomposition of $\mathfrak{p}$ as

$$\mathfrak{p} = \mathfrak{a} + \sum_{\alpha \in \Sigma} \mathfrak{p}_\alpha \tag{1}$$

where p_α is the eigenspace associated with $\alpha \in \Sigma$. In the Cartan classification of irreducible Riemannian globally symmetric spaces (e.g., [14]) there are four types. These are the types I to IV. Types I and II are compact and are dual to the noncompact types III and IV respectively. The Riemannian globally symmetric spaces of type IV are the spaces G/K where G is a connected Lie group whose Lie algebra is $\mathfrak{g}^{\mathbb{R}}$ where $\mathfrak{g}$ is a simple Lie algebra over $\mathbb{C}$ and $\mathfrak{g}^{\mathbb{R}}$ is $\mathfrak{g}$ considered as Lie algebra over $\mathbb{R}$ and K is a maximal compact subgroup of G, (p. 516 [14]). The metric on G/K is G-invariant and is uniquely determined (up to a constant) by this condition. The simple Lie algebras are well known from É. Cartan's classification (e.g., p. 516 [14]). The corresponding group includes the classical groups $SL(n, \mathbb{C})$, $SO(2n+1, \mathbb{C})$, $Sp(n,\mathbb{C})$ and $SO(2n,\mathbb{C})$. Since $\mathfrak{g}$ has a complex structure, each restricted root of a symmetric space of type IV has multiplicity two. In fact the following properties are equivalent: i) $\mathfrak{g}$ has a complex structure, ii) all of the restricted roots have multiplicity two and iii) the compact dual of the symmetric space is a compact Lie group.

A noncompact symmetric space is globally trivial, in fact, the exponential map is a global diffeomorphism so in particular such a space is simply connected and the center is trivial.

The Laplace-Beltrami operator Δ on a Riemannian manifold M with Riemannian metric $g(\cdot,\cdot)$ is defined on a C^∞-function f on M as

$$\Delta f = \text{div grad } f \tag{2}$$

which in terms of local coordinates is

$$\Delta f = \frac{1}{\sqrt{\bar{g}}} \sum \partial_k \left(\sum g^{ik} \sqrt{\bar{g}}\ \partial_i f \right) \tag{3}$$

where

$$g_{ij} = g\left(\frac{\partial}{\partial x_i}, \frac{\partial}{\partial x_j}\right) \tag{4}$$

$$\sum g_{ij}\, g^{jk} = \delta_{ik} \tag{5}$$

$$\bar{g} = |\det(g_{ij})| \tag{6}$$

The radial part of $\Delta_{G/K}$, $R(\Delta_{G/K})$ is given by (p. 267 [15])

$$R(\Delta_{G/K}) = \Delta_A + \sum_{\alpha\in\Sigma^+} m_\alpha(\coth\ \alpha)A_\alpha \tag{7}$$

where Δ_A is the Laplace-Beltrami operator on $A = \exp(\mathfrak{a})$, m_α is the multiplicity of $\alpha \in \Sigma^+$ and $A_\alpha \in \mathfrak{a}$ is dual to α, that is,

$$\langle A_\alpha, H\rangle = \alpha(H) \tag{8}$$

for $H \in \mathfrak{a}$ and A_α is considered as a differential operator on $A^+\cdot o$.

While the radial part of the Laplace-Beltrami operator in (7) is valid for an arbitrary symmetric space for a complex Lie group, that is, a symmetric space of type IV, it has a particularly simple form (e.g., p. 268 [15]) as

$$R(\Delta_{G/K}) = \delta^{-1/2}(\Delta_A - \langle\rho,\rho\rangle) \circ \delta^{1/2} \tag{9}$$

where

$$\delta^{1/2}(a) = \sum_{w\in W}(\det\ w)e^{w\rho(\log\ a)} \tag{10}$$

and $a \in A^+$ and $A^+ = \exp \mathfrak{a}^+$.

Recall that $\mathfrak{a}'$ is the family of regular elements in $\mathfrak{a}$ and $A' = \exp \mathfrak{a}'$. Define $G' = KA'K$. The open dense subset $X' = G'\cdot o$ of $X = G/K$ is diffeomorphic to $(K/M) \times A^+$ under the "polar coordinate map"

$$\Phi : (kM,a) \mapsto kaK \tag{11}$$

where $\Phi : (K/M) \times A^+ \to X'$, M is the centralizer of A in K and o is the origin $\{K\}$ in G/K. This coordinate map can be used to express the Laplace-Beltrami operator on G/K and it is used subsequently to describe the stochastic control problem.

The eigenvalue problem for the spherical functions has been investigated by Harish-Chandra [12]. It is described by the following equation

$$R(\Delta_{G/K})\varphi_\lambda = \lambda_\Delta\varphi_\lambda \tag{12}$$

where $\lambda_\Delta = -(\langle\lambda,\lambda\rangle + \langle\rho,\rho\rangle)$ and $R(\Delta_{G/K})$ is the radial part of $\Delta_{G/K}$. This is a system of differential equations with variable coefficients. Harish-Chandra [12] proved that there is a unique solution of (12) on a positive Weyl chamber of the form

$$\Phi_\lambda(\exp H) = \sum_{\mu \in \Lambda} \Gamma_\mu \exp((\sqrt{-1}\,\lambda - \rho - \mu)H) \tag{13}$$

where $\Gamma_0 = 1$ and Λ is the root lattice. Then the spherical function φ_λ is expressed as

$$\varphi_\lambda(h) = \sum_{w \in W} c(w\lambda)\Phi_{w\lambda}(h) \tag{14}$$

In general (14) is a complicated expression whose appropriateness in a cost functional of a stochastic control problem is not clear. However for special eigenvalues (14) can be simplified and it can be shown that these spherical functions are appropriate for use in a cost functional.

Symmetric spaces of type IV are dual to symmetric spaces of type II and these latter spaces are the simple, compact, connected Lie groups. Furthermore since each restricted root in a symmetric space of type IV has multiplicity two, some properties of root spaces can be used to obtain some other expressions for a spherical function. Berezin [3] and Harish-Chandra (p. 304 [12]) have done this simplification. This expression for a spherical function in a symmetric space of type IV is

$$\varphi_\lambda(h) = \frac{\pi(\rho) \sum_{w \in W} \varepsilon(w) e^{iw\lambda(\log h)}}{\pi(i\lambda) \sum_{w \in W} \varepsilon(w) e^{w\rho(\log h)}} \tag{15}$$

where

$$\pi(\lambda) = \prod_{\alpha \in \Sigma^+} \langle \alpha, \lambda \rangle \tag{16}$$

and Σ^+ are the positive restricted roots. Now special values of λ are chosen to obtain some properties of (15) that are important in the cost functional of a control problem. Let

$$i\lambda = 2^n \rho \tag{17}$$

for $n \in \mathbb{Z}^+$. By a formula of Weyl for the root system $\{2\alpha : \alpha \in \Sigma\}$ we have

$$\sum_{w \in W} \varepsilon(w) e^{w\rho} = \prod_{\alpha \in \Sigma^+} (e^{\alpha} - e^{-\alpha}) \tag{18}$$

For $i\lambda = 2^n\rho$ using (18) in (15) it follows that

$$\varphi_{\frac{2^n\rho}{i}}(\cdot) = 2^{n\ \mathrm{card}(\Sigma+)} \prod_{\alpha \in \Sigma^+} \left(\frac{e^{2^n\alpha} - e^{-2^n\alpha}}{e^{\alpha} - e^{-\alpha}} \right) \tag{19}$$

Fix $\alpha \in \Sigma^+$ and let $x = e^{\alpha}$ so that the corresponding term in the product in (19) can be expressed as

$$\frac{x^{2^n} - x^{-2^n}}{x - x^{-1}}$$

For $n = 1$ we have

$$\frac{x^2 - x^{-2}}{x - x^{-1}} = x + x^{-1} \tag{20}$$

Proceeding by induction we have

$$\frac{x^{2^n} - x^{-2^n}}{x - x^{-1}} = \prod_{k=0}^{n-1} (x^{2^k} + x^{-2^k}) \tag{21}$$

Thus

$$\psi_n(\cdot) \triangleq \varphi_{\frac{2^n\rho}{i}}(\cdot) = \prod_{\alpha \in \Sigma^+} \prod_{k=0}^{n-1} \cosh 2^k\alpha \tag{22}$$

If $(\beta_1, \ldots, \beta_\ell)$ are the simple roots of Σ^+ then each root $\alpha \in \Sigma^+$ can be expressed in this basis as

$$\alpha = \sum_{j=1}^{\ell} n_j(\alpha)\beta_j \tag{23}$$

where $n_j(\alpha) \in \mathbb{Z}^+$. Thus in a positive Weyl chamber ψ_n is an increasing function along rays emanating from the origin. Thus ψ_n has a monotonicity property that is important for its use as a cost functional.

3. The Control Problem.

Let G/K be an irreducible (Riemannian globally) symmetric space of type IV. The Riemannian metric on G/K is induced from the Killing form (or a scaled version of it) on G by restriction and translation. Let $(Z(t),\ t \geq 0)$ be the controlled diffusion process in G/K with the infinitesimal generator

$$\frac{1}{2}\Delta_{G/K} + \sum_{j=1}^{\ell} U_i \frac{\partial}{\partial x_i} \tag{24}$$

where $\frac{\partial}{\partial x_i}$ is the differential operator on $A^+ o$, that is induced from X_i where $\beta_i(H) = \langle X_i, H\rangle$ for all $H \in \mathfrak{a}$ and $i = 1,\ldots,\ell$ and $(\beta_1,\ldots,\beta_\ell)$ are the simple roots for the given Weyl chamber $\mathfrak{a}^+$. This description (24) of the infinitesimal generator for the controlled diffusion is intrinsic using the polar coordinate map (11). Since $A^+ = \exp \mathfrak{a}^+$, A^+ and $\mathfrak{a}^+$ are often identified using the global diffeomorphism exp. For the analysis of the controlled diffusion it suffices to describe the system of (local) stochastic differential equations for the radial part of the infinitesimal generator (24). This system of equations in $\mathfrak{a}^+$ is

$$\begin{aligned} dX_i(t) = \Bigg[\frac{1}{2} & \sum_{\alpha\in\Sigma^+} m_\alpha \coth \alpha(X(t))\alpha(H_i) \\ & + \sum_{j=1}^{\ell} \beta_j(H_i)U_j(t)\Bigg] dt + dB_i(t) \end{aligned} \tag{25}$$

for $i = 1,2,\ldots,\ell$ where $(H_1,\ldots,H_\ell)$ is an orthonormal basis of $\mathfrak{a}$, $(\beta_1,\ldots,\beta_\ell)$ are the simple roots for the fixed positive Weyl chamber $\mathfrak{a}^+$, $X(t) = (X_1(t),\ldots,X_\ell(t))'$, $(B_1(t),\ldots,B_\ell(t),\ t \geq 0)$ is an ℓ-dimensional standard Brownian motion and $X(0) \in \mathfrak{a}^{+\prime}$.

Fix $n \in \mathbb{Z}^+$ and let λ_n be defined as

$$\begin{aligned} \lambda_n &= -\left(\left\langle \frac{2^n\rho}{i}, \frac{2^n\rho}{i}\right\rangle + \langle\rho,\rho\rangle\right) \\ &= -(-2^{2n}\langle\rho,\rho\rangle + \langle\rho,\rho\rangle) \\ &= +(2^{2n} - 1)\langle\rho,\rho\rangle \end{aligned} \tag{26}$$

Clearly λ_n is the eigenvalue associated with the spherical function $\varphi_{\frac{2^n\rho}{i}}$.

Let $T > 0$ be chosen such that the unique positive solution of the Riccati differential equation on $[0,T]$

$$g' + \lambda_n g - \frac{1}{4} g^2 + 1 = 0 \tag{27}$$

$$g(T) = 0 \tag{28}$$

satisfies

$$\sup_{0 \le t \le T} g(t) \le 2^{2n-3} \tag{29}$$

The cost functional for the controlled diffusion $(Z(t),\ t \ge 0)$ with the infinitesimal generator (24) is

$$J_n(U) = E_{X(0)} \int_0^T \psi_n(Z(t)) + \sum_{j=1}^{\ell} f_j(Z(t))U_j^2(t)\ dt \tag{30}$$

where

$$f_i(x) = \frac{(D_i\psi_n(x))^2}{\psi_n(x)} \tag{31}$$

$$D_i = \frac{\partial}{\partial x_i} \tag{32}$$

Since ψ_n is a spherical function and constant on K-orbits, the cost functional can be expressed in terms of the process $(X(t),\ t \ge 0)$ in $\mathfrak{a}^+$, that is, the solution of the system of equations (25), as

$$J_n(U) = E_{X(0)} \int_0^T \tilde{\psi}_n(X(t)) + \sum_{\delta=1}^{\ell} \tilde{f}_j(X(t))U_j^2(t)\ dt \tag{33}$$

where $\tilde{\psi}_n = \bar{\psi}_n \circ \exp^{-1}$, $\bar{\psi}_n = \psi_n|_{A^+}$ and

$$\tilde{f}_j = \frac{D_j \tilde{\psi}_n}{\tilde{\psi}_n}$$

An admissible control vector at time t is a measurable function of Z(t) that is smooth on (G/K)′, that is, the regular elements of G/K, such that the solution of (25) exists and is unique in a sample path sense. By the K-invariance of the cost functional it suffices to consider controls at time t that are measurable functions of X(t) which is the solution of (25).

Initially it is necessary to note that there are admissible controls that give a finite value to the cost $J_n(U)$. In fact, the control $U \equiv 0$ gives a finite cost from the following lemma whose proof is not included.

Lemma. *Let* $n \in \mathbb{Z}^+$ *be fixed and let* $X(t) = (X_1(t),\dots,X_\ell(t))'$ *be the solution of (25) with* $U_j(t) \equiv 0$ *for* $j = 1,2,\dots,\ell$. *Then*

$$E_{X(0)} \int_0^T \tilde{\psi}_n(X(t))dt < \infty \tag{34}$$

To prove this lemma the Brownian motion in G/K can be compared to a Brownian motion in a noncompact symmetric space of rank one as in [16] because G/K has negative (sectional) curvature. In this way the verification of (34) can be reduced to a similar verification for a noncompact symmetric space of rank one [10].

The main result of this paper is the following theorem that solves the stochastic optimal control problem (24-25) and (30) by explicitly exhibiting an optimal control.

Theorem. *The stochastic control problem described by (24-25) and (30) has an optimal control that in the basis* $(A_{\beta_1},\dots,A_{\beta_\ell})$ *determined by the simple roots* $(\beta_1,\dots,\beta_\ell)$ is

$$U_i^*(s,x) = -\frac{1}{2}\,\frac{\tilde{\psi}_n(x)}{D_i\tilde{\psi}_n(x)}\,g(s) \tag{35}$$

where $s \in [0,T]$, $x \in \mathfrak{a}^+$ *and* g *is the unique positive solution of (27).*

Proof. It is well known (e.g., [11]) that the Hamilton-Jacobi or dynamic programming equation for a stochastic optimal control problem of diffusion type is

$$0 = \frac{\partial W}{\partial s} + \min_{v \in U}\,[\mathcal{A}^s(s)W + L(s,x,v)] \tag{36}$$

where $\mathcal{A}^v$ is the infinitesimal generator of the controlled diffusion using the control v and L is the cost function. To apply a verification theorem (p. 159 [11]) to the control problem here it is assumed that the solution W of (36) with the boundary condition $W(x,y) = 0$ for $(s,y) \in \{T\} \times G/K$ is $C^{1,2}[(0,T) \times G/K]$ and continuous on $[0,T] \times G/K$.

Since the integrand of the cost function (30) as a function of the state with the control fixed is K-invariant and the control appears in a K-invariant way, the solution W of (36) for the control problem (24, 30) is K-invariant so (36) can be rewritten using only the radial part of (24) as

$$0 = \frac{\partial W}{\partial s} + \min_{v \in \mathbb{R}^{\ell}} \left[\frac{1}{2} R(\Delta_{G/K}) + \sum_{i=1}^{\ell} v_i \frac{\partial W}{\partial x_i} + \tilde{\psi}_n(x) + \Sigma \tilde{f}_i(x) v_i^2 \right] \tag{37}$$

Performing the minimization in (37), it is clear that the system of controls obtained by this minimization is

$$U_j^*(s,x) = \frac{-1}{2\tilde{f}_j(x)} \frac{\partial W}{\partial x_j}(s,x) \tag{38}$$

Assume a solution W of (37) as

$$W(s,x) = \tilde{\psi}_n(x) g(s) \tag{39}$$

Substitute (39) into (37) to obtain

$$\begin{aligned} 0 &= g'\tilde{\psi}_n + \frac{1}{2}\lambda_n \tilde{\psi}_n g - \frac{1}{4}\tilde{\psi}_n g^2 + \tilde{\psi}_n \\ &= \tilde{\psi}_n (g' + \frac{1}{2}\lambda_n g - \frac{1}{4} g^2 + 1) \end{aligned} \tag{40}$$

If g in (39) satisfies (27) then (39) is a smooth solution to (37) that satisfies the boundary condition $W(s,y) = 0$ for $(s,y) \in \{T\} \times G/K$. Furthermore, the family of controls (38) is an optimal control provided that the stochastic differential equation (25) with the family of controls (38) has one and only one solution in $[0,T]$. Since $\dfrac{\tilde{\psi}_n}{D_i \tilde{\psi}_n}$ is bounded as $|x| \to \infty$ in the interior of the positive Weyl chamber for $i = 1,2,\ldots,\ell$, the drift term for the optimal stochastic differential equation is bounded and globally Lipschitz continuous in the complement of a neighborhood of the boundary of the positive Weyl chamber. Furthermore there is no finite escape time. Thus to verify existence and uniqueness of (25) using the family of controls (38) it suffices to show that

$$P(X(t) \in \partial \mathfrak{a}^+ \text{ for some } t \in [0,T]) = 0 \tag{41}$$

where $\partial \mathfrak{a}^+$ is the boundary of the positive Weyl chamber $\mathfrak{a}^+$.

To verify (41) it is useful to review the lengths of the simple roots for some Lie algebras. For the classical Lie algebras over $\mathbb{C}$ the lengths of the simple roots are well known (e.g., pp. 462-464 [14]). If $\mathfrak{g}$ is such a Lie algebra then the elements in a Cartan subalgebra $\mathfrak{h}$ that are dual to the simple roots of $(\mathfrak{g},\mathfrak{h})$ have length > 1 and a fortiori for the same elements in an associated compact real form of $\mathfrak{g}$. By the duality of the symmetric spaces of types II and IV the length of A_{β_i} is > 1 for $i = 1,\ldots,\ell$ where A_{β_i} is dual to the simple root β_i. For the exceptional Lie algebras the Killing form has to be scaled so that A_{β_i} has length > 1 for $i = 1,2,\ldots,\ell$. For these exceptional Lie algebras it is assumed that the Killing form is suitably scaled.

The boundary of $\mathfrak{a}^+$ can be expressed as a disjoint union of submanifolds as

$$\partial \mathfrak{a}^+ = \bigcup_{i=0}^{\ell-1} \bigcup_{j(i)} \Lambda_{ij} \tag{42}$$

where $\dim \Lambda_{ij} = i$. Fix i and j and consider Λ_{ij}. It is shown that the solution of (25) with the optimal control (38) does not hit Λ_{ij} for all $t \in [0,T]$ a.s. There is a simple root $\beta_{k(i,j)}$ such that this coordinate of the simple root basis is zero on Λ_{ij}. For notational simplicity let

$$\gamma = \beta_{k(i,j)} \tag{43}$$

It follows from (25) that the scalar stochastic differential equation with the basis vector A_γ is

$$dZ(t) = [L \coth LZ(t) + \sum_{\alpha \in \Sigma^+ \setminus \{\gamma\}} L n_k(\alpha) \coth \alpha(Y(t)) \tag{44}$$

$$+ LU_k(t)] + dB(t)$$

where $L = \langle A_\gamma, A_\gamma \rangle^{1/2}$ and $(B(t), t \geq 0)$ is a real-valued standard Brownian motion. Let $(\tilde{Z}(t), t \geq 0)$ be the solution of the stochastic differential equation

$$d\tilde{Z}(t) = \frac{3}{4} \coth \tilde{Z}(t)\, dt + dB(t) \tag{45}$$

The two dimensional Bessel process (e.g., p. 60 [17]) has the infinitesimal generator

$$\frac{1}{2}\left(\frac{d^2}{dr^2} + \frac{1}{r}\frac{d}{dr}\right)$$

and is the solution of the scalar stochastic differential equation

$$dV(t) = \frac{1}{2}\frac{1}{V(t)}\, dt + dB(t) \tag{46}$$

where $(B(t), t \geq 0)$ is a real-valued Brownian motion. It is known that $(V(t), t \geq 0)$ does not hit the origin. If $(\tilde{Z}(t), t \geq 0)$ is compared with $(V(t), t \geq 0)$ then

$$P(\tilde{Z}(t) = 0 \text{ for some } t \in [0,T]) = 0 \tag{47}$$

In the positive Weyl chamber $\mathfrak{a}^+$ near Λ_{ij} the drift term in (46) is a lower bound on the drift term in (44) from (29) so if $(Z(t), t \geq 0)$ is zero then $(\tilde{Z}(t), t \geq 0)$ is zero by a localized version of a comparison theorem (p. 352 [16]) applied to the stochastic differential equations (44) and (46). However $(\tilde{Z}(t), t \geq 0)$ is not zero by (47) so that $(Z(t), t \geq 0)$ is not zero and the solution of (25) with the optimal control (38) does not hit Λ_{ij}. This completes the proof.

REFERENCES

[1] V. E. Benes, L. A. Shepp and H. S. Witsenhausen, Some solvable stochastic control problems, Stochastics 4(1980), 39-83.

[2] A. Bensoussan and J. H. Van Schuppen, Optimal control of partially observable stochastic systems with an exponential-of-integral performance index, SIAM J. Control Optim. 23(1985), 599-613.

[3] F. A. Berezin, Doklady Akad. Nauk SSR N.S. 107 (1956), 9-12 and 110 (1956), 897-900.

[4] T. E. Duncan, Some stochastic systems on manifolds, Lecture Notes in Econ. and Math. Systems 107(1975), 262-270, Springer-Verlag, New York.

[5] T. E. Duncan, Dynamic programming optimality criteria for stochastic systems in Riemannian manifolds, Appl. Math. Optim. 3(1977), 191-208.

[6] T. E. Duncan, Stochastic systems in Riemannian manifolds, J. Optimization Theory Appl. 27(1979), 399-426.

[7] T. E. Duncan, A solvable stochastic control problem in hyperbolic three space, Systems and Control Letters, 8(1987), 435-439.

[8] T. E. Duncan, A solvable stochastic control problem in spheres in Geometry of Random Motion (R. Durrett and M. Pinsky, eds.) Contemporary Mathematics, 73(1988), 49-54, Amer. Math. Soc., Providence.

[9] T. E. Duncan, Some solvable stochastic control problems in compact symmetric spaces of rank one, to appear in Contemporary Mathematics.

[10] T. E. Duncan, Some solvable stochastic control problems in noncompact symmetric spaces of rank one, to appear in Stochastics.

[11] W. H. Fleming and R. W. Rishel, Deterministic and Stochastic Optimal Control, Springer-Verlag, 1975.

[12] Harish-Chandra, Spherical functions on a semi-simple Lie group I., Amer. J. Math. 80(1958), 241-310.

[13] U. G. Haussmann, Some examples of optimal stochastic controls or: the stochastic maximum principle at work, SIAM Review 23(1981), 292-307.

[14] S. Helgason, Differential Geometry, Lie Groups and Symmetric Spaces, Academic Press, 1978.

[15] S. Helgason, Groups and Geometric Analysis, Academic Press, 1984.

[16] N. Ikeda and S. Watanabe, Stochastic Differential Equations and Diffusion Processes, North-Holland, 1981.

[17] K. Itô and H. P. McKean, Jr., Diffusion Processes and Their Sample Paths, Springer-Verlag, 1965.

[18] R. C. Merton, Optimum consumption and portfolio rules in a continuous-time model, J. Economic Theory 3(1971), 373-413.

IMPULSIVE CONTROL OF PIECEWISE-DETERMINISTIC PROCESSES.

Dariusz Gatarek
Systems Research Institute
01-447 Warszawa, Newelska 6
POLAND

The paper deals with value functions for impulsive control for piecewise-deterministic processes . The associated dynamic programming equations are quasi-variational inequalities with integral and first order differential terms. Here we study different regularity properties of the cost function and existence of optimal policies.

Key words : piecewise-deterministic processes, impulsive control, quasi-variational inequality.

1. Introduction.

In this paper we study the problems of impulsive control of piecewise-deterministic processes. The class of piecewise-deterministic Markov processes was first defined by Davis in [3] and covers most non-diffusion stochastic models. Generator of such processes A is an integro-differential operator.

Optimal stopping time problems for piecewise-deterministic processes were studied by Lenhart and Liao in [9] and by Gugerli in [8]. Impulsive control was studied by Lenhart in [11].

The paper is divided into sections. In Section 2 we give stochastic background of studied problems . In Section 3 we characterize the value function for an optimal impulsive control problem with discounted cost as a solution of a quasi-variational inequality. In Section 4 we study the same problem with long run average cost under stronger assumptions.

2. Stochastic background.

To formulate results of the paper denote by $\mathcal{B}(E)$, $C(E)$, $C_b(E)$, $C_o(E)$, $W^{1,\infty}(E)$, $C^1(E)$ and $LC(E)$ respectively the space of all Borel measurable and bounded functions, continuous functions, continuous and bounded, continuous and vanishing at infinity, Lipschitz continuous, continuously differentiable and lower-semi-continuous functions on E. The norm $\|\cdot\|$ is the "sup" or "esssup" norm. The norm in the space $W^{1,\infty}(E)$ will be denoted by $\|\cdot\|_{1,\infty}$. For any function space $\mathcal{V}$ denote by $\mathcal{V}_+$ the space of all nonnegative functions from the space $\mathcal{V}$.

Let $X = (\Omega,\mathcal{F},\mathcal{F}_t,x_t,\theta_t,P_x)$ be a Markov piecewise-deterministic process on a state space E, being a closed subset of $\mathbb{R}^n$. The set of all stopping times with respect to $\mathcal{F}_t$ will be denoted by $\mathcal{M}$.

A piecewise-deterministic process X is characterized by the following quadruple (B,k,μ,Γ), where

1. Γ is a closed subset of E (usually $\Gamma \subseteq \partial E$).
2. B is a vector field on E with its unique integral curves $z^x(t)$,

   such that $z^x(t)=x$ for $x\in\Gamma$ and any $t\geq 0$.
3. k is a nonnegative, bounded function on $E\setminus\Gamma$,
4. $\mu(x,\cdot)$ is a probabilistic measure on $E\setminus\{x\}\setminus\Gamma$ for $x\in E$.

Remark 1.
$z^x(t)$ can be considered as a nonlinear semi-group on E i.e. z satisfies the following semi-group property:

$$z^x(t+s) = z^{z^x(t)}(s).$$

One can construct an adjoint linear semi-group T_t on $\mathcal{B}(E)$ by the following: $T_tw(x) = w(z^x(t))$.
Here B is an extended generator of the semi-group T i.e.

$$w(z^x(t))-w(x) = T_tw(x)-w(x) = \int_0^t T_sBw(x)ds = \int_0^t Bw(z^x(s))ds \tag{1}$$

for any $w \in \mathcal{D} = \{w\in\mathcal{B}(E) : w(z^x(\cdot))\in W^{1,\infty}(\mathbb{R}_+)$ for any $x \in E$ and

$$\sup_{x\in E} \|w(z^x(\cdot))\|_{1,\infty} < +\infty \}. \tag{2}$$

The operator B is determined for any $w \in \mathcal{D}$.
In the points of differentiability of $w(z^x(\cdot))$ holds:

$$\frac{\partial}{\partial t}w(z^x(t)) = Bw(z^x(t)). \tag{3}$$

Denote jump epochs of the process by $T_1,T_2,\ldots,$. The process X is a strong Markov process with the following dynamics:

(i) $x_t = z^x(t)$ for $t < T_1$ P_x-a.s.,

(ii) $P_x(T_1 > t) = \begin{cases} \exp\{-\int_0^t k(z^x(s))ds\} & \text{if } T(x) < t, \\ 0 & \text{if } T(x) \geq t, \end{cases}$ where T is a

function $T : E \to \mathbb{R}_+ \cup \{+\infty\}$ defined by $T(x) = \inf\{t \geq 0 : z^x(t) \in \Gamma\}$.

(iii) $P_x(x_{T_1} \in A \mid \mathcal{F}_{T_1 -}) = \mu(z^x(T_1), A)$.

The following will be assumed in the paper:

A1. $Bw(x) = a(x)\nabla w(x)$ for $w \in C^1(\mathbb{R}^n)$, where a_i are bounded, Borel measurable functions on E for $i=1,\ldots,n$ and $x \in E$.

A2. $\int_E w(y)\mu(\cdot, dy) \in C_o(E)$ for $w \in C_o(E)$.

A3. $k \in C_b(Cl(E \setminus \Gamma))_+$,

A4. $T \in C^*(E)$,

A5. $\rho : \{x \in E : T(x) < \infty\} \to \Gamma$ is continuous, where $\rho(x) = z^x(T(x))$,

A6. $E_x N_t < +\infty$ for any $t < +\infty$, where $N_t = \text{card}\{k : T_k \leq t\}$.

Define an operator A and a function space $\mathcal{U}$ by:

$$Aw(x) = Bw(x) + k(x)\int_E (w(y) - w(x))\mu(x, dy) \text{ for } x \notin \Gamma, \tag{4}$$

$$\mathcal{U} = \Big\{ w \in LC(E) : w(z^x(\cdot)) \in W^{1,\infty}(\mathbb{R}_+) \text{ for any } x \in E ,$$

$$\sup_{x \in E \setminus \Gamma} \|w(z^x(\cdot))\|_{1,\infty} < +\infty \text{ and } w(x) = \int_E w(y)\mu(x,dy) \text{ for } x \in \Gamma \Big\}. \tag{5}$$

By (1) and (3), Aw is well defined for any $w \in \mathcal{U}$.
The following proposition holds:

Proposition 1. [3]
Let A1-A6 be satisfied.
Then the process $w(x_t) - \int_0^t Aw(x_s)ds$ is an $\mathcal{F}_t$-martingale for any $w \in \mathcal{U}$.

Other words A is the extended generator of X and $\mathcal{U} \subseteq \mathcal{D}(A)$ - the domain of the operator A.

□

Remark 2.

In the paper [3] the "reflection" set Γ is assumed to be a subset of the boundary ∂E. This assumption is not necessary, in fact the dynamics of the process on the boundary can be different from immediate jump.

Remark 3.

In general setting [3] a piecewise-deterministic process has two components (N_t, x_t), where N_t is the number of jumps up to time t. If dynamics of the process x_t does not depend of the number of jumps the first component is redundant and we simply use x_t for (N_t, x_t).

Proposition 2.

Let A1-A3 be satisfied. Assume moreover that $\Gamma = \emptyset$.

Then the processes x is a Feller processes i.e. for any $w \in C_o(E)$ holds:

$$P_t w(x) \to w(x) \quad \text{as } t \to 0, \tag{6}$$

$$P_t w \in C_o(E) \quad \text{for any } t \geq 0 \tag{7}$$

where $P_t w(x) = E_x w(x_t)$. (8)

If $\Gamma \neq \emptyset$ then x is not a Feller process.

Proof.

Let $w \in C_o(E)$ and $T_1 = \inf \{t \geq 0 : x_t \neq x_{t-}\}$ be the first jump epoch of the process x_t.

By the strong Markov property:

$$P_t w(x) = E_x I_{\{T_1 > t\}} w(x_t) + E_x I_{\{T_1 \leq t\}} w(x_t) =$$

$$= P_x\{T_1 > t\} w(z^x(t)) + E_x[I_{\{T_1 \leq t\}} P_{t-T_1} w(x_{T_1})] =$$

$$= \exp\{-\int_o^t k(z^x(s))ds\} w(z^x(t)) +$$

$$+ \int_o^t k(z^x(s)) \exp\{-\int_o^s k(z^x(u))du\} \cdot \int_E P_{t-s} w(y) \mu(z^x(s), dy) ds. \tag{9}$$

From this formula (6) follows immediately. In order to prove (7) we view (9) as an integral equation for the function

$$\psi(t, x) = P_t w(x).$$

By the standard technique of successively iterating the equation one can represent ψ as a uniformly convergent series. By A1-A3 each term of this series belongs to $C_o(E)$. Hence the same is true for ψ.

If $\Gamma \neq \emptyset$ then x is not quasi-left-continuous and in consequence not a Feller process.

□

3. Impulsive control problem.

Let α be a positive fixed constant, let f be a measurable function on E and let c be a measurable function on E^2. The aim of this section is to study impulsive control problems for piecewise-deterministic processes. In the case of impulsive control problem an admissible policy is not a single stopping time but a sequence of stopping times and new starting points $\{\sigma_1, \zeta_1, \sigma_2, \zeta_2, \ldots\}$. Start with constructing the probability space associated to given admissible policy.

Let $(\Omega^{\mathbb{N}}, \mathcal{F}^{\otimes\mathbb{N}})$ be the reference probability space and let

$\mathcal{F}_t^n = \sigma(A_1 \times A_2 \times \ldots : A_i \in \mathcal{F}_t$ for $i \leq n$ and $A_i \in \{\emptyset, \Omega\}$ for $i > n$).

The sequence $\pi = (\tau_n, \xi_n)$ will be called *impulsive control policy*, iff

(i) $\tau_n \geq \tau_{n-1}$ and $\tau_n \to \infty$,

(ii) τ_n is an $\mathcal{F}_t^n$-Markov time,

(iii) ξ_n is an $\mathcal{F}_{\tau_n}^n$ -measurable random variable on E .

Denote by Π the set of all impulsive control policies.

Proposition 3. [12]

Let A1-A6 be satisfied and let the operator A be defined by (4). Then for any impulsive policy $\pi \in \Pi$ there exists a sequence of Markov processes $X^n = (\Omega^{\mathbb{N}}, \mathcal{F}^{\otimes\mathbb{N}}, \mathcal{F}_t^n, x_t^n, P_x^\pi)$, such that:

(iv) $(\Omega^{\mathbb{N}}, \mathcal{F}^{\otimes\mathbb{N}}, \mathcal{F}_t^1, x_t^1, P_x^\pi)$ is a piecewise-deterministic Markov process with the extended generator A ,

(v) $P_x^\pi(x_t^n = \xi_{n-1}$ for $t \leq \tau_{n-1}) = 1$ for $n \geq 2$,

(vi) $w(x_t^n) - \int_{\tau_{n-1}\wedge t}^{t} Aw(x_s^n)ds$ is an $\mathcal{F}_t^n$-martingale for any $w \in W^{1,\infty}(E)$ and $n \geq 2$ (other words x_t^n is a Markov process with the generator A with birth time τ_{n-1}).

□

The problem is to characterize the value function:

$$w_\alpha(x) = \inf_{\pi \in \Pi} S_\infty^\alpha(x, \pi) \tag{10}$$

where

$$S_t^{\alpha}(\pi,x) := E_x^{\pi}\left\{ \int_o^{\tau_1 \wedge t} f(x_s^1)e^{-\alpha s}\, ds + c(x_{\tau_1}^1, \xi_1)\cdot I_{\{\tau_1 \le t\}}\cdot e^{-\alpha\tau_1} \right\} +$$

$$+ \sum_{n=2}^{\infty} E_x^{\pi}\left\{ \int_{\tau_{n-1}\wedge t}^{\tau_n \wedge t} f(x_s^n)e^{-\alpha s}\, ds + c(x_{\tau_n}^n, \xi_n)\cdot I_{\{\tau_n \le t\}}\cdot e^{-\alpha\tau_n} \right\} . \qquad (11)$$

and to find a $\pi_o \in \Pi$ such that $w_\alpha(x) = S_\infty^\alpha(x, \pi_o)$ (if possible).

The tool is the following quasi-variational inequality:

$$\begin{cases} (Aw_\alpha(x) - \alpha w_\alpha(x) + f(x)) \wedge (Mw_\alpha(x) - w_\alpha(x)) = 0 & \text{for } x \notin \Gamma, \\ w_\alpha(x) = \int_E w_\alpha(y)\mu(x, dy) & \text{for } x \in \Gamma. \end{cases} \qquad (12)$$

where $Mw(x) = \inf_{y \in U} \{c(x,y) + w(y)\}$. (13)

Remark 9.

Notice that the impulse obstacle is different from the usual

$$\inf_{\substack{y \in U \\ x+y \in E}} \{c(y) + w(x+y)\} \qquad \text{(see [2,11])} .$$

The following theorem holds:

Theorem 1. [6]

Let A1-A6 be satisfied and let:

B1. $U \subseteq E$ be a compact set,

B2. $c(x,z) \ge \int_E c(y,z)\mu(x, dy)$ for $x \in \Gamma$ and $z \in U$,

B3. $c \in C_b(E \times U)_+$,

B4. $c \ge \gamma > 0$,

B5. $c(\cdot,y) \in \mathcal{U}$ and $\|Bc(\cdot,y)\| \le R < \infty$ for any $y \in U$,

B6. $f \in C_b(E)_+$,

B7. $\alpha > 0$.

Let w_α be defined by (10).

Then $w_\alpha \in \mathcal{U}$ and is a unique solution of (12).

Moreover $w_\alpha(x) = S_\infty^\alpha\left(x, \left\{\tau_o, \xi\left(x_{\tau_o}\right)\right\}^\infty\right)$, where $\tau_o = \inf\{t \ge 0 : w_\alpha(x_t) = Mw_\alpha(x_t)\}$

and $c(x, \xi(x)) + w_\alpha(\xi(x)) = Mw_\alpha(x)$.

In addition if $\Gamma = \emptyset$ then $w_\alpha \in C_b(E)$.

□

Under stronger assumptions we have better regularity properties of the value function w_α.

Theorem 2.
Let A1-A6 and B1-B2 be satisfied and let

C1. $Bw(x) = a(x)\cdot\nabla w(x)$, where $a_i \in W^{1,\infty}(E)$ for $i = 1,\dots,n$.

C2. E - bounded with smooth boundary,

C3. $\int_E \phi(y)\mu(\cdot,dy) \in W^{2,\infty}(E)$ for any $\phi \in L^\infty(E)$ and $\|\int_E \phi(y)\mu(\cdot,dy)\|_{2,\infty} \le K\|\phi\|$, where $\|\cdot\|_{2,\infty}$ denotes the norm in the space $W^{2,\infty}(E)$,

C4. $a(x)\cdot\eta(x) > 0$ for any $x \in \partial E$, where η is the exterior normal vector to .

C5. $\Gamma = \partial E$,

C6. $c \in C_b(E^2)$.

C7. $c \ge \gamma > 0$,

C8. $c(\cdot,y) \in W^{1,\infty}(E)$ and $\|c(\cdot,y)\|_{1,\infty} \le K < \infty$ for any $y \in E$,

C9. $f \in W^{1,\infty}(E)$ and $f \ge 0$,

C10. $k \in W^{1,\infty}(E)$ and $k \ge 0$,

Then $w_\alpha \in W^{1,\infty}(E)$ for sufficiently large $\alpha > 0$.

Proof.

Denote $d_\alpha(x) = Mw_\alpha(x)$. Notice that $d_\alpha \in W^{1,\infty}(E)$ and

$d_\alpha(x,z) \ge \int_E d_\alpha(y,z)\mu(x, dy)$ for $x \in \Gamma$ and $z \in U$.

Since w_α satisfies the following variational inequality:

$$\begin{cases} (Aw_\alpha(x)-\alpha w_\alpha(x)+f(x))\wedge(d_\alpha(x)-w_\alpha(x))=0 & \text{for } x \notin \Gamma, \\ w_\alpha(x) = \int_E w_\alpha(y)\mu(x, dy) & \text{for } x \in \Gamma \end{cases} \tag{14}$$

and by [9], $w_\alpha \in W^{1,\infty}(E)$ for sufficiently large $\alpha > 0$.

□

This result will be applied in the next Section.

4. Long run average cost function.

In this section we study the ergodic impulsive control problem. Define the cost function $r(x) = \inf_{\pi\in\Pi} \liminf_{t\to\infty} t^{-1}\cdot S_t^o(\pi,x)$.

The following theorem holds:

Theorem 3. [7]
Let A1-A6, B2 and C1-C10 be satisfied and let
D1. $U = Cl(E)$,
D2. $k \geq \beta$, where β is sufficiently large.

Let $$q = \liminf_{\alpha \to 0} \inf_{x \in E} w_\alpha(x), \tag{15}$$

where w_α is defined by (10).

Then $r(x) \equiv q$ and the policy $\pi^* = \{\tau_n^*, \xi(y_{\tau_n^*})\}_{n=1,2,\ldots}$ is optimal,

where
$$\tau_1^* = \inf\{t \geq 0 : w_\alpha(x_t^1) = Mw_\alpha(x_t^1)\}, \tag{16}$$

$$\tau_n^* = \inf\{t \geq \tau_{n-1}^* : w_\alpha(x_t^n) = Mw_\alpha(x_t^n)\} \quad \text{and} \tag{17}$$

ξ is a function $\xi : E \to E \setminus \Gamma$, such that
$$c(x, \xi(x)) + w_\alpha(\xi(x)) = Mw_\alpha(x) \ , \tag{18}$$

where $w \in W^{1,\infty}(E)$ and is a solution of the following quasi-variational inequality:

$$\begin{cases} (Aw(x)+f(x)-q) \wedge (Mw(x)-w(x))=0 & \text{for } x \notin \Gamma \\ w(x) = \int_E w(y)\mu(x, dy) & \text{for } x \in \Gamma. \end{cases} \tag{19}$$

Proof.
Let $$v_\alpha(x) = w_\alpha(x) - \inf_{x \in E} w_\alpha(x) \geq 0. \tag{20}$$

Notice that:
(i) $0 \leq v_\alpha(x) \leq \|c\|$,
(ii) $\nabla v_\alpha(x) = \nabla w_\alpha(x)$,
(iii) $\|\alpha w_\alpha\| \leq \|f\|$.
Notice that v_α satisfies the following variational inequality:

$$\begin{cases} (a(x)\cdot\nabla v_\alpha(x)-(\alpha+k(x))v_\alpha(x)+f_\alpha(x)) \wedge (d_\alpha(x)-v_\alpha(x))=0 & \text{for } x \notin \Gamma, \\ v_\alpha(x) = \int_E v_\alpha(y)\mu(x, dy) & \text{for } x \in \Gamma, \end{cases}$$

where $d_\alpha(x) = Mv_\alpha(x)$ and (21)
$$f_\alpha(x) = f(x) + k(x)\int_E v_\alpha(y)\mu(x,dy) - \inf_{x \in E} \alpha w_\alpha(x). \tag{22}$$

By assumptions and [9], $\|\nabla v_\alpha\| \leq K_1 + K_2\|v_\alpha\| \leq K_1 + K_2\|c\|$, where the constants K_1 and K_2 are independent of α, if $k \geq \beta$ with β sufficiently large.

By Arzela and Alaoglu theorems there exists a subsequence α_n of α such that :

$$\alpha_n w_{\alpha_n} \to q \quad \text{uniformly,} \tag{23}$$

$$v_{\alpha_n}(x) \to w(x) \quad \text{uniformly ,} \tag{24}$$

$$\nabla v_{\alpha_n} \to \nabla w \quad \text{in } L^\infty(E) \text{ weakly}^*. \tag{25}$$

Hence $Mv_{\alpha_n}(x) \to Mw(x)$ uniformly and $Av_{\alpha_n} \to Aw-q$ in $L^\infty(E)$ weakly*.

Obviously $w(x) = \int_E w(y)\mu(x,dy)$ for $x \in \Gamma$, $w \leq Mw$ and $Aw(x)+f(x)-q \geq 0$ for $x \notin \Gamma$.

To prove that $(Aw(x)+f(x)-q)(w(x)-Mw(x)) = 0$ for $x \notin \Gamma$ assume that $w < Mw$ for $x \in V$ an open subset of E. Thus there exists an open subset $W \subseteq V$ such that $w_\alpha(x) < Mw_\alpha$ in W for α sufficiently small. Therefore $(Aw(x)+f(x)-q)(w(x)-Mw(x))=0$ for $x \notin \Gamma$.

Hence $(Aw(x)-q+f(x))\wedge(Mw(x)-w(x))=0$ for $x \notin \Gamma$.

Notice that

$$M_t = w(x_t^1)-\int_0^t Aw(x_s^1)ds \text{ is an } \mathcal{F}_t^1\text{-martingale since } w \in W^{1,\infty}(E). \tag{26}$$

Let $\pi \in \Pi$ be any admissible policy and let

$$N_t^m = \int_0^{\tau_1 \wedge t} f(x_s^1)ds + c(x_{\tau_1}^1, \xi_1)\cdot I_{\{\tau_1 \leq t\}} - qt\wedge\tau_m + w(x_t^1)\cdot I_{\{\tau_1 > t\}} +$$

$$+ \sum_{n=1}^m \left\{ \int_{\tau_{n-1}\wedge t}^{\tau_n \wedge t} f(x_s^n)ds + c(x_{\tau_n}^n, \xi_n)\cdot I_{\{\tau_n \leq t\}} + w(x_t^n)\cdot I_{\{\tau_{n-1} \leq t < \tau_n\}} \right\}. \tag{27}$$

By (20) and (24) $0 \leq w(x) \leq E_x^\pi\{N_t^1\}$ and by induction $w(x) \leq E_x^\pi\{N_t^n\}$ for any $n \geq 1$ and $t \geq 0$. Hence $w(x) \leq E_x^\pi\{N_t^\infty\}$.

But since w is bounded

$$\liminf_{t\to\infty} t^{-1}\cdot S_t^o(\pi,x) - q = \liminf_{t\to\infty} t^{-1}\cdot E_x^\pi\{N_t^\infty\} \geq 0 . \tag{28}$$

Therefore $q \leq \inf_{\pi\in\Pi} \liminf_{t\to\infty} t^{-1}\cdot S_t^o(\pi,x)$.

Since $w \in \mathcal{U}$ and by assumption B2. there exists a function $\xi : E \to \text{Int}E$, such that $c(x,\xi(x))+w(\xi(x)) = Mw(x)$.

To prove that $q = \inf_{\pi\in\Pi} \liminf_{t\to\infty} t^{-1}\cdot S_t^o(\pi,x)$ define $\pi^* \in \Pi$ by (16)-(18).

Prove that $\pi^* \in \Pi$, i. e. $\tau^*_n \to \infty$.

Suppose that $P^{\pi^*}(\lim_{n\to\infty} \tau^*_n < t) = \beta > 0$.

Let

$$G^m_t = \int_o^{\tau^*_1\wedge t} f(x^1_s)ds + c(x^1_{\tau^*_1},\xi^*_1)\cdot I_{\{\tau^*_1\le t\}} - qt\wedge\tau^*_m + w(x^1_t)\cdot I_{\{\tau^*_1 > t\}} +$$

$$+ \sum_{n=2}^{m}\left\{\int_{\tau^*_{n-1}\wedge t}^{\tau^*_n\wedge t} f(x^n_s)ds + c(x^n_{\tau^*_n},\xi^*_n)\cdot I_{\{\tau^*_n\le t\}} + w(x^n_t)\cdot I_{\{\tau^*_{n-1}\le t < \tau^*_n\}}\right\}.$$

By (26) $0 \le w(x) = E^{\pi^*}_x\{G^1_t\}$ and by induction $w(x) = E^{\pi^*}_x\{G^n_t\}$ for any $n\ge1$ and $t\ge0$.

Hence $w(x) = E^{\pi^*}_x\{G^n_t\} > n\beta\gamma - qt \to \infty$ as $n\to\infty$ a contradiction, since w is bounded. Therefore $\tau^*_n\to\infty$.

Letting n tend to infinity we get $w(x) = E^{\pi^*}_x\{G^{\infty}_t\}$.

Since w is bounded

$$q = \liminf_{t\to\infty} t^{-1}\cdot E^{\pi^*}_x\{G^{\infty}_t\} + q = \liminf_{t\to\infty} t^{-1}\cdot S^o_t(\pi^*,x) =$$

$$= \inf_{\pi\in\Pi} \liminf_{t\to\infty} t^{-1}\cdot S^o_t(\pi,x). \tag{29}$$

This completes the proof.

□

Acknowledgment.

I am indebted to Prof. Jerzy Zabczyk for his very helpful comments.

References.

[1] A. Bensoussan and J.L. Lions, *Applications des Inequations Variationnelles en Controle Stochastique*, Dunod, Paris 1978.

[2] A. Bensoussan and J.L. Lions, *Controle Impulsionnel et Inequations Quasi-Vatiationnelles*, Dunod, Paris 1982

[3] M. H. A. Davis, Piecewise-deterministic Markov processes: a general class of non-diffusion stochastic models, *J.Royal Statist. Soc.* (B) 46 (1984), 353-388.

[4] E. B. Dynkin, *Markov Processes*, Springer, Berlin 1965

[5] A. Friedmann and M. Robin, The free boundary for variational inequalities with nonlocal operators, *SIAM Control Opt.*16 (1978),

347-372.

[6] D. Gatarek, On value functions for impulsive control of piecewise-deterministic processes, submitted for *Stochastics*,

[7] D. Gatarek, Impulsive control of piecewise-deterministic processes with long run average cost, in preparation,

[8] U. S. Gugerli, Optimal stopping of a piecewise-deterministic Markov process, *Stochastics* 19 (1986), 221-236.

[9] S. Lenhart and Y.-C. Liao, Integro-differential equations associated with optimal stopping time of a piecewise-deterministic process, *Stochastics* 15 (1985), 183-207.

[10] S. Lenhart and Y.-C. Liao, Switching control of piecewise-deterministic processes, a preprint.

[11] S. Lenhart, Viscosity solutions associated with impulsive control problems for piecewise-deterministic process, a preprint

[12] M. Robin, *Controle Impulsionnel des Processus de Markov*, Thesis, Universite Paris IX, 1978

[13] Ł. Stettner and J. Zabczyk, Optimal stopping for Feller processes, Preprint Nr 284, Institute of Mathematics PAS, Warsaw 1983

[14] A. A. Yushkevich, Continuous-time Markov decision processes with interventions, *Stochastics 9* (1983), 235-274.

SYNTHESIS OF OPTIMAL CONTROLS

U.G. Haussmann

Mathematics Department
University of British Columbia
121 - 1984 Mathematics Road
Vancouver, Canada, V6T 1Y4

1. Introduction

In [2] we consider the stochastic control problem

$$\inf \{J(u,S;s,x) \; : \; (u,S) \text{ admissible}\}$$

where

$$J(u,S;s,x) = E \int_s^{S\wedge\rho} f_0(t,X(t),u(t))\, dt + h_0(S\wedge\rho, X(S\wedge\rho)),$$

(1.1) $$X(t) = x + \int_s^t b(\theta,X(\theta),u(\theta))\, d\theta + \int_s^t \sigma(\theta,X(\theta),u(\theta))\, dw(\theta),$$

ρ is the first exit time of $(t,X(t))$ from $D \subset \mathbb{R}^{d+1}$, and (u,S) is admissible if (i) S is a stopping time and (ii) u is a suitable control process such that (1.1) has a solution on $[s,S]$ which satisfies certain hard and soft constraints. The point of [2] is to establish a very general existence result and then to prove the existence of an optimal control in Markov or feedback form, i.e. existence of an optimal control law. In this note we elaborate on these points in the deterministic case, i.e. $\sigma = 0$ above. We show that the existence result in [2] gives a result as general as just about any in the literature and furthermore we give conditions on the data such that in this case ($\sigma = 0$) the synthesis problem has a solution, i.e. there exists an optimal control law.

In the past, [1] Ch. IV, Theorem 7.2, the existence of optimal control laws was established by showing that the Hamilton-Jacobi-Bellman equation has a (piecewise) smooth solution and then using a measurable selection theorem. But if the data are not sufficiently regular to guarantee a smooth solution then one could say nothing about optimal control laws. The present result remedies this situation. We emphasize that for the proof we consider the deterministic problem as a degenerate stochastic problem.

In section two we formulate the problem precisely and give the existence result for optimal controls. Then in section three we give conditions such that the sythesis problem has a solution.

We continue with some notation.

- $x \wedge y = \min\{x,y\}$.
- $\mathbf{R}_+ = [0,\infty)$, $\overline{\mathbf{R}}_+ = [0,\infty]$, $\mathbf{R}_+^m = (\mathbf{R}_+)^m$, $\overline{\mathbf{R}}_+^m = (\overline{\mathbf{R}}_+)^m$.
- If $D \subset \mathbf{R}_+ \times \mathbf{R}^d$ then ∂D is the relative boundary of D, i.e. ∂D consists of all $(t,x) \in \mathbf{R}_+ \times \mathbf{R}^d$ such that any ϵ-ball about (t,x) intersects both D and $(\mathbf{R}_+ \times \mathbf{R}^d)\setminus D$. $\bar{D}$ is the closure of D, and

$$\bar{D}_\infty = \{(\infty,x) \in \overline{\mathbf{R}}_+ \times \mathbf{R}^d : \text{there exist } x_n \to x,\ t_n \to \infty,\ (t_n,x_n) \in D\}$$

$$\bar{\bar{D}} = \bar{D} \cup \bar{D}_\infty .$$

2. The Existence of Optimal Controls

Let us formulate the control problem to be considered. We are given D, a subset of $\mathbf{R}_+ \times \mathbf{R}^d$, open in the relative topology of $\mathbf{R}_+ \times \mathbf{R}^d$, U, a closed subset of a Euclidean space (or more generally a closed, σ-compact subset of a Banach space), $\lambda \in \mathbf{R}^m$ and functions b, f, h satisfying

(2.1) $b : D \times U \to \mathbf{R}^d$, measurable, such that $(x,u) \to b(t,x,u)$ is continuous for each t, and for some k

$$|b(t,x,u)| \le k\,(1 + |x| + |u|),$$

(2.2) $f : D \times U \to \overline{\mathbf{R}}_+^{1+m}$, measurable, such that each component of f is lower semi-continuous, i.e. ℓ.s.c., in (x,u) for each t,

(2.3) $h : \bar{\bar{D}} \to \overline{\mathbf{R}}_+^{1+m}$, such that each component of h is ℓ.s.c. and constant on $\bar{D}_\infty$.

For $(s,x) \in D$ fixed we say that $\alpha := (X,u,S)$ is <u>admissible</u>, i.e. $\alpha \in \underline{\underline{U}}(s,x)$, if $S \ge s$, X is absolutely continuous on $[s, S\wedge\rho]$ where ρ is the first exit time of $(t,X(t))$ from D, $u \in L^1_{\ell oc}[s,S\wedge\rho]$, and the pair (X,u) satisfies

(2.4) $X(t) = x + \int_s^t b(\theta,X(\theta),u(\theta))\,d\theta, \quad s \le t \le S\wedge\rho.$

Note that S and/or ρ may be $+\infty$. For $\alpha = (X,u,S) \in \underline{\underline{U}}(s,x)$ we define

(2.5) $J(\alpha,s,x) = \int_s^{S\wedge\rho} f(t,X(t),u(t))\,dt + h(S\wedge\rho, X(S\wedge\rho)).$

Then $\alpha = (X,u,S)$ is said to be <u>feasible</u>, i.e. $\alpha \in \underline{\underline{U}}^f(s,x)$, if $J_0(\alpha,s,x) < \infty$ and

$J_i(\alpha,s,x) \le \lambda_i, i=1,\ldots,m$. The control problem is

(2.6) $\quad \inf \{J_0(\alpha,s,x) : \alpha \in \underline{\underline{U}}^f(s,x)\}$.

Observe that J in (2.5) is well-defined because $f_i \ge 0$ implies that the integral is defined, and h_i = constant (possibly $+\infty$) on $\bar{D}_\infty$ implies that $h(S\wedge\rho, X(S\wedge\rho))$ is defined even when $S\wedge\rho = +\infty$. Usually h is taken to have the form : $h(t,x) = e^{-\delta t}\,\bar{h}(x)$ with $\bar{h}$ bounded and $\delta > 0$. Then indeed h is constant (=0) on $\bar{D}_\infty$.

We point out that in the deterministic literature, [1] Ch. III, the existence theorems usually take $D = \mathbf{R}_+ \times \mathbf{R}^d$ but allow S to be a terminal time chosen so that $(S,X(S))$ lies in a specified target set B (the free terminal time problem). This can be achieved by setting $h_0(t,x) = +\infty$ if $(t,x) \notin B$. If B is closed and h is ℓ.s.c. on B then it is ℓ.s.c. on $\bar{\bar{D}}$. On the other hand when dynamic programming is treated, [1], Ch. IV, then the controller <u>must</u> stop when the boundary of some region is reached, and earlier stopping is not allowed, i.e. the problem is formulated with the set D but without the stopping time S. In the above model we need only set $h_0(t,x) = +\infty$ on D. Hence our problem formulation encompasses both possibilities.

We mention next that the hypotheses $h_i \ge 0$, $f_i \ge 0$ can be relaxed as follows. If $h_i(t,x) \ge -M > -\infty$ and $f_i(t,x,u) \ge -\bar{f}_i(t)$ where $\bar{f}_i \ge 0$, integrable, then set

$$\tilde{h}_i(t,x) = h_i(t,x) + M + \int_t^\infty \bar{f}_i(\theta)d\theta$$

$$\tilde{f}_i(t,x,u) = f_i(t,x,u) + \bar{f}_i(t)$$

$$\tilde{\lambda}_i = \lambda_i + M + \int_s^\infty \bar{f}_i(t)dt.$$

Then $\tilde{h}_i \ge 0$, $\tilde{f}_i \ge 0$ and they have the same regularity as h_i, f_i. Moreover $J_i \ge \lambda_i$ if and only if $\tilde{J}_i := \int_s^{\rho\wedge S} \tilde{f}_i\, dt + \tilde{h}_i(\rho\wedge S, \tilde{X}(\rho\wedge S)) \le \tilde{\lambda}_i$.

As pointed out above, since h_0 can assume the value $+\infty$ then contraints at the terminal time can be included. Since f_0 may also assume the value $+\infty$ then the generic problem includes problems with constraints of the form

$$X(t) \in A_o(t) \ , \ u(t) \in U(t,X(t))$$

for given sets $A_o(t)$ and $U(t,x) \in U$, or more generally of the form

$$(X(t),u(t)) \in A(t) \quad \text{a.e.}(t)$$

provided the sets A(t) are closed and the graph of A

$$G(A) = \{(t,x,u) : (x,u) \in A(t)\}$$

is measurable. We simply set $f_0(t,x,u) = +\infty$ if $(x,u) \notin A(t)$. If $f_0(t,\cdot,\cdot)$ is ℓ.s.c. on the closed set A(t), then it is ℓ.s.c. Moreover it is measurable if it is measurable when restricted to the measurable set G(A).

Let us add that $\underline{\underline{U}}(s,x) \ne \emptyset$ since for any u in $L^1_{\ell oc}[s,\infty)$ $b(t,x,u(t))$ satisfies the Caratheodory condition on any "rectangle" $\{|x| < N\} \times (s,T)$ so that solutions of (2.4) exist and are defined up to the boundary of this rectangle. Hence for S sufficiently small (but $S\ge s$) , $(X,u,S) \in \underline{\underline{U}}(s,x)$. On the other hand we do not claim that X is the unique solution of (2.4); hence for some u there may exist (X^1,S^1) and (X^2,S^2) such that $(X^i,u,S^i) \in \underline{\underline{U}}(s,x)$ for i = 1,2 with $X^1 \neq X^2$.

For any continuous map $\xi : [0,\infty) \to \mathbf{R}^d$ we define

(2.7) $\rho_o(\xi) = \inf\{t\geq 0 : (t,\xi(t)) \notin D\}$.

If $(X,u,S) \in \underline{\underline{U}}(s,x)$ then X(t) is defined for $s\leq t\leq \rho\wedge S$ where ρ is the first exit time of $(t,X(t))$ from D, so $\rho = \rho_o(X(\cdot))$ as defined by (2.7) if we extend X to $[0,\infty)$ by defining

$$X(t) = \begin{cases} X(\rho\wedge S), & t > \rho\wedge S \\ x, & t < s \end{cases}.$$

Since D is (relatively) open then $\xi \to \rho_o(\xi)$ is ℓ.s.c. Unfortunately it is not always continuous.

<u>Theorem 1.</u> Assume (2.1) - (2.3) and

(2.8) $\{((t,x,u),z) \in \mathbf{R}^d \times \mathbf{R}^{1+m} : z_i \geq f_i(t,x,u), i = 0,1,\ldots,m,\ u \in U\}$

is convex for all $(t,x) \in D$,

(2.9) there exists $\ell \in\{0,1,\cdots,m\}$ and $\tilde{f} : U \to \mathbf{R}_+$, continuous, such that for all $(t,x,u) \in D \times U$, $f_\ell(t,x,u) \geq \tilde{f}(u)$ and

$$\lim_{\substack{|u|\to\infty \\ u\in U}} \frac{\tilde{f}(u)}{|u|} = +\infty,$$

(2.10) either (i) $X(\cdot) \to \rho_o(X(\cdot))$ is continuous for each $(X,u,S) \in \underline{\underline{U}}^f(s,x)$, or (ii) $D = [0,T) \times O$, O open, $h(t,x) = \tilde{h}(t)$ for $x \in \partial O$ and each component of $\tilde{h}$ is left-continuous, non-decreasing,

(2.11) $\underline{\underline{U}}^f(s,x) \neq \emptyset$,

then there exists $\alpha^* \in \underline{\underline{U}}^f(s,x)$ such that $J(\alpha^*,s,x) \leq J(\alpha,s,x)$ for all $\alpha \in \underline{\underline{U}}^f(s,x)$.

<u>Remark.</u> The conditions (2.1) - (2.3) are rather minimal and innocuous, but (2.8) is not. This convexity is required to move from an optimal control wich is relaxed (or a chattering control) to one which is not. The coercivity condition (2.9) is trivially satisfied if U is compact. Moreover nothing is gained by taking $p \geq 1$ and assuming $u \in L^p_{\ell oc}[s,\infty)$,

$$|b(t,x,u)| \leq k(1+|x|+|u|^p)$$

$$\frac{\tilde{f}(u)}{|u|^p} \to \infty$$

since the transformation $v_i = u_i^p$ $\forall i$ maps U into a closed set V, and if $\tilde{g}(v) = \tilde{f}(u)$, then

$$\frac{\tilde{g}(v)}{|v|} \geq \frac{\tilde{f}(u)}{|u|^p} \to \infty.$$

Hence by rewriting the problem in terms of v and V we are back to the framework of the theorem. Conditions (2.8) and (2.9) are standard for existence. Condition (2.10) is satisfied if $D = [0,T) \times \mathbf{R}^d$ (with $T = +\infty$ possibly) - the usual situation for existence theorems in the literature. Below we shall give conditions such that (2.10i) holds. Finally (2.11) simply requires that the minimization problem not be trivial.

Proof: For any $\alpha = (X,u,S) \in \underline{\underline{U}}^f(s,x)$ we can extend X to be defined on $[0,\infty]$ by setting

$$X(t) = \begin{cases} x & \text{if } t \leq s \\ X(S \wedge \rho) & \text{if } t \geq S \wedge \rho \end{cases}$$

Then $X \in C(\mathbf{R}_+;\mathbf{R}^d) := C$, the space of continuous functions : $\mathbf{R}_+ \to \mathbf{R}^d$.

Let P be the probability measure on C which assigns unit mass to $\{X(\cdot)\}$. If we denote the Borel σ-algebra on C by $\underline{\underline{C}}$ and the canonic Borel filtration by $\{\underline{\underline{C}}_t\}$, then $(C,\underline{\underline{C}},P,\{\underline{\underline{C}}_t\},X(\cdot),u(\cdot),S)$ is a strict control in the sense of [2] Proposition 3.1, with a = 0 (in the notation of [2]); in fact it is a natural control, cf. [2] Definition 3.8. Thus α can be considered to be a feasible control for a stochastic control problem as considered in [2] and the minimum cost for this problem will be no greater than the inf in (2.6). According to [2] Remark 4.3, Lemma 4.4 and Corollary 4.8, there exists an optimal control for the stochastic problem, i.e. there exists a probability space with filtration $(\tilde{\Omega},\tilde{\underline{\underline{F}}},\tilde{P},\{\tilde{\underline{\underline{F}}}_t\})$, two stochastic processes $\{\tilde{X}_t\},\{\tilde{u}_t\}$ and a stopping time $\tilde{S}$ such that (E denotes expectation)

$$E \int_s^{T\wedge\rho\wedge\tilde{S}} |\tilde{u}_t| \, dt < \infty \quad \forall\, T < \infty,$$

$\tilde{X}_t(\omega),\tilde{u}_t(\omega)$ satisfy (2.4) on $s \leq t \leq \tilde{S}\wedge\rho$ almost surely,

$$E\, J_i((\tilde{X},\tilde{u},\tilde{S}),s,x) \leq \lambda_i, \quad i = 1,2,\ldots,m,$$

with J_i given as in (2.5) and $E\, J_0((X,u,S),s,x)$ attains its minimum at $(\tilde{X},\tilde{u},\tilde{S})$ when the min is taken over all of $\Omega,\underline{\underline{F}},P,\{\underline{\underline{F}}_t\},\{X_t\},\{u_t\},S$ which satisfy the above restrictions. Define

$$\lambda_0 = E\, J_0((\tilde{X},\tilde{u},\tilde{S}),s,x).$$

Then

$$\tilde{P}\{\omega\in\tilde{\Omega} : J_i((\tilde{X}(\omega),\tilde{u}(\omega),\tilde{S}(\omega)),s,x)) \leq \lambda_i, i=0,1,\ldots,m\} \neq \emptyset,$$

hence there exists ω_o such that if $\alpha_o = (\tilde{X}(\omega_o),\tilde{u}(\omega_o),\tilde{S}(\omega_o))$ then $\alpha_o \in \underline{\underline{U}}^f(s,x)$. Considering this α_o as a feasible strict control for the stochastic problem, cf. above for the imbedding, then by construction of α_o and by the optimality of $(\tilde{X},\tilde{u},\tilde{S})$

$$\lambda_0 \leq J_0(\alpha_o,s,x) = E\, J_0(\alpha_o,s,x) \leq \lambda_0$$

if we recall that E is expectation with respect to P_o, a Dirac measure on C, with unit mass at $\tilde{X}(\omega_o)$. It follows that α_o solves (2.6). ∎

<u>Theorem 2.</u> (2.10 i) holds if $D = [0,T) \times O$ with $T \le +\infty$, O open with smooth boundary $\partial O = \Gamma_0 \cup \Gamma_1$, Γ_0, Γ_1 disjoint. Moreover if $n(x)$ is the unit outward normal to O at $x \in \partial O$ then there exist neighbourhoods N_i of Γ_i,
$i = 0,1$, and $\nu > 0$, $K < \infty$ such that

i) $b(t,x,u)\cdot n(x) \ge \nu \qquad \forall t \ge s,\ x \in \Gamma_1,\ u \in U,$
and b is continuous in x uniformly in $(t,x,u) \in [s,\infty) \times (N_1 \cap O) \times U$

ii) $b(t,x,u)\cdot n(x) \le 0 \qquad \forall\ t \ge s,\ x \in \Gamma_o,\ u \in U$
$|b(t,x,u) - b(t,x',u)| \le K(1+|u|)|x - x'| \quad \forall t \ge s,\ x,x' \in N_0 \cap O,$
$u \in U$.

Proof: We first show that $X(t)$ cannot exit O on Γ_o. Define $\delta(x) = \text{dist}(x,\partial O)$. For $x \in N_0 \cap O$ there exists $\tilde{x} \in \Gamma_0$ such that $\nabla\delta(x) = -n(\tilde{x})$ and $|x-\tilde{x}| = \delta(x)$. Hence

$$\begin{aligned} b(t,x,u) \cdot \nabla\delta(x) &= -b(t,x,u) \cdot n(\tilde{x}) \\ &= [b(t,\tilde{x},u) - b(t,x,u)] \cdot n(\tilde{x}) - b(t,\tilde{x},u) \cdot n(\tilde{x}) \\ &\ge -K(1 + |u|)\ \delta(x) \end{aligned}$$

since $b(t,\tilde{x},u) \cdot n(\tilde{x}) \ge 0$; hence for any solution $X(t)$ of (2.4)

$$\begin{aligned} \frac{d}{dt}\ \delta[X(t)] &= b(t,X(t),u(t)) \cdot \nabla\delta[X(t)] \\ &\ge -K(1 + |u(t)|)\ \delta[X(t)] \end{aligned}$$

i.e. $$\delta[X(t)] \ge \delta[X(t')] \exp\{-K \int_{t'}^{t} (1 + |u(\theta)|)\, d\theta\}$$

for $t \ge t'$, $X(\theta) \in N_0$ for all $\theta \in [t',t]$. Since the integral in the exponent is finite for all $t < \infty$, then the exponential decay for $X \in N_0$ implies $X(\cdot)$ cannot reach Γ_0 in finite time.

Let us now show that ρ_o is continuous if $X(\rho) \in \Gamma_1$. Since $b(t,x,u)\cdot\nabla\delta(x) \le -\nu$ for $x \in \Gamma_1$ (n.b. $\nabla\delta(x) = -n(x)$ for $x \in \partial O$) then by continuity of b uniformly in (t,u) we may assume that

$$b(t,x,u)\cdot\nabla\delta(x) \le -\nu/2\ ,\ x \in N_1.$$

We now require a certain uniformity of N_1. Choose $T' < \infty$ and let $\sigma = \rho \wedge T'$.
Then (2.1) implies

$$\sup\{|X(t)| : s \le t \le \sigma\} \le k_o(1 + |x| + \int_s^\sigma |u(t)|dt) \exp k(T'-s)$$

for any solution of (2.4). The coercivity (2.9) implies that for some m

$$\begin{aligned} \int_s^\sigma |u(t)|dt &\le m(T'-s) + J_\ell(\alpha,s,x) \\ &\le m(T'-s) + \lambda_\ell \end{aligned}$$

if $\alpha \in \underline{\underline{U}}^f(s,x)$. Note if $\ell = 0$ we take $\lambda_0 = J_0(\alpha_o,s,x)$ where α_o is some element of $\underline{\underline{U}}^f(s,x)$ ($\ne \emptyset$ by (2.11)) and we consider the problem with the added constraint

$J_0(\alpha,s,x) \le \lambda_0$ - clearly the optimal solutions are unchanged. Hence

$$\sup\{|X(t)| : s \le t \le \sigma\} \le k_1(1 + |x| + (T'-s))\, e^{k(T'-s)} := k_o.$$

Now choose $\epsilon > 0$ such that

$$N_\epsilon := \{x \epsilon O : \mathrm{dist}(x,\Gamma_1) < \epsilon,\ |x| \le k_o\} \subset N_1.$$

Suppose that $X(\rho) \in \Gamma_1$ and $X^n \to X$. We write $\rho = \rho_o(X)$, $\rho^n = \rho_o(X^n)$ and we assume that we have chosen $T' > \rho = \rho_o(X)$. If $\rho^n < \rho$ then $\lim_n \rho^n = \rho$ since $\rho_o(\cdot)$ is ℓ.s.c. If $\rho < \rho^n$ then for n sufficiently large $X^n(\rho) \in N_\epsilon$, and if $X^n(t) \in N_\epsilon$ then

$$\frac{d}{dt}\,\delta[X^n(t)] \le -\nu/2$$

so that X^n will continue to stay in N_ϵ as long as $|X^n| \le k_o$. Hence $X^n(t) \in N_\epsilon$ for $\rho \le t \le \sigma^n$ if $\sigma^n = \min\{\rho^n, T'\}$. It follows that

$$\begin{aligned}\sigma^n - \rho &\le 2\nu^{-1}(\delta[X^n(\rho)] - \delta[X^n(\sigma^n)])\\ &\le 2\nu^{-1}\delta[X^n(\rho)]\\ &\le 2\nu^{-1}|X^n(\rho) - X(\rho)|\end{aligned}$$

so $\sigma^n \to \rho$, i.e. $\rho^n \to \rho$ since $T' > \rho$.

The remaining possibility is that $\rho = T$ (if $T < \infty$). But clearly $\rho(\cdot)$ is continuous at X such that $\rho(X) = T$. ∎

We observe that we can dispense with "$D = [0,T) \times O$" provided we work with ∂D rather than ∂O and assume the required continuity of b is in (t,x) not just x. The above proof goes through with the normal now n(t,x) and b replaced by $\tilde{b} = (1,b)$. We also point out that if $D = [0,T) \times R^d$ then Γ_0 and Γ_1 are empty and so ρ_o is continuous.

3. The Synthesis Problem

We begin by defining the optimal Markov controls or optimal control laws. An <u>optimal control law</u> consists of a set $D' \subset D$ and a Borel function $u' : D' \to U$ such that for any $(s,x) \in D$ there exists a control (X,u,S) optimal in $\underline{\underline{U}}(s,x)$ and

(i) $S = \inf\{t \ge s : (t,X(t)) \notin D'\}$

(ii) $u(t) = u'(t,X(t))$.

It follows that for $(s,x) \notin D'$, immediate stopping is optimal, and for $(s,x) \in D'$, $u(t) = u'(t,X(t))$ is optimal for the problem where D' replaces D and $S = +\infty$, i.e. the terminal time is ρ, the first exit time of $(t,X(t))$ from D'.

We cannot allow integral constraints, so we take $m = 0$ and $J(\alpha,s,x) = J_0(\alpha,s,x)$. We also simplify the hard constraints by assuming that $f(t,x,u) < \infty$ if and only if $u \in U(t)$ for some closed sets $U(t) \subset U$.

We now add further hypotheses:

(3.1) either U is compact or (2.10i) holds;

(3.2) there exists $\lambda : D \to R_+$, locally bounded and upper semi-continuous, such that

$\{\alpha \in \underline{\underline{U}}^f(s,x) \ : \ J(\alpha,s,x) \le \lambda(s,x)\} \ne \emptyset$

for each $(s,x) \in D$;

(3.3) for any $(s,x) \in D$ there exists $\delta > 0$, $u^\delta \in \bigcap_{s \le t < s+\delta} U(t)$ and

$g : R_+ \to R_+$ locally integrable such that for all $t \in [s,s+\delta)$, $|x - x'| < \delta$

$$f(t,x',u^\delta) \le g(t).$$

If we have no stopping, i.e. if $S \ge \rho$, i.e. if $h = +\infty$ on D, then [2] Theorem 5.15 gives us

<u>Theorem 3</u> Assume (2.1) - (2.3), (2.8) - (2.10), (3.1) - (3.3) and $h = +\infty$ on D. Then there exists an optimal control law with $D' = D$.

<u>Remark</u> The above case of no stopping is the one to which dynamic programming is usually applied in the literature. Observe that (3.3) is satisfied provided

(3.4) $\quad f(t,x,u) \le k\ (1 + |x|^p + |u|^p)$

for some $p \in [1,\infty)$ and some constant k.

<u>Theorem 4</u> Assume (3.4), and

(3.5) h is continuous on $\partial D \cup \bar{D}_\infty$, f is continuous in x for all u and almost all t,

$|b(t,x,u) - b(t,x',u)| \le K\ (1+|u|)\ |x-x'|$

and with p as in (3.4) there exists a function $\overset{\circ}{u}$ in $L^p(R_+;U)$ such that $\overset{\circ}{u}(t) \in U(t)$ for all t.

Let $\overset{\circ}{X}_{sx}$ be the unique solution of (2.4) corresponding to $\overset{\circ}{u}$. If $X(\cdot) \to \rho_o(X(\cdot))$ is continuous at $X = \overset{\circ}{X}_{sx}$ for all $(s,x) \in D$ then (3.2) holds.

Proof: It follows easily that for $\overset{\circ}{S}_{sx} := \rho_o(\overset{\circ}{X}_{sx})$

$$(s,x) \to J((\overset{\circ}{X}_{sx},\overset{\circ}{u},\overset{\circ}{S}_{sx}),s,x)$$

is continuous on D. Hence (3.2) holds with

$$\lambda(s,x) = J((\overset{\circ}{X}_{sx},\overset{\circ}{u},\overset{\circ}{S}_{sx}),s,x).$$ ∎

<u>Remark</u> It now follows that in the standard situation without stopping where $D = [0,T) \times O$ there exists an optimal Markov control provided we assume (2.1) - (2.3), (2.8), (2.9), (3.4), (3.5) and $\partial O = \Gamma_0 \cup \Gamma_1$ as in Theorem 2.

In case stopping is allowed we have the same result.

<u>Theorem 5</u> Assume (2.1) - (2.3), (2.8), (2.9), (2.10i), (3.4), (3.5). Then there exists an optimal control law.

Proof: From [2] Remark 5.10 and Theorem 4 above we deduce that

$$(s,x) \to v(s,x) := \inf \{J(\alpha,s,x) \ : \ \alpha \in \underline{\underline{U}}^f(s,x)\}$$

is continuous, hence again by [2] Remark 5.10 it follows that we may set

$$D' = \{(s,x) : v(s,x) < h(s,x)\}$$

and solve the synthesis problem without stopping on D', i.e. in Theorem 3 we set $D = D'$ to obtain a control u'. Then (u',D') constitutes the optimal control law.

∎

References

[1] W.H. Fleming and R.W. Rishel, Deterministic and Stochastic Optimal Control, Springer Verlag, New York, 1975.

[2] U.G. Haussmann, On the existence of optimal controls, submitted to SIAM J. Control Optimization.

Consistent ML estimator for drift parameters of both ergodic and nonergodic diffusions

PIOTR KAZIMIERCZYK

Institute of Fundamental Technological Research, Polish Academy of Sciences
Świętokrzyska 21, 00-049 Warsaw, Poland

Consistency of the maximum-likelihood estimator (MLE) of a (multidimensional) parameter entering the drift term of an Itô equation is investigated. The results concern both ergodic and transient and even exploding solutions. The datum for the estimation is one sample-path (s-p). Accordingly, the definition of consistency is localised and the (sufficient) consistency condition is formulated in terms of a given s-p. In this way the consistency inference is based on the datum exclusively, and the convergence of the estimator can be analysed independently of whether the MLE is globally consistent (convergent for all possible data) or not. The theoretical considerations are illustrated by the numerical simulations of estimation of parameters of stochastic Duffing oscillator in ergodic, transient and exploding cases. Some essential computational aspects are clarified.

1. INTRODUCTION.

Efficiency of the stochastic calculus and models of the form of Itô equations attracts attention of rapidly increasing number of researchers interested in effective descriptions of dynamical systems. Accordingly, parametric identification of models formed of Itô equations becomes the more and more important scientific problem. Simultaneously, only a small part of models resulting from various applications satisfies the assumptions which due to the existing mathematical theory assure desirable properties of estimators proposed in context of such a parametric identification.

Let us focus the attention on the case, where the model is known up to the parameter $\theta \in R^m$ entering the drift term exclusively:

$$dX_t = \mathbf{m}_t(X,\theta)\,dt + \mathbf{b}_t(X)\,dW_t, \tag{1-1}$$

(here: $X_t \in R^n$ —is a state vector of a system, $W_t \in R^k$ —is an instantaneous value of a standard Wiener process W), and the datum for the identification is the function ϕ modelled as a sample-path (s-p) $X_\bullet(\omega_0)$ of a solution process X of equation (1-1)

$$\phi : [0,T] \ni t \to \phi(t) \simeq X_t(\omega_0) \in R^n, \tag{1-2}$$

($\omega_0 \in \Omega$ —is fixed, $\{\Omega, \mathcal{F}, P\}$ denotes a complete probability space). In such cases, usually, the maximum-likelihood estimator (*MLE*) $\hat{\theta}(X_\bullet(\omega))$ was proposed, see e.g. [1-7] and references there. The property which was investigated in most of the above mentioned articles was the consistency of MLE. All proofs (except of few connected with one-dimensional systems: $n = m = 1$) make use of the assumption that X is an ergodic process. To avoid problems with ergodicity most authors assume that the diffusion matrix $\mathbf{bb}^*$ ('$*$' denotes the transpose) is nondegenerate. Such assumptions are extremally limiting as to the applications. Mechanical oscillators require diffusion matrices degenerate everywhere. Modelling dynamical systems one is often interested mainly in nonsteady states. Failure analysis requires descriptions of systems with exploding solutions.

In this paper we give the conditions which are suitable for infering from s-ps $X_\bullet(\omega)$ about the convergence of the MLE $\hat{\theta}(X_\bullet(\omega))$ to θ in all above mentioned situations, not only for diffusions but for diffusion-type processes as well.

Our approach is innovative in one more apsect. The question which is commonly analyzed is whether $P(\hat{\theta}(X_\bullet(\omega)) \to \theta) \to 1$ (the weak consistency). However, during the identification we are given one particular s-p $X_\bullet(\omega_0)$. Therefore it is much better to know whether for almost *all* sample-paths (s-ps) the MLE tends to θ (the strong consistency). But the question which is *really* of primary importance is whether just for *this* very s-p $X_\bullet(\omega_0)$ which we deal with it holds true that $\hat{\theta}(X_\bullet(\omega_0)) \to \theta$. And that is still the essential question also where it is *not* true that $P(\hat{\theta}(X_\bullet(\omega)) \to \theta) = 1$. Accordingly, we modify the definition of consistency to be able to handle such situations—our inference is based on individual s-ps.

Results of numerical simulations (we simulate diffusions—that is the reason for the restriction in the title) illustrate the theory as well as the above mentioned and some more aspects of practical importance. In particular some problems connected with the approximation (1-2) are considered.

2. Consistency Conditions

It was proven in [8](see also [9]) that, if coordinate functionals of drift and diffusion coefficients $\mathbf{m}$ and $\mathbf{b}$ satisfy local Lipschitz condition and if the instant of estimation $T = \tau(\omega)$ is the Markov stopping time not exceeding the explosion time $\bar{\tau}(X_\bullet(\omega))$ of the solution process X and satisfies some technical assumptions (assuring existence of the likelihood function) then, the MLE has the following form (similar to that known from the literature devoted to the case of globally Lipschitzian coefficients and constant T).

$$\hat{\theta}_{\tau(\omega)}(X_\bullet(\omega)) = \arg\max_{\theta \in \Theta} \Big\{ \int_0^{\tau(\omega)} \mathbf{m}_s^*(X_\bullet(\omega),\theta)\, [\mathbf{b}_s \mathbf{b}_s^*]^+ (X_\bullet(\omega))\, \mathbf{m}_s(X_\bullet(\omega),\theta)\, ds$$

$$+\frac{1}{2} \int_0^{\tau(\omega)} \mathbf{m}_s^*(X_\bullet(\omega),\theta)\, [\mathbf{b}_s \mathbf{b}_s^*]^+ (X_\bullet(\omega))\, dX_s(\omega) \Big\} \qquad (2\text{-}1)$$

The sign '+' in the above formula denotes the pseudoinverse matrix (see [1]).

In the case where $\mathbf{m}$ depends on θ linearly:

$$\mathbf{m}(x) = \mathbf{m}^0(x) + \theta_1 \mathbf{m}^1(x) + \ldots + \theta_m \mathbf{m}^m(x) \tag{2-2}$$

the embraced expression in (2-1) becomes the square function of parameter θ wich yields the following closed formula for the MLE.

$$\hat{\theta}(\omega) = \mathbf{G}^{-1}(X_\bullet(\omega))\,\mathbf{H}(X_\bullet(\omega)). \tag{2-3}$$

Here:

$$\begin{aligned}
\mathbf{H} &= [h_1 - g_{01}, \ldots, h_m - g_{0m}]^*, \qquad \mathbf{G} = \{g_{ij}\}_{i,j=1,\ldots,m}, \\
h_i &= h_i(\omega, \tau) \\
&= \int_0^{\tau(\omega)} \tilde{m}_s^{i*}(X_\bullet(\omega)) \quad [\mathbf{b}_s \mathbf{b}_s^*]^+(X_\bullet(\omega))\, dX_s(\omega), \qquad (i = 1, \ldots, m), \qquad (2\text{-}4) \\
\tilde{m}^i &= \mathbf{b}^+ \mathbf{m}^i, \qquad (i = 1, \ldots, m), \qquad \mathbf{M} = [\tilde{m}^1, \ldots, \tilde{m}^m]^* \\
g_{ij} &= g_{ij}(\omega, \tau) \\
&= \int_0^{\tau(\omega)} [\tilde{m}_s^{i*} \tilde{m}_s^j](X_\bullet(\omega))\, ds, \qquad (i = 0, 1, \ldots, m; \quad j = 1, \ldots, m). \qquad (2\text{-}5)
\end{aligned}$$

Definition (see [9], cf. [10])

We shall say that the stochastic process $\hat{\theta}_t(\omega)$ defined by the formula (2-3) for $t \le \tau(\omega) < \bar{\tau}(\omega)$ is strongly consistent at τ on the set $A \in \mathcal{F}$ iff $\hat{\theta}_t(\omega) \to \theta$ as $t \nearrow \tau(\omega)$, P-almost everywhere on A. If moreover $P(A) = 1$ then $\hat{\theta}_t(\omega)$ is called globally consistent.

In what follows we state the main theoretical result of this paper. Let $\lambda_t^1(\omega) \le \lambda_t^2(\omega \le \ldots \le \lambda_t^m(\omega)$ be the eigenvalues of the matrix $\mathbf{G}(X_\bullet(\omega))$.

Theorem

Suppose, that for all $\omega \in A \in \mathcal{F}$ the following conditions hold true as $t \nearrow \tau(\omega)$:

(*i*) $\lambda_t^1(\omega) \to \infty$,

(*ii*) $(\lambda_t^m(\omega) \ln \ln \lambda_t^m(\omega))^{\frac{1}{2}} (\lambda_t^1(\omega))^{-1} \to 0.$

Then, the MLE defined by the formula (2-3) is strongly consistent at τ on the set A.

Sketch of the proof. Note that due to (*i*) if t is sufficiently close to $\tau(\omega)$, then the matrix $\mathbf{G}_t$ is positive definite so the MLE is well defined. Moreover, $\mathbf{G}_t$ as a real symmetric matrix can be represented in the form:

$$G_t = U_t^* \operatorname{diag}\{\lambda_t^i\} U_t,$$

where $U_t = \{u_{ij}(\omega, t)\}$ is an unitary matrix. Let $\tilde{h}_t^i$ be defined alike h_t^i with dW_s instead of dX_s. Define also

$$g_t^i = \sqrt{g_t^{ii} \ln \ln g_t^{ii}}, \qquad \tilde{G}_t = \operatorname{diag}\{g_t^i\},$$

$$z_t^i = \frac{h_t^i}{g_t^i}, \qquad Z_t = \left[z_t^1, \dots, z_t^m\right]^*.$$

Using the identity $(A^+)^* A^+ A = (A^+)^*$ (see [1]) and the theorem on the integration by substitution for stochastic integrals (see [11]) we obtain for $\omega \in A$ and $t \nearrow \tau(\omega)$:

$$\begin{aligned} H_t &= \int_0^t M_s b_s^+ \, dX_s - \int_0^t M_s \tilde{m}_s^0 \, ds \\ &= \int_0^t M_s b_s^+ \Big[\Big(m_s^0 + \sum_{i=1}^m m_s^i \theta_i\Big) ds + b_s \, dW_s\Big] - \int_0^t M_s \tilde{m}_s^0 \, ds \\ &= \int_0^t M_s M_s^* \, ds \, \theta + \int_0^t M_s \, dW_s \\ &= G_t \theta + \left[\tilde{h}_t^1, \dots, \tilde{h}_t^m\right]^*. \end{aligned}$$

Thus,

$$U_t(\hat{\theta}_t - \theta) = \operatorname{diag}\left\{(\lambda_f^i)^{-1}\right\} U_t \tilde{G}_t Z_t. \tag{2-6}$$

Thus, the error of MLE splits into Z_t and $\operatorname{diag}\{(\lambda_f^i)^{-1}\} U_t \tilde{G}_t Z_t$. The remaining two steps concern bounds for these factors.

For Z_t we use the following.

LEMMA

Under the assumptions of the theorem P-a.e. on A for any $i = 1, \dots, m$

$$\limsup_{t \nearrow \tau} |z_t^i| \le \sqrt{2}. \tag{2-7}$$

As the second factor is concerned the following estimate cannot be improved without additional assumptions.

$$g_t^{ii} = \sum_{j=1}^m \lambda_t^j (u_t^{ij})^2 \le \lambda_t^m. \tag{2-8}$$

Equality (2-6) together with estimates (2-7) and (2-8) yield the thesis of the theorem.

In what follows we outline the proof of the lemma. The idea is to observe that the process $\tilde{h}_t$ under the appropriate time change admits on $\mathcal{A}$ the use of the law of iterated logarithm for local martingales (see e.g. [12]).

Note that *(i)* together with the equality from (2-8) yield that $g_t^{ii} \nearrow \infty$ as $t \nearrow \tau$ on $\mathcal{A}$. Let $f : [0,\infty) \to [0,\infty)$ be any increasing, continuous function such that $f(s) \nearrow \infty$ as $s \nearrow \infty$. Let us define

$$\mathrm{A}_f = \{\omega \mid \liminf_{s\to\infty} g_s^{ii}\,(f(s))^{-1} \geq 2\},$$

$$\tau_s(\omega) = \inf\{r \geq 0 \mid (\forall q \geq r)\ g_q^{ii} \geq f(s)\} \wedge s,$$

$$\tilde{W}_{f(s)}(\omega) = \int_0^{\tau_s(\omega)} \left(\tilde{m}_s^i(X_\bullet(\omega))\right)^* dW_s(\omega).$$

It follows from these definitions that

$$\tilde{W}_{f(s)} = \tilde{h}^i_{\tau_s} \qquad \text{and} \qquad \langle \tilde{W}_{f(s)} \rangle^2 = g^{ii}_{\tau_s}.$$

Simultaneously, $\tau_s(\omega) \nearrow \tau(\omega)$ on $\mathcal{A}_t$. Thus, the law of iterated logarithm can be applied to $\tilde{W}$ on $\mathcal{A}_f$ yielding the tesis of the lemma on this set. As the above can be mimiced for any function f with the above required properties then the thesis holds true P-a.e. on $\mathcal{A}$.

Remark 1. From the proof one can easily notice that in addition to the above statement any measurable set on which *(i)* and *(ii)* hold true but the MLE does not converge to θ as $t \nearrow \tau(\omega)$ must be of probability zero. This fact ancourages to the use of the above conditions as a test for the convergence of the estimator independently of wheather it is globally consistent or not. This is especially important in such cases, where the character of behaviour of the solution X depends on chance: some of s-ps $X_\bullet(\omega)$ may fall into ergodic sets (see [13]), where the MLE is consistent, but some of them may wander in such a way that $\hat{\theta}_t(X_\bullet(\omega))$ never approaches θ. The above test can be applied to infer about the convergence from each s-p individually.

Remark 2. It is also easilly seen that the proof remains valid for general coefficients and W beeing a generalized Wiener process (defined as in [1]).

Remark 3. It is clear that the limiting condition *(ii)* is required because of the very general but not so flexible estimate (2-7). In the case where the matrix U_t tends to a diagonal limit (as $t \nearrow \tau$) quick enough in comparison with the ratio of eigenvalues then, the condition *(ii)* can be weaken or even discarded.

3. Results of Digital Simulations

To complete the pattern we present the results of digital simulations of the action of the MLE for ergodic, transient and exploding systems. The object of estimation was the parameter $\theta = [\theta_1, \theta_2, \theta_3]^*$ characterizing the stochastic Duffing oscillator

$$\ddot{y} + \theta_1 \dot{y} + (\theta_2 + \xi_1)\, y + \theta_3\, y^3 = \xi_1. \tag{3-1}$$

Parameters $\theta_1, \theta_2, \theta_3$ are called—damping coefficient, linear stiffness coefficient and cubic stiffness coefficient, respectively. $\xi = [\xi_1, \xi_2]^*$ is the "physical" Gaussian white noise. The "pre-equation" (3-1) is given the precise meaning through the equation (1-1) with $X_0^1 = 1, X_0^2 = -1$ (any other initial condition could have been used equally well), $X_t = [X_t^1, X_t^2]^*$ modelling $[y, \dot{y}]^*$ and with $\mathbf{m}_t^0 = [X_t^2, 0]^*$, $\mathbf{m}_t^0 = [0, -X_t^2]^*, \mathbf{m}_t^0 = [0, -X_t^1]^*$, $\mathbf{m}_t^0 = [0, -(X_t^1)^3]^*, \mathbf{b}_t = [\mathbf{b}_{1t}^*, \mathbf{b}_{2t}^*]^*$, $\mathbf{b}_{1t}^* = [0, 0]^*$, $\mathbf{b}_{2t}^* = [-1, -X_t^1]\, D^{\frac{1}{2}}$ where D is the intensity matrix of the noise. More deteils concerning modelling together with the ergodicity conditions for (3-1) can be found in [14].

Obviously, during the simulation the solution process X can be at most approximated. We used the polygonal approximation of the Wiener process with the diameter of the time-axis partition equal to Δ. The s-p of solution X was approximated by the functiiob ϕ obtained through one-step forward difference integration (Euler method) with integration step δ (see below for the particular values of Δ and δ). Although the form of the model (3-1) does not require the Stratonovič (or Wong-Zakai, say) corrections in the governing equation, another corrections (in formulae (2-4)) are necessary for proper approximation of stochastic integrals by finite Cauchy integral sums (they are evaluated with respect to the approximate solution which is not an Itô process—its s-ps ϕ are piece-wise linear). The problem of corrections requiered for the sake of ML estimation is studied in detail in [15] (the general methodology of [11] is employed there). The results of simulations are shown on Figs. 1–3. The paths obtained through the direct numerical integration of formulae for MLE are referred as *classical*. Those which resulted from the application of corrections described in [15] (see also [14] as to the case of ergodic Duffing oscillator) are referred as *modified*. Note that the modified MLE works really effectively.

(a) Ergodic case (Fig. 1)

For positive θ_i the solution process X is ergodic (see [14]). To trace the s-ps of MLE in Fig. 1 we used $\theta_1 = \theta_2 = \theta_3 = 1$ (it makes the error analysis straightforward). Conditions *(i)* and *(ii)* are satisfied in this case. Results match the theory pretty well already for $\delta = 0.005, \Delta = 0.01$.

(b) Transient case (Fig. 2)

For $\theta_3 = 0, \theta_2 < 0$ we obtain a bilinear system with transient but still regular solution wandering exponentially to infinity. We used $\theta_1 = 1, \theta_2 = -1$. In this case the condition *(ii)* is not satisfied but the diagonal elements of the matrix $\mathbf{G}$ are

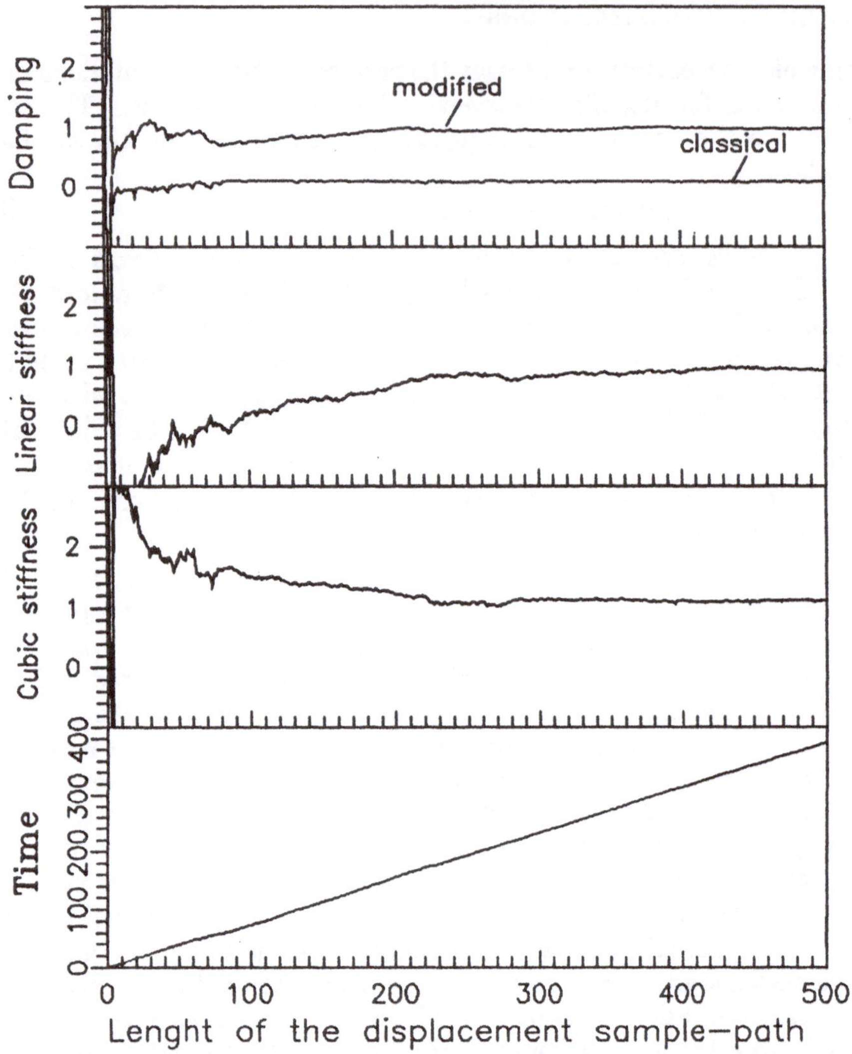

Figure 1

really dominant and remark 3 applies leading to consistency of MLE. However, to gain quality of estimation comparable to that in the case (a) the time-axis partition had to be refined to $\delta = 0.001$ (Δ remaining equal to 0.01).

(c) Explosive case (Fig. 3)

For negative θ_2, θ_3 the solution explodes in finite time. We used $\theta_1 = 1$ (for $\theta_1 = -1$ no qualitative differences were observed), $\theta_2 = \theta_3 = -1$. The situation was alike in the transient case—remark 3 applied. However, the differences between the rates of growth of g^{3i} and g^{2j} were so drastic that the condition *(ii)* returned under the form of numerical instability until the order of integration precision

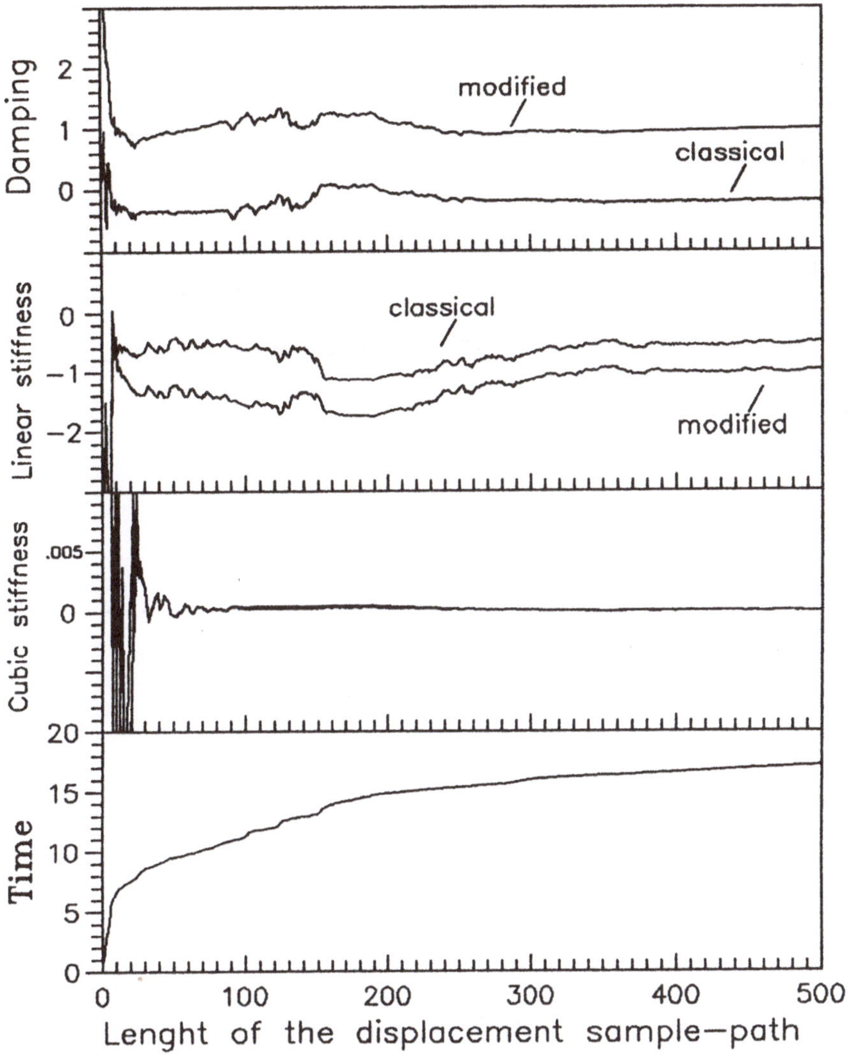

Figure 2

had been increased once more. To arrive at the results plotted in Fig.3 we were forced to use $\Delta = 0.0005, \delta = 0.00001$.

4. Conclusions

To conclude let us underline that it is *not ergodicity* but a kind of law of large numbers for martingales which is the mechanism responsible for consistency of the MLE under consideration. It was evident from the proof of the theorem of

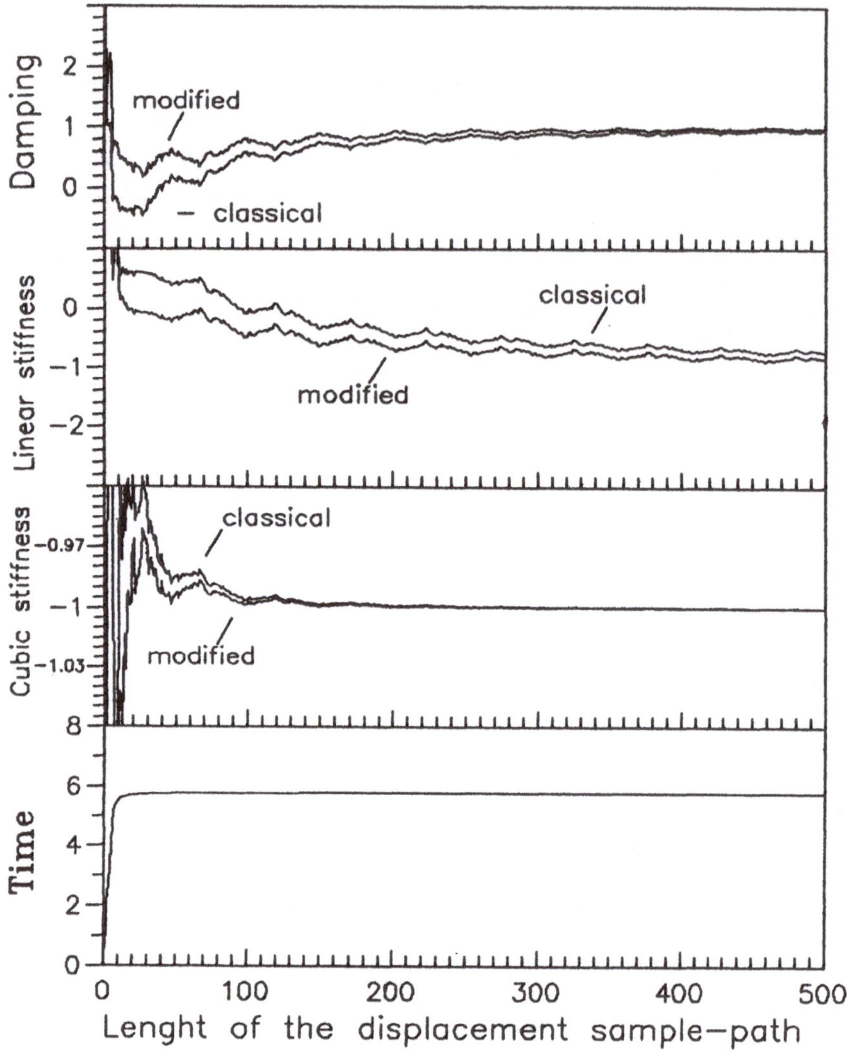

Figure 3

sec. 2 as well as from the simulations described in sec. 3. Accordingly, not only for ergodic systems the MLE can be effectively employed.

Our consistency conditions *(i)* and *(ii)* together with remark 3 are suitable for analysis of ergodic, transient and even exploding systems.

The use of our corrections for MLE (see [15]) is suggested not only for simulated data but for real dynamical systems as well. Their answers to real loadings can be approximated by Itô processes but their s-ps, alike our functions ϕ resulting from simulation, cannot be treated as functions with infinite variation. They do not undergo the rules of stochastic Itô calculus. Therefore, to employ formulae resulting from this efficient theory one has to adopt them to realistic data. Our simulations show that this approach leads to really applicable estimators.

5. References

1. Liptser R. S. and Shiryayev A. N. "Statistics of Random Processes",Springer Vlg., N.Y., 1977—vol. 1, 1978—vol. 2.
2. Borkar V. and Arunbha Bagchi "Parameter estimation in continuous-time stochastic processes", Stochastics, vol. 8, pp. 193-212, 1982.
3. Prakasa-Rao B. L. S. "Estimation of the drift for diffusion processes", Proc. 6-th Int. Summer School, Sellin/Rügen, Nov. 1983, Semminarbericht no. 31.
4. Bellach B. "Parameter estimators in linear stochastic differential equations and their assymptotic properties", Math. Operationsforsch. Stat. Ser. Statistics, vol. 14, no. 1, pp. 141-191, 1983.
5. Liňkov Ju.N. "On asymptotic behaviour of likelihood function for certain problems for semimartingales", Teor. Sluč. Proc. no. 12, pp. 40-47, 1984.
6. Kozin F. and Natke H.G. "System identification techniques", Structural Safety, vol. 3, nos. 3+4, pp. 269-317, 1986.
7. Musiela M. "Processus de Diffusion; Aspects Probabilistes et Statistiques", PhD thesis, L'Universite Scientifique et Medicale de Grenoble, June 1984.
8. Kazimierczyk P. "Parametric Identification of Dynamical Systems Modelled by Stochastic Diffusion Markov Processes", Ph.D.-thesis, I.F.T.R. Polish A.S., IFTR Reports 88/17, 320 pp. (in Polish)
9. Kazimierczyk P. "ML estimation of stochastic differential systems; the case of general coefficients", IFTR Reports 89/6, (in print).
10. Ibragimov I.A. and Has'minskii R.Z. "Statistical Estimation. Asymptotic Theory", Springer Vlg., N.Y. 1981.
11. Mc Shane E.J. "Stochastic Calculus and Stochastic Models", Acad. Press, N.Y. 1974.
12. Ikeda N. and Watanabe Sh. "Stochastic Differential Equations and Diffusion Processes", North-Holland/Kodansha, 1981.
13. Arnold L. and Kliemann W. "Qualitative theory of stochastic systems", in "Probabilistic Analysis and Related Topics", vol. 3. (ed. Bharucha Reid), pp. 1-79, Acad. Press 1983.
14. Kazimierczyk P. "Consistent maximum-likelihood estimators for nonlinear oscillators; ergodic case", IFTR Reports 89/5, (in print).
15. Kazimierczyk P. "Correction formulas for estimators of drift parameters of an Itô equation", IFTR Reports 88/39.

ON ADAPTIVE CONTROL OF CONTINUOUS TIME LINEAR STOCHASTIC SYSTEMS*

BOZENNA PASIK-DUNCAN
Department of Mathematics
University of Kansas
Lawrence, Kansas 66045

The solution of adaptive control of continuous time linear stochastic systems without and with delays is presented. The asymptotic distribution of the cost in the adaptive control of a continuous time linear stochastic system without and with delays is given. The consistency of a least squares identification procedure for continuous time linear stochastic systems without and with delays is discussed with the aid of control theory methods.

1. Introduction.

The problem of adaptive control of stochastic systems has been studied extensively. While the majority of the work has been for discrete time stochastic systems, the problem that is studied here is for continuous time stochastic systems. The study of continuous time stochastic systems is necessary for some problems that evolve naturally in continuous time. The continuous time theory is in fact relevant for discrete time schemes using high sampling rates. In other words it is important in discrete time systems for the case of small sampling times and for the analysis of computational questions. The continuous time stochastic theory is one way of addressing the question of robustness in continuous time adaptive schemes designed for deterministic signal models. The stochastic systems are modelled by linear stochastic differential equations where the unknown parameters appear affinely in the stochastic differential equation that describes the state. We shall investigate the problem of the adaptive control of continuous time linear stochastic systems and we shall obtain a solution of it. The problem of adaptive control is to identify the system and simultaneously to control it. A solution of the adaptive control problem means that the family of parameter estimates are strongly consistent and that the average costs for the

* Research partially supported by NSF Grant ECS-8718026.

control problem using the parameter estimates converge almost surely to the optimal average cost.

There is a large body of work on adaptive control problems. For discrete time linear systems the work that is most closely associated with our presented results is that of P. Mandl [8] and P. Hall and C. C. Heyde [6].

Mandl's solution imposed some conditions that seem to be difficult to verify and are not natural in terms of the control problem. Hall and Heyde simplified Mandl's assumptions but restricted the model to one unknown parameter. The conditions that will be imposed here are natural and an improvement over both Mandl's conditions and Hall-Heyde's conditions. The adaptive control problem for continuous time linear stochastic systems with only one unknown parameter is solved in [12].

It is probably useful to describe briefly the adaptive control scheme that will be used here. The control as feedback gains will be obtained from the solution of the infinite time, quadratic cost, control problem [7] by replacing the correct values of the parameters in this solution by the estimates of the parameters at time t to obtain the feedback gain at time t. The estimates of the parameters are the maximum likelihood estimates. The maximum likelihood estimate based on the observations of the state until time t is used as the correct value of the parameter to solve the infinite time control problem by solving the algebraic Riccati equation. This method gives a feedback gain for each t and therefore a control policy.

2. Preliminaries.

The stochastic system for the adaptive control problem that we shall consider first is

$$dX_t = \left(F_0 X_t + \sum_{i=1}^{p} \alpha_i F_i X_t + BU_t\right) dt + dW_t \tag{1}$$

where $\alpha_i \in I_i \subset R$ is an unknown parameter, the true value α_0 of which is to be estimated from the observation of X and U and I_i is bounded, connected, open interval for $i = 1,2,\ldots,p$. $(W_t, t \geq 0)$ is a standard $\mathbb{R}^n$-valued Wiener process, $X_t \in R^n$, $U_t \in R^m$, $(U_t, t \geq 0)$ is a random process nonanticipative with respect to Wiener process, $F_i \in L(R^n, R^n)$ $i = 0,1,\ldots,p$. $B \in L(R^m, R^n)$, and $X_0 \equiv a \in R^n$. It is assumed that $(F_i, i = 1,2,\ldots,p)$ are linearly independent.

The probability space $(\Omega, \mathcal{F}, P)$ can be chosen such that Ω is the Fréchet space of R^n-valued continuous functions on $R_+ = [0,\infty)$ with the seminorms of local uniform convergence, P is the Wiener measure on Ω and $\mathcal{F}$ is the P-completion of the Borel σ-algebra of Ω.

The cost functional C_t is defined as

(2)
$$C_t = \int_0^t (\langle QX_s, X_s\rangle + \langle PU_s, U_s\rangle)\, ds$$

where $Q \in L(R^n, R^n)$ and $P \in L(R^m, R^m)$ are symmetric and positive definite.

The control $(U_t, t \geq 0)$ will be given in a feedback form, e.g. $U_t = K_t X_t$, $t \geq 0$. Specifically it will be in terms of the feedback gains $(K_t, t \geq 0)$. The feedback gain K_t is obtained by solving the algebraic Riccati equation for the infinite time, deterministic, quadratic cost control problem replacing the correct values of the parameters by their maximum likelihood estimates based on the observations of the state until time t. Since $(K_t, t \geq 0)$ will be a stochastic process that is adapted to $(\sigma(X_u), u \leq t)$ it is necessary to define the solution of (1).

It is known [3] that if $(K_t, t \geq 0)$ is uniformly bounded almost surely then there is a probability measure that is (mutually) absolutely continuous with respect to Wiener measure such that Brownian motion for the Wiener measure satisfies the stochastic equation (1) under the other probability measure. This method for defining solutions of stochastic equations has been used extensively in stochastic control problems [2].

The question of the existence of the solution of (1) using the feedback gains that depend on the estimates can be eliminated by computing K_t from the estimate at time $t - \delta$ where $\delta > 0$. In this way K_t is adapted to $\sigma(X_u, u \leq t - \delta)$ and it becomes a known stochastic process for (1) at time t.

For notational convenience let $R = BP^{-1}B'$ and $A = A(\alpha) = F_0 + \sum_{i=1}^{p} \alpha_i F_i$. It is well known that if (A,B) is reachable then the deterministic system

(3)
$$\frac{dx}{dt} = Ax + Bu$$

is stabilizable by state feedback.

A stabilizing feedback gain can be obtained from the unique, symmetric positive definite solution of the algebraic Riccati equation

(4)
$$0 = \Pi A + A'\Pi - \Pi R \Pi + Q.$$

The optimal feedback control $u^*(t)$ for the infinite time, deterministic optimal control problem with state equation (3) and cost function C_∞ from (2) is $-P^{-1}B'\Pi x(t)$.

Let $(X_t, t \geq 0)$ be the solution of (1) using the feedback control policy $(K_t, t \geq 0)$. Then $(K_t, t \geq 0)$ is said to be a stable control policy or $(X_t, t \geq 0)$ is said to be stable if

$$\text{(5)} \qquad \limsup_{t\to\infty} \frac{1}{t} \int_0^t E\langle X_s, X_s\rangle \, ds \leq C < \infty.$$

Let $(X_t, t \geq 0)$ be the solution of (1) with the feedback control policy $(K_t, t \geq 0)$. Define $(a_{mn}(t), m, n = 1,\ldots,p)$ as

$$a_{mn}(t) := \int_0^t \langle F_m X_s, F_n X_s\rangle \, ds.$$

Let $\tilde{A}(t) = \{\tilde{a}_{ij}(t)\}$ where $\tilde{a}_{ij}(t) = \dfrac{a_{ij}(t)}{a_{ii}(t)}$.

For the computation of the maximum likelihood estimate it will be assumed that

$$\text{(6)} \qquad \liminf_{t\to\infty} |\det \tilde{A}(t)| \geq c > 0 \qquad \text{a.s.}$$

where $c \in \mathbb{R}$.

In specific cases it is often easy to verify (6). For example, let $(f_k^i, k = 1,2,\ldots,n)$ be the columns of F_i for $i = 1,2,\ldots,p$. If $\langle f_k^i, f_\ell^j\rangle = 0$ for all $i \neq j$ and $k,\ell \in \{1,2,\ldots,n\}$, then the condition (6) is trivially satisfied. More generally one can often use well known inequalities to verify (6). For an elementary example of this let $n = 3$ and $p = 2$ where $F_1 = E_{11} + E_{22}$, $F_2 = E_{12} + E_{31}$ and E_{ij} is the elementary 3×3 matrix with 1 in the (i,j) position and zeroes elsewhere.

3. Main results.

The first problem of adaptive control to be presented here is the solution of adaptive control for continuous time linear stochastic systems.

The first verification that is necessary in the solution of the adaptive control problem is to show that the family of maximum likelihood estimates is strongly consistent. The likelihood function is obtained from the mutual absolute continuity of the probability measure of (1) and the Wiener measure for $(W(t), t \geq 0)$ as

$$L_t(\alpha) = \exp\Big(\int_0^t \langle(A(\alpha)X_s + BU_s), dX_s\rangle - \frac{1}{2}\int_0^t \langle A(\alpha)X_s + BU_s, A(\alpha)X_s + BU_s\rangle\, ds\Big). \tag{7}$$

To maximize L_t it is necessary (and in fact sufficient) that $DL_t = 0$. As usual, $\ell n\, L_t$ is maximized.

$$\frac{\partial L_t}{\partial \alpha_j} = \int_0^t \langle F_j X_s, dX_s\rangle - \int_0^t \langle F_j X_s, F_0 X_s + \sum_{k=1}^{p} \hat{\alpha}_k F_k X_s + BU_s\rangle\, ds = 0, \tag{8}$$

$j = 1,2,\dots,p.$

Now use

$$dX_t = A(\alpha_0)X_t\, dt + BU_t\, dt + dW_t$$

and rewrite (8) to obtain

$$\begin{aligned} \hat{\alpha}_j \int_0^t \langle F_j X_s, F_j X_s\rangle\, ds & \\ = \alpha_{oj} \int_0^t \langle F_j X_s, F_j X_s\rangle\, ds &+ \int_0^t \langle F_j X_s, dW_s\rangle \\ &+ \int_0^t \langle F_j X_s, \sum_{\substack{k \\ k\neq j}} \alpha_k F_k X_s\rangle\, ds \\ &- \int_0^t \langle F_j X_s, \sum_{\substack{k \\ k\neq j}} \hat{\alpha}_k F_k X_s\rangle\, ds. \end{aligned} \tag{9}$$

Define

$$b_j(t) := \int_0^t \langle F_j X_s, dW_s \rangle$$

$$\tilde{b}_j(t) = \frac{b_j(t)}{a_{jj}(t)}.$$

Thus the family of linear equations (9) can be rewritten as

(10) $$\tilde{A}(t)\hat{\alpha}(t) = \tilde{A}(t)\alpha_0 + \tilde{b}(t)$$

where $\tilde{A}(t) = (\tilde{a}_{jk}(t))$, $\hat{\alpha}(t) = (\hat{\alpha}_1(t),\ldots,\hat{\alpha}_p(t))'$, $\alpha_0 = (\alpha_{01},\ldots,\alpha_{0p})'$ and $\tilde{b}(t) = (\tilde{b}_1(t),\ldots,\tilde{b}_p(t))'$. From (10) we get

(11) $$\hat{\alpha}(t) = \alpha_0 + \tilde{A}^{-1}(t)\tilde{b}(t).$$

The following theorem [5] gives the strong consistency of parameter estimates.

Theorem 1. *Let* $(K_t, t \geq 0)$ *be a control policy that is uniformly bounded almost surely. Assume that (6) is satisfied. Then the family of maximum likelihood estimates* $(\hat{\alpha}_t)$ *of* α_0 *is strongly consistent, that is*

(12) $$P(\lim_{t\to\infty} \hat{\alpha}_t = \alpha_0) = 1.$$

To verify this theorem the following result is important. If (W_t,F_t) is a Wiener process on (Ω,F,P) and (f_t) is (F_t)-adapted such that

$$\int_0^t \langle f_s,f_s\rangle\, ds < \infty \qquad \text{a.s. for} \quad 0 \leq t < +\infty$$

and

$$\int_0^\infty \langle f_s,f_s\rangle\, ds = +\infty \quad \text{a.s.}$$

then

$$\lim_{t\to\infty} \frac{\int_0^t \langle f_s,dW_s\rangle}{\int_0^t \langle f_s,f_s\rangle ds} = 0 \qquad \text{a.s.}$$

For the applications of adaptive control and identification it is important to have recursive estimates of the unknown parameters. Thus it has been shown [5] how to obtain a stochastic differential equation for the maximum likelihood estimates $(\hat{\alpha}_t, t > 0)$. Rewriting (9) as a matrix equation we have

$$A(t)\hat{\alpha}_t = A(t)\alpha_0 + b(t)$$

where $b(t) = (b_1(t),\dots,b_p(t))'$. Let $\langle \tilde{F}X, Y\rangle$ be the vector whose i-th component is $\langle F_iX, Y\rangle$. With this notation

$$\hat{\alpha}_t = A^{-1}(t)\int_0^t \langle \tilde{F}X_s, dX_s - F_0X_s ds - BU_s ds\rangle. \tag{12}$$

$A^{-1}(t)$ satisfies the following differential equation

$$dA^{-1}(t) = -A^{-1}(t)dA(t)A^{-1}(t)$$

so

$$d\hat{\alpha}_t = A^{-1}(t)\langle \tilde{F}X_t, dX_t - A(\hat{\alpha}_t)X_t dt - BU_t dt\rangle.$$

The following theorem [5] shows that the average costs using the adaptive control policy converge almost surely to the optimal average cost as $t \to \infty$.

Theorem 2. *Let* $(K_t, t \geq 0)$ *be the feedback gains determined from the algebraic Riccati equation using the maximum likelihood estimates* $(\hat{\alpha}_t)$ *as the parameter values. Assume that (6) is satisfied for the control policy* $(K_t, t \geq 0)$ *and that* (A,B) *is reachable. The feedback gains* $(K_t, t \geq 0)$ *form a stable control policy and*

$$\lim_{t\to\infty} \frac{1}{t} C_t = \operatorname{tr} V \quad \text{a.s.} \tag{13}$$

where V *is the unique, symmetric, positive definite solution of Riccati equation using* α_0 *and* C_t *is the cost at time* t *given by (2) using the feedback gains* $(K_t, t \geq 0)$.

This self-optimizing property of the adaptive control is proved for a time averaged cost functional of the type $t^{-1}C_t$.

A trivial consequence of uniform boundedness of K_t is that the adaptive control scheme presented here is also self-optimizing w.r.t. the mean time averaged cost functional $t^{-1}EC_t$ where the expectation is taken w.r.t. the measure induced by the trajectories of the true system controlled adaptively by means of $(K_t, t \geq 0)$, i.e., we have

(14) $$\lim_{t\to\infty} t^{-1}EC_t = \mathrm{tr}\, V.$$

This simple extension is very important in application.

Similar results were obtained ([12], [4]) for the stochastic system of the following type

(15) $$dX_t = \left(\sum_{i=1}^{k} F_0^i X_{t-\tau_i} + \sum_{i=1}^{k} \sum_{j=1}^{n_i} \alpha_{ij} F_j^i X_{t-\tau_i} + BU_t \right) dt + dW_t$$

where $\alpha_{ij} \in I_{ij} \subset R$ is an unknown parameter and I_{ij} is a bounded, connected, open interval for all (i,j), $X_t \in R^n$, $U_t \in R^m$, $F_j^i \in L(R^n, R^n)$, $B \in L(R^m, R^n)$ $(W_t, t \geq 0)$ is a standard Wiener process and $X_0 = h$ is a fixed function. It is assumed that for each $i \in \{1,2,\ldots,k\}$ the family of linear transformations $(F_j^i, j = 1,2,\ldots,n_i)$ are linearly independent.

The cost functional, C_t, is defined as

(16) $$C_t = \int_0^t (\langle QX_s, X_s\rangle + \langle PU_s, U_s\rangle)\, ds$$

where $Q \in L(R^n, R^n)$ and $P \in L(R^m, R^m)$ are symmetric and positive definite. The controls as feedback gains are obtained from the solutions of the infinite time, quadratic cost problem that was given by Delfour, McCalla and Mitter.

The second problem of adaptive control to be studied here is the asymptotic distribution of the cost in the adaptive control of a continuous time linear stochastic system without and with delays that was considered above. The following theorem shows that the cost is asymptotically normally distributed.

Theorem 3. *Let* $(C(t), t \geq 0)$ *be the family of costs ((2), (16)) associated with the feedback gains* $(K_t, t \geq 0)$ *that are the optimal stationary feedback gains using the maximum likelihood estimates for the unknown parameters. Then*

(17) $$\frac{C_t - t\, \mathrm{tr}\, V}{t^{1/2}} \to Z$$

where Z *is* $N(0,\mathrm{tr}\Gamma)$, *the convergence is in distribution.* V *is the unique, positive definite solution of (4) with the true parameters and* Γ *is the unique, positive definite solution of*

$$\Gamma(A + Bk_0) + (A + Bk_0)'\Gamma + 4V^2 = 0 \tag{18}$$

where the true parameters are used and $k_0 = -P^{-1}B'V$.

The proof of this theorem is not given here but the basic ideas of the proof are outlined. The cost C_t can be rewritten using (4) as

$$\begin{aligned} C_t = 2 \int_0^t \langle VX_s, dW_s\rangle + \int_0^t \mathrm{tr}\, V\, ds - \langle VX_t, X_t\rangle + \langle VX_0, X_0\rangle \\ + \int_0^t \langle P(k_0 - K_s)X_s, (k_0 - K_s)X_s\rangle\, ds. \end{aligned} \tag{19}$$

To show that $\frac{1}{t^{1/2}}\langle VX_t, X_t\rangle$ is negligible as $t \to \infty$ the following two stochastic equations are used

$$\begin{aligned} dX_t &= (A + BK_t)X_t dt + dW_t \\ dY_t &= (A + Bk_0)Y_t dt + dW_t \\ X_0 &\equiv a, \qquad Y_0 \equiv a. \end{aligned}$$

Let $Z_t = X_t - Y_t$ for $t \geq 0$. From the above two stochastic differential equations it follows directly that $(Z_t, t \geq 0)$ satisfies the stochastic differential equation

$$\begin{aligned} dZ_t &= [(A + Bk_0)Z_t + B(K_t - k_0)X_t]dt \\ Z_0 &\equiv 0. \end{aligned} \tag{20}$$

The solution $(Z_t, t \geq 0)$ can be explicitly expressed by the variation of parameters formula. Since the homogeneous differential equation associated with (20) is uniformly asymptotically stable, the "fundamental matrix" Φ for this differential equation is bounded in operator norm

$$\|\Phi(t)\| \leq \beta e^{-at}$$

where $a > 0$. From this inequality and stability of $(X_t, t \geq 0)$ it follows that

$$(21) \qquad \lim_{t \to 0} \frac{\langle Z_t, Z_t \rangle}{t^{\alpha}} = 0 \quad \text{in probability}$$

for each $\alpha > 0$. Furthermore it is elementary to verify that

$$(22) \qquad \lim \frac{1}{t^{\alpha}} \langle Y_t, Y_t \rangle = 0 \text{ in probability} .$$

Using the above two limits it follows that

$$(23) \qquad \lim \frac{1}{t^{\alpha}} \langle X_t, X_t \rangle = 0 \text{ in probability}$$

for each $\alpha > 0$.

By the stability property of $(X_t, t \geq 0)$, the convergence of $(K_t - k_0, t \geq 0)$ to 0 as $t \to \infty$ and Chebyshev's inequality it easily follows that

$$(24) \qquad \lim_{t \to \infty} \frac{1}{t^{1/2}} \int_0^t \langle P(k_0 - K_s)X_s, \; (k_0 - K_s)X_s \rangle \, ds$$

$$= 0 \quad \text{in probability.}$$

By a time change we have

$$2 \int_0^t \langle VX_s, dW_s \rangle = W_{A_t}$$

where

$$A_t = 4 \int_0^t \langle VX_s, VX_s \rangle \, ds.$$

Applying Ito's formula to $\langle \Gamma X_t, X_t \rangle$ where Γ is the solution of (18) it can be shown that

$$(25) \qquad \lim_{t \to \infty} \frac{A_t}{t} = \mathrm{tr}\Gamma \quad \text{a.s.}$$

Returning to the equation (19) for $(C_t, t \geq 0)$ we have

$$\frac{C_t - t \,\mathrm{tr}\, V}{t^{1/2}} + \frac{1}{t^{1/2}} \left(\langle VX_t, X_t \rangle - \langle VX_0, X_0 \rangle \right)$$

$$(26) \qquad - \frac{1}{t^{1/2}} \int_0^t \langle P(k_0 - K_s)X_s, (k_0 - K_s)X_s \rangle \, ds$$

$$= \frac{W_{A_t}}{t^{1/2}}.$$

From (23) it follows that

$$(27) \qquad \lim_{t \to \infty} \frac{1}{t^{1/2}} \left(\langle VX_t, X_t \rangle - \langle VX_0, X_0 \rangle \right) = 0 \quad \text{in probability.}$$

By the stability property (5) of $(X_t, t \geq 0)$, the convergence of $(K_t - k_0, t \geq 0)$ to zero and Chebyshev's inequality we obtain

$$(28) \qquad \lim_{t \to \infty} \frac{1}{t^{1/2}} \int_0^t \langle P(k_0 - K_s)X_s, (k_0 - K_s)X_s \rangle \, ds = 0 \quad \text{in probability.}$$

Using (25, 27-28) in (26) verifies (17).

This asymptotic normality of the cost is similar to the results of Mandl [8] and Hall-Heyde [6] for the adaptive control of discrete time linear systems. However as we mentioned earlier, they imposed additional restrictive conditions on the problem by comparison with our results. Furthermore our techniques of proof are different from their methods. Mandl [11] also discusses the asymptotic normality of the cost for continuous time linear systems but he imposes some restrictive conditions on the estimates that are not satisfied for the maximum likelihood estimators.

The third problem of adaptive control to be discussed here is the consistency of a least squares identification procedure for continuous time linear stochastic systems without and with delays with the aid of control theory methods.

We have been continuing the research of parameter estimation in linear systems initiated above and have showed that the applications of control theory methods to the consistency presented in M. Boschková [1] and P. Mandl [9] can be developed to obtain explicit results.

Let us again deal with random processes the trajectory of which fulfills

$$dX_t = F(\alpha)X_t dt + U_t dt + dW_t \qquad t \geq 0.$$

$W = (W_t, t \geq 0)$ is an n-dimensional Wiener process with incremental variance matrix h

$$dW_t dW_t' = h dt.$$

$U = (U_t, t \geq 0)$ is a random process nonanticipative with respect to W. $F(\alpha)$ denotes and n×n matrix of the form

$$F(\alpha) = F_0 + \alpha_1 F_1 + ... + \alpha_m F_m \quad \alpha = (\alpha_1,...,\alpha_m) \in R^m.$$

$F_0,...,F_m$ are given matrices, α is a parameter, the true value α_0 of which is to be estimated from the observation of X and U.

The least squares estimate of α_0 on the basis of $(X_t, t \leq T)$, $(U_t, t \leq T)$ is denoted by $\hat{\alpha}_T$. It is defined as follows.

Let ℓ be a nonnegative definite symmetric matrix. Heuristically, $\hat{\alpha}_T$ is the minimizer of the quadratic functional

$$(29) \qquad \int_0^T (\dot{X}_t - F(\alpha)X_t - U_t)' \ell (\dot{X}_t - F(\alpha)X_t - U_t)dt$$

where $\dot{X}_t$ denotes derivative of X_t which in fact does not exist. To improve this we subtract

$$\int_0^T \dot{X}_t \ell \dot{X}_t \, dt$$

from (29) which does not depend on α and rewrite the remaining terms as

$$(30) \qquad \int_0^T (F(\alpha)X_t + U_t)' \ell (F(\alpha)X_t + U_t)dt - 2\int_0^T (F(\alpha)X_t + U_t)' \ell \, dX_t.$$

Equating derivatives of it with respect to α_i to 0 one obtains the system of linear equations

$$(31) \qquad \sum_j \int_0^T X' F_i' \ell F_j X \, dt \, \hat{\alpha}_{jT} = \int_0^T X' F_i' \ell (dX - U \, dt - F_0 X \, dt).$$

We remark that this is a recursive estimation procedure.

We proved in [10] the following

Theorem 4. *Let* h *be nonsingular and let the matrices*

$$\sqrt{\ell}\,F_i \qquad i = 1,\dots,m$$

be linearly independent where $\sqrt{\ell}$ *is the symmetric square root of* ℓ.
If

$$\frac{1}{T}\int_0^T \left(|X_t|^2 + |U_t|^2 \right) dt \qquad T>0$$

is bounded in probability (resp. a.s.) and

$$\lim_{T\to\infty} \frac{|X_T|^2}{T} = 0 \qquad \textit{in probability} \qquad \textit{(a.s.)}$$

then

$$\lim_{T\to\infty} \hat{\alpha}_T = \alpha_0 \qquad \textit{in probability} \qquad \textit{(a.s.)}.$$

Again we shall not give a proof here but only sketch of it.
Inserting (17) with $\alpha = \alpha_0$ into (31) we get

$$\sum_j \int_0^T X' F_i' \ell F_j X \, dt(\hat{\alpha}_T^j - \alpha_0^j) = \int_0^T X' F_i' \ell \, dW,$$

and hence

$$\sum_{i,j} \frac{1}{T} \int_0^T X' F_i' \ell F_j X \, dt(\hat{\alpha}_T^i - \alpha_0^i)(\hat{\alpha}_T^j - \alpha_0^j) = \sum_i \frac{1}{T} \int_0^T X' F_i' \ell \, dW(\hat{\alpha}_T^i - \alpha_0^i). \tag{32}$$

To investigate the left-hand side of (32) take $\mu \in R^m$, $|\mu| = 1$ and denote

$$p(\mu) = \sum_i \mu^i \sqrt{\ell}\, F_i, \quad q(\mu) = p(\mu)' p(\mu). \tag{33}$$

Consequently,

$$\sum_{i,j} \frac{1}{T} \int_0^T X' F_i' \ell F_j X \, dt \, \mu^i \mu^j = \frac{1}{T} \int_0^T X' q(\mu) X \, dt.$$

Set $F = F(\alpha_0)$. It can be assumed that F is a stable matrix because without loss of generality it can be replaced by F - aI where I is the unit matrix. Introduce the quadratic functional

$$(34) \qquad Q_T(\mu) = \int_0^T X'q(\mu)X\,dt + c\int_0^T |U|^2 dt$$

where $c > 0$. Consider U as a control process and $Q_T(\mu)$ as a cost functional. The minimum of EQ_T over all U nonanticipative is obtained by solving a Riccati equation. Using it we estimate the left-hand side of (32).

Let $\varepsilon > 0$ and $\delta > 0$. We get

$$(35) \qquad P\left(\sum_{i,j} \frac{1}{T}\int_0^T X'F_i' \,\ell\, F_j X\,dt\left(\hat{\alpha}_T^i - \alpha_0^i\right)\left(\hat{\alpha}_T^j - \alpha_0^j\right) \ge \delta|\hat{\alpha}_T - \alpha_0|^2\right) \ge 1 - 2\varepsilon.$$

Regarding the right side of (32) we have for $T > T_0$

$$(36) \qquad P\left(\sum_{i} \frac{1}{T}\int_0^T X'F_i \,\ell\, dW\left(\hat{\alpha}_T^i - \alpha_0^i\right) \le \delta^2|\hat{\alpha}_T - \alpha_0|\right) \ge 1 - \varepsilon.$$

From (32), (35), (36) it follows that

$$P\left(|\hat{\alpha}_T - \alpha_0| < \delta\right) \ge 1 - 2\varepsilon, \quad T > T_0.$$

Since δ can be chosen arbitrarily small, the validity of Theorem 4 is thus established.

Example. A self-tuning model is described by the equation

$$dX_t = F(\alpha_0)X_t dt + k(\hat{\alpha}_t)X_t dt + dW_t \quad t \ge 0$$

where $K(\alpha)$ are given feedback gain matrices. Assume that

$$\mathcal{K} = \{K(\alpha), \alpha \in R^m\}$$

is a bounded set and that the following Liapunov type assumption (see: [12]) is fulfilled. There exists an nxn matrix G and a symmetric matrix $Z > 0$ s.t.

$$Z(F + GK) + (F + GK)'Z + I \le 0, \qquad K \in \mathcal{K}.$$

We continue the research of parameter estimation in linear systems initiated above and show that the applications of control theory methods to the consistency problems presented above can be developed to obtain explicit results also for linear stochastic delay-time systems. The situation is more complicated because not only is the system described by a stochastic differential equation with delays (15) but also the cost functional (34) contains delays.

To investigate a strong consistency result that is analogous to Theorem 4 it is necessary to have a result for a linear quadratic optimization problem for linear systems with discrete delays where delays can also occur in the quadratic cost functional. We obtain such a result by considering the Hamiltonian equations associated with the optimization problem and obtaining smoothness of the solution of the "adjoint" system of equations because the adjoint solution defines the optimal control. The adjoint variable is related to the state variable by a linear transformation and the optimal cost is described by a quadratic form in the starting point. Formal differentiation of this description of the optimal cost gives an operator-valued Riccati differential equation. Some results of DaPrato can be used to show that this equation has a unique global mild solution. The infinite time control problem is well posed by results of Delfour and Mitter. Thus one can obtain an algebraic operator-valued equation for the optimal cost that is the steady state solution of the Riccati differential equation. For strong consistency of the maximum likelihood estimate of the unknown parameters in (15) one proceeds in analogy with [10] now. In this way we obtain the following result.

Theorem 5. *Let the matrices* F_i *be linearly independent. If*

$$\frac{1}{T}\int_0^T \left(|X_t|^2 + |U_t|^2 \right) dt, \qquad T > 0,$$

is bounded in probability (respectively a.s.), and

$$\lim_{T\to\infty} \frac{\langle X_T, X_T\rangle}{T} = 0 \qquad \text{in probability (respectively a. s.)}$$

then

$$\lim_{T\to\infty} \hat{\alpha}_T = \alpha_0 \qquad \text{in probability (respectively a.s.).}$$

We shall omit the proof of this result because it is rather complicated.

REFERENCES

[1] M. Boschková: Self-tuning control of stochastic linear systems in presence of drift, Kybernetika 24 (1988).

[2] T. Duncan and P. Varaiya: On the solutions of a stochastic control system, SIAM J. Control, 9 (1971).

[3] T. Duncan and P. Varaiya: On the solutions of a stochastic control system II, SIAM J. Control, 13 (1973).

[4] T. Duncan and B. Pasik-Duncan: Adaptive control of linear delay time systems, Stochastics, Vol. 24 (1988), pp. 45-74.

[5] T. Duncan and B. Pasik-Duncan: Adaptive control of continuous time linear systems, MCSS (1989).

[6] P. Hall and C. C. Heyde: Martingale Limit Theory and Its Applications, Academic Press (1980).

[7] R. E. Kalman: Contributions to the theory of optimal control, Bol. Soc. Math. Mexicana, 5 (1960), pp. 102-119.

[8] P. Mandl: The use of optimal stationary policies in the adaptive control of linear systems, Proc. Sym. to Honour J. Neyman, Warsaw (1974).

[9] P. Mandl: On evaluating the performance of self-tuning regulators. In Proc. of 2nd International Symposium on Numerical Analysis, Prague 1987, B. E. Feubner, Leibzig.

[10] P. Mandl, T. Duncan, B. Pasik-Duncan: On the consistency of a least squares identification procedure, Kybernetika 24 (1988).

[11] P. Mandl: Asymptotic ordering of probability distributions for linear controlled systems with quadratic cost. Lecture Notes in Control and Infor. Sciences 78 (1986), pp. 277-283.

[12] B. Pasik-Duncan: On Adaptive Control, Habilitation, Central School of Planning and Statistics Publishers, Warsaw 1986.

A MINIMAX CONTROL OF LINEAR SYSTEMS

Z.Porosiński, K.Szajowski
Institute of Mathematics, Technical University of Wrocław
Wybrzeże Wyspiańskiego 27, PL-50-370 Wrocław

1. Introduction

Let us consider an m-dimensional linear system defined by the equation

$$x_{n+1} = a_n x_n + b_n u_n + c_n v_n, \quad n = 0,1,\dots,M-1, \tag{1.1}$$

where x_n is a state of the system, u_n is a control, v_n is a disturbance at time n; x_n, u_n, v_n are the m-dimensional vectors; a_n, b_n, c_n are given m×m-matrices (m is a fixed integer number, $m\geq 2$).

A horizon N of a control is the random variable independent of the disturbances with the known distribution given by

$$P(N=k) = \gamma_k, \quad k=0,1,\dots,M, \quad \sum_{k=0}^{M}\gamma_k = 1, \quad \gamma_M > 0. \tag{1.2}$$

The vectors v_n, $n=0,1,\dots,M$ are independent identically distributed random vectors with distribution dependent on some unknown vector parameter μ.

The following notations are used: $\mathcal{X}_n = (x_1, x_2, \dots, x_n)$, $\mathcal{U}_n = (u_0, u_1, \dots, u_n)$, $\mathcal{U}^n = (u_n, u_{n+1}, \dots, u_M)$, $v_n = (v_{n1}, v_{n2}, \dots, v_{nm})^T$, if $A = \left[a_{ij}\right]_{m\times m}$ then $di(A) = diag((a_{11}, a_{22}, \dots, a_{mm})^T)$, $\rho_n = \sum_{i=n}^{M}\gamma_i$, $\vartheta_n^i = \rho_i/\rho_n$, $e = (1,1,\dots,1)^T$.

It is assumed that the control u_n is based on $\mathcal{X}_n$ and $\mathcal{U}_{n-1}$, then before any data are obtained, the control u_n is a random vector determined by the random disturbances $v_0, v_1, \dots, v_{n-1}$. For convenience we denote $\mathcal{U}_M$ by $\mathcal{U}$ and it will be called a control policy.

Let us assume that the loss function (the cost criterion) is given by

$$L(\mathcal{U},\mathcal{X}_N) = \sum_{i=0}^{N}(y_i^T s_i y_i + u_i^T k_i u_i) = \tag{1.3}$$

$$\sum_{i=0}^{N}(x_i^T s_i^1 x_i + 2\mu^T s_i^3 x_i + \mu^T s_i^2 \mu + u_i^T k_i u_i)$$

where $\mathcal{U}$ is a control policy, $y_i = (x_i^T, \mu^T)^T$, $s_i = \begin{bmatrix} s_n^1 & s_n^3 \\ s_n^{3T} & s_n^2 \end{bmatrix}$ is a non-negative definite 2m×2m-matrix and k_i is a non-negative definite m×m-matrix.

The risk connected with the control policy $\mathcal{U}$ when the parameter p is given and the state x_0 is known is defined as

$$R(p,\mathcal{U}) = E_N\left[E_p\left(L(\mathcal{U},x_N)\,|x_0\right)\right]$$

where E_N denotes the expectation with respect to the distribution (1.2) of the random variable N and E_p denotes the expectation with respect to the distribution of the disturbances $\upsilon_0,\upsilon_1,\ldots,\upsilon_{N-1}$ when p is fixed.

Definition 1. A policy $\hat{\mathcal{U}}$ such that $\sup_{p\in\mathcal{P}} R(p,\hat{\mathcal{U}}) = \inf_{\mathcal{U}\in\mathcal{D}} \sup_{p\in\mathcal{P}} R(p,\mathcal{U})$, where $\mathcal{D}$ is a set of all control policies, is called a minimax policy.

Previous observations of the disturbances frequently give an information about the parameter p. We can used them to reduce the cost of the control. Suppose the information about p is given by some prior distribution π of the parameter p. Denote by Π the set of all these distributions. The expected risk, when π is given and $\mathcal{U}$ is chosen, is defined as $r(\pi,\mathcal{U}) = E_\pi R(p,\mathcal{U})$, where E_π denotes the expectation with respect to the distribution π. Let $\Pi^0\subset\Pi$.

Definition 2. A policy $\tilde{\mathcal{U}}$ such that $\sup_{\pi\in\Pi^0} r(\pi,\tilde{\mathcal{U}}) = \inf_{\mathcal{U}\in\mathcal{D}} \sup_{\pi\in\Pi^0} r(\pi,\mathcal{U})$ is called a Π^0-minimax policy. When Π^0 contains only one distribution the control policy $\tilde{\mathcal{U}}$ is denoted by $\mathcal{U}^*$ and it is called a Bayesian control (BC) policy.

BC policies play an important role in procuring of minimax policies. Since $\sup_{p\in\mathcal{P}} R(p,\mathcal{U}) = \sup_{\pi\in\Pi} r(\pi,\mathcal{U})$ for $\mathcal{U}\in\mathcal{D}$ then

$$\inf_{\mathcal{U}\in\mathcal{D}} \sup_{p\in\mathcal{P}} R(p,\mathcal{U}) \le \sup_{p\in\mathcal{P}} R(p,\mathcal{U}) \le \sup_{\pi\in\Pi} r(\pi,\mathcal{U}) =$$

$$\inf_{\mathcal{U}\in\mathcal{D}} \sup_{\pi\in\Pi} r(\pi,\mathcal{U}) \le \inf_{\mathcal{U}\in\mathcal{D}} \sup_{p\in\mathcal{P}} R(p,\mathcal{U}),$$

therefore every Π-minimax policy is a minimax one.

The control of discrete-time linear system with additive disturbances has already been investigated extensively during the past decade. The development of digital control systems increases interest in investigation systems with discrete state space. Interesting results in this field were obtained by De Koning [4]. In such systems we ought to use as a model of disturbances the sequence of discrete random variables. On the other hand, the description of the disturbances are often incomplete in practice. Under the circumstances an adaptive approach can be used. Models of adaptive stochastic control theory were described by Aoki [1]. These models were generalized to a non-stationary Bayesian dynamic decision model by Rieder [7] and to a model of the Bayesian control of Markov chains by van Hee [3]. In the cited papers the existence of a Bayesian control was shown. Kumar [5] has given review of some recent results in stochastic adaptive control.

A minimax control problem of stochastic system was formulated by Aoki [1,p.298]. Some comparison between the minimax and Bayesian criterion was given by van Hee [3,p.179]. The minimax control problem of an one-dimensional linear system with quadratic costs and disturbances belonging to the exponential class with the variance being the quadratic function of the mean has been considered by Trybuła [9]. It follows from results of this paper that for the most of disturbances the risk as a function of the parameter of the distribution of disturbances is unbounded for all controls. Then there are no possibilities to compare the controls on this way , unless we restrict the states of nature i.e. the values of the parameter. The restriction of the value of the parameter can be given in many ways. We can assume that the value of the parameter is random and it is known that the second moments of the parameter is bounded or that the distribution of the parameter belongs to some class of distributions. The existence of the minimax control when some restrictions on the moments of disturbances are given, the distributions are one-dimensional and they belongs to the exponential family is proved in [6]. In the present paper it is shown that minimax controls exist for the considered system when disturbances have the multinomial distribution. Two cases are considered. First one, when the distribution of the parameter belongs to the class of the Dirichlet distributions and the second, when there is no restriction on the unknown parameter of the disturbances. We generalize results of Sworder [8] and Aoki [1,p.18] in several ways: first we consider the multidimensional case, secondly we allow the costs to be a quadratic form of controls and the unknown parameter. Thirdly we allow the plant to be non-stationary.

The main result of the paper is based on the following fact from the decision theory.

Lemma 1 (cf [10],[9]) Suppose that there is a sequence $\{\pi_k\}$ of prior distributions belonging to Π^o for which the corresponding sequence $\{\mathcal{U}^*_{\pi_k}\}$ of the BC policies satisfies the condition $\lim_{k\to\infty} r(\pi_k, \mathcal{U}^*_{\pi_k}) = c$ and there is a control policy $\hat{\mathcal{U}} \in \mathcal{D}$ such that $r(\pi, \hat{\mathcal{U}}) \le c$ for each $\pi\in\Pi^o$, then $\hat{\mathcal{U}}$ is a Π^o-minimax control policy.

In application of Lemma 1 the properties of prior distribution are used. The conjugate priors form an important class of such distributions. In the next section information about the conjugate priors to the multinomial distribution are given.

2. The multinomial distribution and his conjugate priors.

The vector v has the multinomial distribution with a known parameter K and an unknown one $p = (p_1, p_2, \dots, p_m)^T$ when the density function with respect to a certain σ-finite measure on R^m has the form

$$p(v|K,p) = \frac{K!}{v_1! \dots v_m!} p_1^{v_1} \dots p_m^{v_m} \tag{2.1}$$

where $v = (v_1, \dots, v_m)^T \in \mathscr{V} = \left\{ v \in \{0,1,\dots,K\}^m : \sum_{i=1}^m v_i = K \right\}$ and

$p \in \mathscr{P} = \left\{ p : 0 < p_i < 1,\ i=1,2,\dots,m,\ \sum_{i=1}^m p_i = 1 \right\}$.

If $v_0, v_1, \dots, v_M$ have the multinomial distribution with the parameters K and $p = (p_1, p_2, \dots, p_m)^T$ then

$$E_p v_n = Kp,\quad E_p(v_n^T A v_n) = K(K-1)p^T A p + K e^T di(A) p \tag{2.2}$$

for any m×m-matrix A (cf [2] p.48).

Suppose that p is an unknown parameter and the prior distribution of p is conjugate to that given by (1.4) (cf[2] p.174), i.e. it is the Dirichlet distribution with the density function

$$f(p|q) = \frac{\Gamma(q_1 + q_2 + \dots + q_m)}{\Gamma(q_1) \dots \Gamma(q_m)} p_1^{q_1 - 1} \dots p_m^{q_m - 1} \tag{2.3}$$

where $q \in \mathscr{S} = \{ q \in R^m : q_i > 0,\ i=1,2,\dots,m \}$ and $p \in \mathscr{P}$ (cf [2] p. 49-51). We denote this distribution by π_q.

To determine BC we must obtain the posterior density for p after any new observation. It is possible only if the matrices c_n in (1.1) are such that the equation $c_n v_n = x_{n+1} - a_n x_n - b_n u_n$ has the unique solution v_n for $n=0,1,\dots,M-1$ (x_0 is given). We have

$$g(p|\mathcal{X}_n, \mathcal{U}_{n-1}) = g(p|v_0, \dots, v_{n-1}) = f(p|q_n), \tag{2.4}$$

where

$$q_n = q_{n-1} + v_{n-1} = q + \sum_{i=0}^{n-1} v_i . \tag{2.5}$$

Notice that the distribution (2.4) does not depend on $\mathcal{U}_{n-1}$. Let $\alpha_n = \sum_{i=1}^m q_{ni} = \alpha + nK$ where $\alpha = \sum_{i=1}^m q_i$. From (2.4) and (2.5) we have

$$\begin{aligned} E(p|\mathcal{X}_n, \mathcal{U}_{n-1}) &= \alpha_n^{-1} q_n, \\ E(p^T A p|\mathcal{X}_n, \mathcal{U}_{n-1}) &= \alpha_n^{-1}(\alpha_n + 1)^{-1}\left[q_n^T A q_n + e^T di(A) q_n \right] \end{aligned} \tag{2.6}$$

for any m×m-matrix A, where E denotes the expectation with respect to the joint distribution of v_n and p.

The moments of v_n for given $\mathcal{X}_n$ and $\mathcal{U}_{n-1}$ are then

$$\begin{aligned} E(v|\mathcal{X}_n, \mathcal{U}_{n-1}) &= K\alpha_n^{-1} q_n, \\ E(v^T A v|\mathcal{X}_n, \mathcal{U}_{n-1}) &= K\alpha_n^{-1}(\alpha_n + 1)^{-1}\left[(K-1) q_n^T A q_n + \alpha_{n+1} e^T di(A) q_n \right] \end{aligned} \tag{2.7}$$

for any m×m-matrix A.

3. The Bayesian control.

Suppose the disturbances $v_0, v_1, \ldots$ have the distribution (2.1), the prior distribution π_q of the parameter p is given by (2.3), the control horizon N has the distribution (1.2) and the initial state x_0 is known. Consider the BC problem for the system (1.1) with a starting point at the moment n, when x_n, u_{n-1} are given. The expected risk is then

$$r_n(\pi_q, \mathcal{U}^n) = E_N\left[E\left(\sum_{i=n}^{N}(y_i^T s_i y_i + u_i^T k_i u_i) \mid x_n, u_{n-1}\right) \mid N \geq n\right] =$$

$$E\left[\sum_{i=n}^{M}\vartheta_n^i (y_i^T s_i y_i + u_i^T k_i u_i) \mid x_n, u_{n-1}\right].$$

Let $W_n = \inf_{\mathcal{U}^n} r_n(\pi_q, \mathcal{U}^n)$ denotes the Bayesian risk for the above truncated problem . If there exists $\mathcal{U}^{n*} = (u_n^*, u_{n+1}, \ldots u_M^*)$ such that $r(\pi_q, \mathcal{U}^{n*}) = W_n$ then it is called the BC policy for the truncated problem and u_i^* is called the BC, i=n,...,M. Obviously $r(\pi_q, \mathcal{U}_q^0) = r(\pi_q, \mathcal{U})$ and $W_0 = r(\pi_q, \mathcal{U}^*)$.

For the solution of BC problem we derive BC u_n^* for n=M,...,0 recursively. Then $\mathcal{U}^* = \mathcal{U}^{0*}$ is the solution of the above problem. Using the backward induction and the properties of the multinomial distribution and his conjugate prior we can prove

Lemma 2. For the system (1.1) and the disturbances having the multinomial distribution (2.1) with the parameters K and p, where p has the prior distribution given by (2.3), and for the random, independent of the disturbances, bounded horizon N having the distribution (1.2), the BC u_n^* and the Bayesian risk W_n for the cost criterion (1.3) have the form:

$$u_M^* = 0, \qquad u_n^* = -\xi_n x_n - \alpha_n^{-1}\eta_n q_n, \qquad n=0,\ldots,M-1 \tag{3.1}$$

where

$$\begin{aligned} \xi_n &= \vartheta_n^{n+1}\left[k_n + \vartheta_n^{n+1} b_n^T A_{n+1} b_n\right]^+ b_n^T A_{n+1} a_n, \\ \eta_n &= \vartheta_n^{n+1}\left[k_n + \vartheta_n^{n+1} b_n^T A_{n+1} b_n\right]^+ b_n^T (K A_{n+1} c_n + B_{n+1}^T), \end{aligned} \tag{3.2}$$

(A^+ denotes the Moore-Penrose pseudoinverse matrix of A [1]) and

$$W_n = x_n^T A_n x_n + 2\alpha_n^{-1} q_n^T B_n x_n + q_n^T C_n q_n + \alpha_n^{-1} d_n^T q_n$$

where the matrices A_n, B_n, C_n and the vector d_n satisfy the equations

$$A_n = s_n^1 + \xi_n^T k_n \xi_n + \vartheta_n^{n+1}(a_n - b_n\xi_n)^T A_{n+1}(a_n - b_n\xi_n),$$

$$B_n = s_n^3 + \vartheta_n^{n+1}\Big[(Kc_n - b_n\eta_n)^T A_{n+1}a_n + B_{n+1}a_n\Big],$$

$$\begin{aligned} C_n = {} & \alpha_n^{-1}(\alpha_n+1)^{-1}s_n^2 + \vartheta_n^{n+1}\Big[K(K-1)\alpha_n^{-1}(\alpha_n+1)^{-1}(c_n^T A_{n+1}c_n + \\ & 2\alpha_{n+1}^{-1}B_{n+1}c_n + C_{n+1}) - \alpha_n^{-2}\eta_n^T(k_n + \vartheta_n^{n+1}b_n^T A_{n+1}b_n)\eta_n + \\ & K\alpha_n^{-1}(2\alpha_n^{-1}B_{n+1}c_n + C_{n+1}^T + C_{n+1}) + C_{n+1}\Big], \\ d_n = {} & (\alpha_n+1)^{-1}di\Big(s_n^2 + K\vartheta_n^{n+1}(\alpha_{n+1}c_n^T A_{n+1}c_n + 2B_{n+1}c_n + \\ & \alpha_{n+1}C_{n+1})\Big)e + \vartheta_n^{n+1}d_{n+1} \end{aligned} \tag{3.3}$$

for $n=M-1,\ldots,0$, with conditions

$$A_M = s_M^1,\ B_M = s_M^3,\ C_M = \alpha_M^{-1}(\alpha_M+1)^{-1}s_M^2,\ d_M = (\alpha_M+1)^{-1}di(s_M^2)e.$$

The matrices ξ_n, η_n, A_n, B_n do not depend on q, $n=0,\ldots,M$.

4 The risk for the Bayesian control.

Now, for further considerations we calculate the risk function for the BC policy $\mathcal{U}_q^*$ given by (3.1) (i.e. we assume that the parameter p has the prior distribution π_q given by (2.3)). To calculate the risk function the truncated problem is considered i.e. the problem of the control for the system (1.1) with the starting point at the moment n, when $\mathcal{X}_n, \mathcal{U}_{n-1}$ are given. The truncated risk is then:

$$R_n(p,\mathcal{U}) = E_N\Big[E_p\Big(\sum_{i=n}^M (y_i^T s_i y_i + u_i^T k_i u_i)\,|\mathcal{X}_n, \mathcal{U}_{n-1}\Big)\,|\ N\geq n\Big].$$

It may be transformed to

$$R_n(p,\mathcal{U}) = E_p\Big(\sum_{i=n}^M \vartheta_n^i (y_i^T s_i y_i + u_i^T k_i u_i)\,|\mathcal{X}_n, \mathcal{U}_{n-1}\Big).$$

For the BC policy $\mathcal{U}_q^*$ the risk will be denoted by R_n. R_n fulfils the recurrence relation

$$R_n = y_n^T s_n y_n + u_n^{*T} k_n u_n^* + \vartheta_n^{n+1} E_p(R_{n+1}|\mathcal{X}_n, \mathcal{U}_{n-1})$$

for $n=M-1,\ldots,0$ and $R_M=0$, since

$$\begin{aligned} R_n = {} & E_p\Big[(y_n^T s_n y_n + u_n^{*T} k_n u_n^*)\,|\mathcal{X}_n, \mathcal{U}_{n-1}\Big] + \\ & \vartheta_n^{n+1} E_p\Big[E_p\Big(\sum_{i=n+1}^M \vartheta_{n+1}^i (y_i^T s_i y_i + u_i^{*T} k_i u_i^*)\,|\mathcal{X}_{n+1}, \mathcal{U}_n\Big)\,|\mathcal{X}_n, \mathcal{U}_{n-1}\Big]. \end{aligned} \tag{4.1}$$

Using backward induction we obtain that for the BC policy $\mathcal{U}_q^*$ the truncated risk has the form given by

Lemma 3. The truncated risk R_n has the form:

$$R_n = x_n^T A_n x_n + q_n^T f_n q_n + 2p^T B_n x_n + p^T h_n p + 2p^T i_n q_n + j_n^T p \tag{4.2}$$

where the matrices f_n, h_n, i_n and the vector j_n satisfy

$$\begin{aligned} f_n &= \rho_n^{-1} \sum_{k=n}^{M-1} \alpha_k^{-2} P_k, \\ i_n &= -\rho_n^{-1} \sum_{k=n}^{M-1} \alpha_k^{-1} P_k + K\rho_n^{-1} \sum_{k=n+1}^{M-1} (k-n)\alpha_k^{-2} P_k, \\ h_n &= \rho_n^{-1}\Big[Z_n - 2K\sum_{k=n+1}^{M-1}(k-n)\alpha_k^{-1}P_k + K\sum_{k=n+1}^{M-1}(k-n)\big[K(k-n)-1\big]\alpha_k^{-2}P_k\Big], \\ j_n &= K\rho_n^{-1}\Big[t_n + \sum_{k=n+1}^{M-1}(k-n)\alpha_k^{-2} di(P_k)e\Big], \end{aligned} \tag{4.3}$$

with the conditions $f_M=0$, $h_M=s_M^2$, $i_M=0$, $j_M=0$. The matrices A_n, B_n are given by (3.3), the symmetric non-negative definite matrix P_n, the matrix Z_n and the vector t_n do not depend on q and

$$\begin{aligned} P_n &= \eta_n^T(\rho_n k_n + \rho_{n+1} b_n^T A_{n+1} b_n)\eta_n, \\ Z_n &= \sum_{k=n}^{M} \rho_k s_k^2 + K\sum_{k=n}^{M-1}\rho_{k+1}\Big[(K-1)c_k^T A_{k+1} c_k + 2B_{k+1}c_k\Big], \\ t_n &= \sum_{k=n}^{M-1}\rho_{k+1} di(c_k^T A_{k+1} c_k)e . \end{aligned} \tag{4.4}$$

5. The minimax control.

From Lemma 2 we obtain an explicit form of the risk function for the BC policy. It can also be written as

$$R(p, \mathcal{U}_q^*) = p^T h_o p + p^T(2B_o x_o + 2i_o q + j_o) + q^T f_o q + x_o^T A_o x_o .$$

The expected risk is then

$$r(\pi_q, \mathcal{U}_q^*) = \alpha^{-1}(\alpha+1)^{-1}\Big[q^T h_o q + e^T di(h_o) q\Big] + \alpha^{-1} q^T z(q) + t(q),$$

where

$$\begin{aligned} h_o &= h_o(\alpha), \\ z(q) &= 2B_o x_o + 2i_o q + j_o = 2B_o x_o + 2i_o(\alpha) q + j_o(\alpha), \\ t(q) &= q^T f_o q + x_o^T A_o x_o = q^T f_o(\alpha) q + x_o^T A_o x_o . \end{aligned}$$

Hence we obtain

$$\begin{aligned} r(\pi_q, \mathcal{U}_q^*) = {} & q^T \alpha^{-1}\Big[\alpha(\alpha+1)^{-1} h_o(\alpha) + 2\alpha i_o(\alpha) + \alpha^1 f_o(\alpha)\Big] q\alpha^{-1} + \\ & \alpha^{-1} q^T\Big[2B_o x_o + j_o + (\alpha+1)^{-1} di(h_o(\alpha))e\Big] + x_o^T A_o x_o . \end{aligned} \tag{5.1}$$

Let us define for $m \in \mathcal{P}^c = \{p: 0 \le p_i \le 1,\ i=1,\dots,m,\ \sum_{i=1}^{m} p_i = 1\}$ two control policies $\mathcal{U}_m^+$ and $\mathcal{U}_m^-$, which can be obtained as a limit of the BC policies with probability one. Denote $\mathcal{U}_m^+ = (u_0^+, u_1^+, \dots, u_M^+)$ the control policy defined by

$$u^+_M=0, \quad u^+_n=-\xi_n x_n-\eta_n m, \quad n=0,\dots M-1.$$

Since $\sum_{i=0}^{n-1}\upsilon_i$ is bounded a.s., taking $q=q(m)$ such that $\alpha^{-1}q(m)\to m$ when $\alpha\to\infty$ we see that $\alpha_n^{-1}q_n = (\alpha+nK)^{-1}(q(m)+\sum_{i=0}^{n-1}\upsilon_i)$ is convergent a.s. to m when α tends to infinity. Thus $\mathcal{U}^+_m$ may be treated as the a.s. limit of $\mathcal{U}^*_q$, for $\alpha\to\infty$ and $\alpha^{-1}q\to m$, $q\in\mathcal{P}$.

From (4.2) and (4.3) we have

$$R(p,\mathcal{U}^+_m) = \text{a.s.-}\lim_{\substack{\alpha\to\infty\\ q/\alpha\to m}} R(p,\mathcal{U}^*_q) = p^T(Z_o-\sum_{k=1}^{M-1}P_k)p + \tag{5.2}$$
$$p^T(2B_ox_o+K\ell_o) + x_o^TA_ox_o + (p-m)^T(\sum_{k=1}^{M-1}P_k)(p-m).$$

Notice also

$$\varphi(m) \overset{df}{=} \lim_{\substack{\alpha\to\infty\\ q/\alpha\to m}} r(\pi_q,\mathcal{U}^*_q) = m^T(Z_o-\sum_{k=1}^{M-1}P_k)m + m^T(2B_ox_o+K\ell_o) + x_o^TA_ox_o$$

Denote $\mathcal{U}^-_m=(u^-_o,u^-_1,\dots,u^-_M)$ the following control policy:

$$u^-_M=0,\quad u^-_n=-\xi_nx_n-(nK)^{-1}\eta_n\sum_{i=0}^{n-1}\upsilon_i,\quad n=1,\dots,M-1,\quad u_o=-\bar{\xi}_ox_o-\eta_om.$$

The policy $\mathcal{U}^-_m$ can be obtained as the a.s.-limit of the BC policies $\mathcal{U}^*_q$ when $\alpha\to 0+$ and $\alpha^{-1}q\to m$. By the form of $R(p,\mathcal{U}^*_q)$ we have

$$R(p,\mathcal{U}^-_m) = \text{a.s.-}\lim_{\substack{\alpha\to 0+\\ q/\alpha\to m}} R(p,\mathcal{U}^*_q) = p^T\Big[Z_o-\sum_{i=1}^{M-1}(1+K^{-1}i^{-1})P_i\Big]p +$$
$$p^T\Big[2B_ox_o-2P_om+K\ell_o+K^{-1}\sum_{i=1}^{M-1}i^{-1}di(P_i)e\Big] + m^TP_om + x_o^TA_ox_o.$$

Notice also that

$$\psi(m) \overset{df}{=} \lim_{\substack{\alpha\to 0+\\ q/\alpha\to m}} r(\pi_q,\mathcal{U}_q)^* = -m\,P_o^Tm + x_oA_o^Tx_o +$$
$$m^T\Big[2B_ox_o+K\ell_o+di(Z_o-\sum_{i=1}^{M-1}P_i)e\Big].$$

<u>Theorem 1.</u> For the system (1.1) with the disturbances having the multinomial distribution (2.1), the random, independent of the disturbances, bounded horizon N having the distribution (1.2) and the cost criterion (1.3)

(a) if $h_o(\alpha)$ is a non-positive definite matrix for each $\alpha>0$ then the policy $\mathcal{U}^+_{m^*}$, where $m^*\in\mathcal{P}^c$ fulfils the equality $\varphi(m^*) = \sup_{m\in\mathcal{P}}\varphi(m)$, is the minimax one;

(b) if $h_o(\alpha)$ is a non-negative definite matrix for each $\alpha>0$ then the policy $\mathcal{U}^-_{m^*}$, where $m^*\in\mathcal{P}^c$ fulfils the equality $\psi(m^*) = \sup_{m\in\mathcal{P}} \psi(m)$, is the minimax one;

(c) if there exists $\alpha^*>0$ such that $h_o(\alpha^*)=0$ then the policy $\hat{\mathcal{U}}_{m^*} = \text{a.s.-}\lim_{q/\alpha^*\to m^*} \mathcal{U}^*_q$, where m^* fulfils the equality $\lim_{q/\alpha^*\to m^*} r(\pi_q,\mathcal{U}^*_q) = \sup_{q\in\mathcal{P}} r(\pi_q,\mathcal{U}^*_q)$, is the minimax one.

Proof. Consider the case (a). The assumption of $h_o(\alpha)$ implies that the matrices Z_o and $Z_o-\sum_{k=1}^{M-1}P_k$ are non-positive definite, where Z_o and P_k are given by (4.4) (to see this compute the limit of $h_o(\alpha)$ as α tends to ∞ or 0+, respectively). Let $m^*\in\mathcal{P}^c$ be such that $\sup_{m\in\mathcal{P}} \varphi(m)=\varphi(m^*)$. Obviously m^* always exists. Since $Z_o-\sum_{k=1}^{M-1}P_k$ is non-positive definite matrix then $R(p,\mathcal{U}^+_{m^*})$ given by (5.2) is the concave quadratic form of p. Since it is a sum of a concave func- tion which has the supremum at $p=m^*$ and a convex function which has the infimum at $p=m^*$ then the supremum of $R(p,\mathcal{U}^+_{m^*})$ is attained at $p=m^*$ and it is equal to $\varphi(m^*)$. Therefore $R(p,\mathcal{U}^+_{m^*})\le\varphi(m^*)$ for each $p\in\mathcal{P}$. Using Lemma 1 we obtain that $\mathcal{U}^+_{m^*}$ is the minimax policy.

In the case (b) from the assumption of $h_o(\alpha)$ we have that the matrix $Z_o-\sum_{i=1}^{M-1}(1+K^{-1}i^{-1})P_i$ is non-negative definite. Let $m^*\in\mathcal{P}^c$ fulfils the equality $\psi(m^*)=\sup_{m\in\mathcal{P}}\psi(m)$. Since $p^TAp \le p^T di(A)e$ for a non-negative definite matrix A (because the left-hand side of this inequality is a convex function) then

$$\begin{aligned}R(p,\mathcal{U}^-_{m^*}) &= (p-m^*)^TP_o(p-m^*) + p^T\Big[Z_o-\sum_{i=1}^{M-1}(1+K^{-1}i^{-1})P_i\Big]p - \\ &-p^TP_op + p^T\Big[2B_ox_o+Kt_o+K^{-1}\sum_{i=1}^{M-1}i^{-1}di(P_i)e\Big] + x_o^TA_ox_o \le \\ &(p-m^*)^TP_o(p-m^*) - p^TP_op + p^T\Big[2B_ox_o+Kt_o+ di(Z_o-\sum_{i=1}^{M-1}P_i)e\Big] + x_o^TA_ox_o.\end{aligned}$$

Analogously as in the case (a) we can show that the right-hand side of the above inequality attains its supremum at $p=m^*$ and it is equal to $\psi(m^*)$. Therefore $R(p,\mathcal{U}^-_{m^*}) \le \psi(m^*)$ for each $p\in\mathcal{P}$. Lemma 1 implies that $\mathcal{U}^-_{m^*}$ is the minimax policy in the case (b).

Now consider the case (c). Let $\alpha^*>0$ be such that $h_o(\alpha^*)=0$. Let $m^*\in\mathcal{P}^c$ fulfils the equality $\beta(m^*) = \sup_{m\in\mathcal{P}}\beta(m)$, where $\beta(m)= \lim_{q/\alpha^*\to m} r(\pi_q,\mathcal{U}^*_q)$. We have

$$\begin{aligned}\beta(m) &= m^T\Big[2B_ox_o + Kt_o + \sum_{i=1}^{M-1}\alpha_i^{*-2}di(P_i)e\Big] + x_o^TA_ox_o + \\ &m^T\Big[-2P_o-2\sum_{i=1}^{M-1}\alpha^*\alpha_i^{*-1}P_i+2K\sum_{i=1}^{M-1}\alpha^*\alpha_i^{*-2}P_i+\sum_{i=1}^{M-1}\alpha^{*2}\alpha_i^{*-2}P_i+P_o\Big]m\end{aligned}$$

Let $\hat{\mathcal{U}}_{m^*} = \lim_{q/\alpha^* \to m^*} \mathcal{U}^*_q$. Thus

$$R(p,\hat{\mathcal{U}}_{m^*}) = p^T\left[-2P_o-2\sum_{i=1}^{M-1}\alpha^*\alpha_i^{*-1}P_i+2K\sum_{i=1}^{M-1}\alpha^*\alpha_i^{*-2}iP_i\right]m^* + x_o^T A_o x_o +$$
$$m^{*T}\left[\sum_{i=1}^{M-1}\alpha^{*2}\alpha_i^{*-2}P_i+P_o\right]m^* + p^T\left[2B_o x_o+K\ell_o+\sum_{i=1}^{M-1}\alpha_i^{*-2}i\,di(P_i)e\right] \le \beta(m^*).$$

Since $R(p,\hat{\mathcal{U}}_{m^*}) \le \beta(m^*)$ for each $p\in\mathcal{P}$ then $\hat{\mathcal{U}}_{m^*}$ is the minimax policy in this case. The theorem is proved.

Remark. In the case (c) if $m^*\in\mathcal{P}$ then $\hat{\mathcal{U}}_{m^*} = \mathcal{U}^*_{\alpha^* m^*}$.

Theorem 2. For the considered system, if the unknown parameter of disturbances has distribution belonging to the class of the Dirichlet distributions (2.3) then the minimax control exists.

Proof. Let us consider a prior distribution given by (2.3). We have from (5.1) that the expected risk for the Bayes control has the form

$$r(\pi_q,\mathcal{U}^*_q) = \alpha^{-1}q^T\left[\alpha(\alpha+1)^{-1}h_o(\alpha)+2\alpha i_o(\alpha)+\alpha^2 f_o(\alpha)\right]\alpha^{-1}q +$$
$$\alpha^{-1}q^T\left[2B_o x_o+\ell_o+(\alpha+1)^{-1}di(h_o(\alpha))e\right] + x_o^T A_o x_o.$$

Since $\alpha^{-1}q = r\in\mathcal{P}$, the expected risk can be treated as a function of two variables $r_1(\alpha,r)$, where $\alpha>0$, $r\in\mathcal{P}$ and

$$\sup_{q\in\mathcal{P}} r(\pi_q,\mathcal{U}^*_q) = \sup_{\alpha>0}\sup_{r\in\mathcal{P}} r_1(\alpha,r)$$

where $r\in\mathcal{P}$. From (4.3) and the definition of α, α_n we obtain that for $\alpha\in(0,\infty)$ the risk $r(\pi_q,\mathcal{U}^*_q)=r_1(\alpha,r)$ is bounded by some quadratic form of $r\in\mathcal{P}$. Let $c = \sup_{\alpha>0}\sup_{r\in\mathcal{P}} r_1(\alpha,r)$. Since $\mathcal{P}$ is compact, there exists r^* such that $\sup_{\alpha>0} r_1(\alpha,r^*) = c$. Let $\{\alpha_k\}$ and $\{r_k\}$ be such that $\lim_{k\to\infty} r_1(\alpha_k,r_k) = c$. Denote by π_k the Dirichlet prior with parameter $q = r_k\alpha_k$ and $\hat{\mathcal{U}} = \text{a.s.-}\lim_{k\to\infty} \mathcal{U}_{\pi_k}$. By Lemma 1, $\hat{\mathcal{U}}$ is the minimax control when the possible priors of p are the Dirichlet ones.

6. Examples.

Let us consider the 3-dimensional linear system (1.1) with the matrices $a_n = diag((a_1,a_2,a_3)^T)$, $b_n=c_n= I$ (the unit matrix) and $x_o=x = (x_1,x_2,x_3)^T$. Let the parameter K=1, M=2 and P(N=2)=1. The co-efficients of the loss function (1.3) are: $s_i^1=s_i^2=0$ for i=0,1, $s_i^1=s_i^2=I$, $s_i^3=0$, $k_i=I$ for i=0,1,2. Denote $di(a_i) = diag((a_1,a_2,a_3)^T)$. Then:

$$\xi_1 = di(a_i/2), \qquad \eta_1 = 2^{-1}I,$$
$$\xi_o = di(a_i^3/(a_i^2+2)), \qquad \eta_o = di((a_i^2+a_i)/(a_i^2+2)),$$

$$A_o = di(a_i^4/(a_i^2+2)),$$

$$h_o(\alpha) = di(1+a_i-1/(\alpha+1)).$$

The expected risk for the BC policy $\mathcal{U}_q^*$ has the following open form:

$$r_1(\alpha,r) = \sum_{i=1}^{3}\left[\alpha^2/2(\alpha+1)^2+a_i\alpha/(\alpha+1)-(a_i+1)^2a_i^2/2(a_i^2+2)\right]r_i^2$$
$$+\sum_{i=1}^{3}\left[a_i/(\alpha+1) + a_i^2/2 + x_i(a_i+1)a_i^2/(a_i^2+2)\right]r_i$$
$$+ 1 - 1/2(\alpha+1)^2 + 1/(\alpha+1) + x_o^T A_o x_o.$$

Let $m=(1/3,1/3,1/3)^T$, $m_1=(1,0,0)^T$, $m_2=(0,1,0)^T$, $m_3=(0,0,1)^T$.

1. Let $a_1=a_2=-1$, $a_3=0$. Then for each $\alpha>0$ $\sup_{r\in\mathcal{P}} r_1(\alpha,r)$ is attained at the point $m_3\in\mathcal{P}^c$ and it does not depend on α. Thus a minimax policy is the policy $\hat{\mathcal{U}}_{m_3}$, which is the a.s.-limit of BC policies $\mathcal{U}_q^*$ when $\alpha^{-1}q\to m_3$ for any $\alpha>0$. It has the form:

$$\hat{u}_2=0,\quad \hat{u}_1=-\xi_1x_1-\eta_1(\alpha+1)^{-1}(v_o+\alpha m_3),\quad \hat{u}_o=-\xi_o x_o-\eta_o m.$$

It is interesting in this case that a minimax policy is not unique.

In the cases below it is assumed that $a_1=a_2=a_3=a$, $x_1=x_2=x_3=x$ and that there is no restriction on a class of prior distributions of disturbances.

2. Let $a\le-1$. Then the matrix $h(\alpha)$ is non-positive and the policy $\mathcal{U}_m^+$ is minimax. It has the form:

$$u_2^+=0,\qquad u_i^+=-\xi_i x_i-\eta_i m,\qquad i=0,1.$$

3. Let $a\in(-1,0)$. Then $h_o(\alpha_1)=0$ for $\alpha_1=-a/(a+1)$ and the BC policy $\mathcal{U}_q^*$ for $q=\alpha_1 m$ is a minimax one. Its form is given by (3.1).

4. Let $a=0$. Then $h_o(\alpha)$ is non-negative definite, but for each $\alpha>0$ $\sup_{r\in\mathcal{P}} r_1(\alpha,r)$ is attained at the points $m_i\in\mathcal{P}^c$, $i=1,2,3$, and it does not depend on α. Thus every policy $\hat{\mathcal{U}}_{m_i}$, which is the a.s.-limit of BC policies $\mathcal{U}_q^*$ when $\alpha^{-1}q\to m_i$ for any $\alpha>0$, is minimax. It has the form:

$$\hat{u}_2=0,\quad \hat{u}_1=-\xi_1x_1-\eta_1(\alpha+1)^{-1}(v_o+\alpha m_i),\quad \hat{u}_o=-\xi_o x_o-\eta_o m_i.$$

5. Let $a>0$. Then $h_o(\alpha)$ is non-negative definite and the policy $\mathcal{U}_m^-$ is minimax. It has the form:

$$u_2^-=0,\quad u_1^-=-\xi_1x_1-\eta_1v_o,\quad u_o^-=-\xi_o x_o-\eta_o m.$$

References

[1] Aoki, M.: Optimization of Stochastic Systems. Topics in Discrete Time Systems, Academic Press, New York (1967)

[2] De Groot,M.H.: Optimal Statistical Decision. New York, McGraw-Hill Book Company (1970)

[3] Van Hee,K.M.: Bayesian Control of Markov Chains, Mathem. Center Tracts 95, Amsterdam (1978)

[4] De Koning,W.L.: Digital optimal control systems with stochastic parameters. Delft Univ. Press 1984

[5] Kumar,P.R.: A survey of some results in stochastic adaptive control, SIAM J.Control and Optimization 23 (1985), 329-380

[6] Porosiński,Z.;Szajowski,K.;Trybuła,S.: Minimax control of a second order linear system. Opsearch 23 (1986), 215-228.

[7] Rieder,U.: Bayesian dynamic programming, Adv.Appl.Probab. 7, (1975), 330-348

[8] Sworder,D.D.: Minimax control of discrete time stochastic systems, Soc.Ind.Appl.Math., Ser.A, Control 2 (1964), 433-449

[9] Trybuła,S.: Control with use of previous experience, Zastos. Matem. 19 (1986), 1-12

[10] Zacks,S.: The Theory of Statistical Inference, Wiley, New York (1971)

ON A PACKING PROBLEM

Andrzej Sierociński
Institute of Mathematics, Warsaw Technical University
Pl.Jedności Robotniczej 1, Warsaw, Poland

Jerzy Zabczyk
Institute of Mathematics, Polish Academy of Sciences
Śniadeckich 8, Warsaw, Poland

1. Introduction.

Let $\xi_1,\xi_2,\dots,\xi_n$ be a sequence of random variables defined on a probability space $(\Omega,\mathcal{F},\mathbb{P})$. Packing scheme is a procedure which allocates each of the consecutively observed values $\xi_1,\xi_2,\dots$ in one of the n cells marked with numbers $1,2,\dots,n$, called ranks.
Each cell can contain only one value and bigger values should be allocated in cells with higher ranks. Primary above problem was posed by Guzicki in connection with preliminary sorting in data base. It is well known that sorting a large quantity of items is time-consuming, but this time needed for sorting can be substantially reduced if we know the distribution of the characteristic according to which we want to sort. Namely we can try to put coming observations straightforward in proper order.

The packing problem we are interested in here is to construct a packing scheme which maximizes the probability that all n values $\xi_1,\xi_2,\dots,\xi_n$ will be properly located in all n cells. We solve the problem under the additional assumption, made also in Guzicki's original formulation, that the random variables $\xi_1,\xi_2,\dots,\xi_n$ are independent with the uniform distribution on the interval $(0,1)$. We reformulate the packing problem as a discrete-time stochastic control problem and solve it using the dynamic programming method.

2. Formulation.

To formulate the packing problem mathematically and formalize the concept of strategy it is convenient to introduce notation :

$$I = (0,1)\ , \quad J = \{1,2,\dots,n\}$$

$$E_1 = I \quad , \quad E_{2k-1} = I^k \times J^{k-1},\ k = 2,\dots,n$$

$$E_{2k} = I^k \times J^k\ ,\ k = 1,2,\dots,n$$

Elements of E_{2k-1} and E_{2k} will be denoted respectively as $((x_1,\dots,x_k);(m_1,\dots,m_{k-1}))$ and $((x_1,\dots,x_k);(m_1,\dots,m_k))$ and numbers m_j will be called ranks of x_j , $j = 1,2,\dots$. The transition mechanism of the controlled process we are constructing depends on whether the stage is even or odd. Namely at an odd stage $2k-1$ the transition from E_{2k-1} to E_{2k} is deterministic and depends only on choice of the control parameter $m \in J$ according to the formula : $((x_1,\dots,x_k);(m_1,\dots,m_{k-1})) \longrightarrow ((x_1,\dots,x_k);(m_1,\dots,m_{k-1},m))$, $k = 1,2,\dots,n$. At an even stage $2k$ the transition from E_{2k} to E_{2k+1} is completely stochastic : $((x_1,\dots,x_k);(m_1,\dots,m_k)) \longrightarrow ((x_1,\dots x_k,\xi_{k+1});(m_1,\dots,m_k))$ $k = 1,2,\dots,n-1$, where $\xi_1,\dots,\xi_n$ is a sequence of independent random variables uniformly distributed on I .

A packing strategy Π is a sequence of Borel functions $f_k : E_{2k-1} \longrightarrow J$, $k = 1,2,\dots,n$. If a strategy is given then the position $X_k = (Y_k;Z_k) \in E_k$, $k = 2,\dots,n$ of a controlled system at stage k is defined as follows :

$$X_{2k+1} = ((Y_{2k},\xi_{k+1});Z_{2k}) \qquad k = 1,2,\dots,n-1$$

$$X_{2k} = (Y_{2k-1},(Z_{2k-1},f_k(X_{2k-1})), \quad k = 1,2,\dots,n$$

and

$$X_1 = (\xi_1).$$

Let $r : E_{2n} \longrightarrow R$ be the following function :

$$r((x_1,\dots,x_n);(m_1,\dots,m_n)) = \begin{cases} 1 & \text{if } x_i \neq x_j \text{ for } i\neq j \text{ and} \\ & \text{if } x_i < x_j \text{ then } m_i < m_j \\ 0 & \text{otherwise} \end{cases}$$

The packing problem can be formulated as follows :
Find a strategy Π for which the expected value $E(r(X_{2n}))$ attains its maximal value.

The existence of an optimal strategy it is possible to prove by straightforward checking the Bellman's equations. (See e.g. [1] or [3,Thm.1.1]). The corresponding value sequence and the optimal strategy can be calculated explicitly.

3. The optimal strategy.

The optimal strategy can be defined recursively. Define first a sequence of numbers $p_0, p_1, \ldots$ and a sequence of functions as follows

$$v_j^l : (0,1) \longrightarrow R \;, \quad l = 1,2,\ldots, \; j = 1,\ldots,l$$

$$p_0 = p_1 = 1, \quad v_1^1(x) = 1 \;, \quad x \in (0,1).$$

Assume that $p_0, p_1, \ldots, p_{l-1}$ are known, $l > 1$ and define :

$$v_j^l(x) = \binom{l-1}{j-1} x^{j-1}(1-x)^{l-j} \, p_{j-1} p_{l-1} \;, \quad x \in (0,1) \;, \; j = 1,2,\ldots,l$$

$$p_l = \int_0^1 \max_j \{ \, v_j^l(x) \, \} dx$$

For arbitrary $l = 1,2,\ldots$, let $(A_1^l, \ldots, A_l^l)$ be a partition of the interval $I = (0,1)$ into disjoint sets such that

$$\max_j \{ \, v_j^l(x) \, \} = v_k^l(x) \;, \qquad \text{for} \quad x \in A_k^l \;, \quad k = 1,2\ldots \,.$$

If $(x_1, \ldots, x_l) \in I^l$, $(m_1, \ldots, m_l) \in J^l$ then one writes $(x_1, \ldots, x_l) \sim (m_1, \ldots, m_l)$ if for all $i \neq j$ either $x_i > x_j$ and then $m_i > m_j$ or $x_i < x_j$ and then $m_i < m_j$.

We can define now functions $\hat{f}_1, \ldots, \hat{f}_n$:

$$\hat{f}_1(x_1) = j \quad \text{if and only if} \quad x_1 \in A_j^n \;, \; j = 1,\ldots,n.$$

More generally, if $((x_1, \ldots, x_k); (m_1, \ldots, m_{k-1})) \in E_{2k-1}$ and there are no $m \in J$ such that

$$(x_1, \ldots, x_k) \sim (m_1, \ldots, m_{k-1}, m) \quad \text{then} \quad \hat{f}_k((x_1, \ldots, x_k); (m_1, \ldots, m_{k-1})) = 1.$$

Assume that $(x_1, \ldots, x_k) \sim (m_1, \ldots, m_{k-1}, m)$ for some $m \in J$ then $(x_1, \ldots, x_{k-1}) \sim (m_1, \ldots, m_{k-1})$ and numbers $x_1, \ldots, x_{k-1}$ partition the interval I into disjoint subintervals. A subinterval $\bar{I} = (\alpha, \beta)$ from the partition is called accessible if it has positive capacity where the capacity c is the maximal natural number such that for arbitrary but different $\bar{x}_1, \bar{x}_2, \ldots, \bar{x}_c \in \bar{I}$ there exist natural numbers $\bar{m}_1, \ldots, \bar{m}_c \in J$ for which

$$(x_1, \ldots, x_{k-1}, \bar{x}_1, \ldots, \bar{x}_c) \sim (m_1, \ldots, m_{k-1}, \bar{m}_1, \ldots, \bar{m}_c) \,.$$

We can assume that $x_k \in \bar{I} = (\alpha,\beta)$ for some accessible interval $\bar{I}$ with capacity $c \geq 1$. Then

$$x_k \in \alpha + \frac{1}{\beta-\alpha} A^c_s \quad \text{for some} \quad s = 1,2,\dots,c \quad \text{and we define}$$

$$\hat{f}_k(x_1,\dots,x_{k-1},x_k) = s + 1 \quad \text{where} \quad 1 \quad \text{is the rank of} \quad \alpha .$$

Theorem.

The strategy $\hat{\Pi} = (\hat{f}_1,\dots,\hat{f}_n)$ is an optimal packing strategy and the optimal probability is equal to p_n .

The optimal probabilities p_n tend to 0. Since sets $(A^n_1,\dots,A^n_n)$ are disjoint intervals for all n it is easy to compute the optimal probabilities p_n . Numerical analysis shows that p_n converge to 0 with geometric rate ($p_{n+1}/p_n \longrightarrow q$). Under the assumption of geometric rate of convergence one can show that for sufficiently large n , $|A^n_1| \simeq \frac{q}{n-1}$ and $|A^n_k| \simeq \frac{1}{n-1}$, for all k which are not too small and not too large ($k/n \longrightarrow r \in (0,1)$). So the optimal strategy differs substantially from the trivial strategy based on the subdivision of the (0,1) interval into n subintervals of the same length. For example it has been checked numerically that for $n = 50$ the probability of the proper allocation for the trivial strategy is equal to 62% of the optimal probability p_{50} .

References :

[1] K. Hinderer, Foundations of non-stationary dynamic programming with discrete time parameter. Berlin, Heidelberg, 1970.

[2] E. Lanery, Optimisation en control impulsionnel evolutif (to appear).

[3] J. Zabczyk, Stochastic Control of Discrete-Time Systems, in "Control Theory and Topics in Functional Analysis", Vol. 3, pp. 187-224. International Energy Agency, Vienna 1976.

QUADRATIC CONTROL FOR LINEAR STOCHASTIC
EQUATIONS WITH PATHWISE COST

C. TUDOR
Faculty of Mathematics, University of Bucharest,
14 Academy St., 70109 Bucharest, ROMANIA

1. INTRODUCTION

In the paper we consider an infinite dimensional controlled stochastic system of Ito type with bounded and continuous coefficients on $(-\infty,\infty)$ or $[t_0,\infty)$. We associate two optimization problems with a pathwise cost of average form. We obtain the optimal control as an affine function of the state with the aid of the bounded solution of an infinite dimensional Riccati equation. We consider also the periodic case.

2. PRELIMINARIES

Let H, K, U be real separable Hilbert spaces. We shall denote by $\langle\cdot,\cdot\rangle$, $|\cdot|$ scalar products and norms in Hilbert spaces. L(K,H) denotes the family of bounded linear operators mapping K into H with the operator norm denoted by $\|\cdot\|_{L(K,H)}$. We shall denote by Tr A the trace of the operator A (if it exists) and by $L^+(H)$ the class of all nonnegative bounded operators on H. Next J denotes R or $[t_0,\infty)$, $t_0\in R$, and we define $C_b(J,L(K,H))$ to be the set of all strongly continuous bounded mappings from J into L(K,H).
Let $(\Omega,\mathcal{F},P,(\mathcal{F}_t)_{t\in R})$ be a filtered probability space on which is defined a K-valued Wiener process w with W as covariance operator.
We consider the following hypotheses.

(H1) a) For every $t\in R$, A(t) is a linear operator on H which generates an evolution operator $\{U(t,s)\}_{s\leq t}$ on H, i.e.,

a_1) $U(t,t)=I$ for all t,

a_2) $U(.)h$ is continuous for all $h\in H$,

a_3) For $t>s$, $U(t,s):\mathcal{D}(A(s))\longrightarrow\mathcal{D}(A(t))$ and

$$\frac{\partial}{\partial t}U(t,s)h=A(t)U(t,s)h \quad \text{for } h\in\mathcal{D}(A(s)).$$

b) For n large enough the resolvent operator $R(n,A(t))$ is well defined and if $\{U_n(t,s)\}$ is the evolution operator generated by the Yosida operator $A_n(t)=n^2R(n,A(t))-nI$ then for all $h\in H$ we have

$U_n(t,s)h \longrightarrow U(t,s)h$ uniformly on bounded sets in (s,t).

Remark 2.1. The hypothesis (H1) is fulfiled under the usual hypotheses of Tanabe and Kato-Tanabe (see [9],[1o]).

(H2) $B\in C_b(J,L(U,H))$, $f\in C_b(J,H)$, $G\in C_b(J,L(K,H))$, $M\in C_b(J,L(H))$, $N\in C_b(J,L^+(H))$, $N(t)\geq \varepsilon I>0$ for all $t\in J$.

Definition. 1) We say that the pair (A,B) is stabilizable if there exists $K\in C_b(J,L(H,U))$ such that the evolution operator U_{A-BK} generated by $A-BK$ is (exponentially) stable. i.e.,

$$\|U_{A-BK}(t,s)\|_{L(H)}\leq \alpha \exp\{-\beta(t-s)\} \quad \text{for some } \alpha,\beta>0 \text{ and all } t\geq s.$$

2) We say that the pair (A,M) is detectable if there exists $K_1\in C_b(J,L(H))$ such that U_{A-K_1M} is stable.

The following Riccati equation is useful

$$(2.1)\quad Q' + A^*Q + QA + M^*M - QBN^{-1}B^*Q=0$$

Definition. Q is a bounded solution of (2.1) if $Q\in C_b(J,L^+(H))$ and

$$(2.2)\quad Q(t)h = U^*(s,t)Q(s)U(s,t)h + \int_t^s U^*(r,t)\left[M^*M-QBN^{-1}B^*Q\right]U(r,t)h\,ds$$

for all $h\in H$, $s\geq t$, $s,t\in J$.

The main result concerning the Riccati equation is the following

THEOREM 2.1 [3]. Assume (H1),(H2) and that (A,B) is stabilizable and (A,M) is detectable. Then there exists a unique bounded solution Q of (2.1) and $U_{A-BN^{-1}B^*Q}$ is stable. Moreover, if A,B,M,N are θ-periodic then Q is θ-periodic.

3. THE OPTIMAL CONTROL PROBLEM WITH ONE-SIDED AVERAGE PATHWISE COST

We assume (H1),(H2) hold for $J=[t_0,\infty)$. Consider the following controlled stochastic equation

$$(3.1)\begin{cases} dx(t) = [A(t)x(t) + B(t)u(t) + f(t)]dt + G(t)dw(t)\ ; \ t\geq t_0 \\ x(t_0) = x_0 \end{cases}$$

where u is some progressively measurable U-valued process and x_o is an $\mathcal{F}_o$-measurable random element with values in H.

Definition. The progressively measurable process

(3.2) $$x(t)= U(t,t_o)x_o+ \int_{t_o}^{t} U(t,s)\left[B(s)u(s)+f(s)\right]ds+\int_{t_o}^{t}U(t,s)G(s)dw(s)$$

is called the mild solution of (3.1).

Remark 3.1. From [8] it follows that the process $\int_{t_o}^{\cdot} U(\cdot,s)G(s)dw(s)$ has a progressively measurable modification and if U^o is of contraction type it follows from [5] that the above process has a continuous modification.

Definition. A process $\{x(t)\}_t$ is θ-periodic if it has the same distribution as the shifted process $\{x(t+\theta)\}_t$.

Next we define

(3.3) $$J(u)=\overline{\lim_{T\to\infty}} \frac{1}{T}\int_{t_o}^{t_o+T} \left[|M(t)x(t)|^2 + \langle N(t)u(t),u(t)\rangle\right] dt$$

U_1 is the class of all progressively measurable U-valued processes $\{u(t)\}_{t\geq t_o}$ such that

(3.4) $$\sup_{t\geq t_o} E(|u(t)|^4) < \infty$$

and its response given by (3.2) satisfies

(3.5) $$\sup_{t\geq t_o} E(|x(t)|^4) < \infty .$$

U_2 is the class of all progressively measurable U-valued processes $\{u(t)\}_{t\geq t_o}$ such that u,x are θ-periodic and

(3.6) $$\sup_{t\geq t_o} E(|u(t)|^2) < \infty$$

(3.7) $$\sup_{t\geq t_o} E(|x(t)|^2) < \infty.$$

We introduce the following optimization problems:

(P_i) $$\min_{u\in U_i} J(u)$$

where minimum is taken in the sense of the almost surely inequality.

Consider the hypothesis:

(H3) $\mathcal{D}(A(t))=D$ for all t and

$$\sup_t |A(t)h| < \infty \text{ for all } h\in D.$$

THEOREM 3.1. Assume (H1)-(H3) and that

1) (A,B) is stabilizable,
2) (A,M) is detectable,
3) $E(|x_o|^4) < \infty$.

Then the optimal control for (P_1) is given by the feedback law

$$(3.8)\quad \bar{u} = -N^{-1}B^*(Qx+r)$$

where Q is the unique bounded solution of (2.1) and r is the unique bounded solution of

$$(3.9)\quad r' + L^*r + Qf = o\ ,\quad L = A - BN^{-1}B^*Q\ ,\ \text{that is}$$

$$(3.10)\quad r(t) = \int_t^{\infty} U_L(s,t)Q(s)f(s)ds\ ,\quad t \geq t_o.$$

The optimal cost is

$$(3.11)\quad J(\bar{u}) = \overline{\lim_{n\to\infty}}\ \frac{1}{n}\int_{t_o}^{t_o+n} \left[2\langle r,f\rangle - |N^{-1/2}B^*r|^2 + Tr(GWG^*Q)\right]dt$$

and the optimal trajectory is globally asymptotically stable in mean square, i.e., any solution y of (3.1) with $E(|y(o)|^2) < \infty$ satisfies

$$(3.12)\quad \lim_{t\to\infty} E(|x(t)-y(t)|^2) = o.$$

THEOREM 3.2. Assume (H1)-(H3) and that (A,B) is stabilizable and (A,M) is detectable. Moreover, suppose that A,B,f,G,M,N are θ-periodic. Then $\bar{u}$ given by (3.8) is also optimal for (P_2) and the optimal cost is

$$(3.13)\quad J(\bar{u}) = \frac{1}{\theta}\int_{t_o}^{t_o+\theta} \left[2\langle r,f\rangle - |N^{-\frac{1}{2}}B^*r|^2 + Tr(GWG^*Q)\right]dt\ .$$

We need the following results.

LEMMA 3.1. a) Let $\{y(t)\}_{t\geq t_o}$ be a real process such that

$$(3.14)\quad \sup_{t\geq t_o} E(|y(t)|^2) < \infty.$$

Then $y(t_o+n)/n \xrightarrow{a.s.} o.$

b) Let $\{y(t)\}_{t\geq t_o}$ be a θ-periodic process such that

$$(3.15)\quad \sup_{t\geq t_o} E(|y(t)|) < \infty.$$

Then $y(t_o+n\theta)/n \xrightarrow{a.s.} o.$

Proof. For $\varepsilon > o$ we have

$$\sum_n P(|y(t_o+n)| > n\varepsilon) \leq \frac{1}{\varepsilon^2}\sup_t E(|y(t)|^2)\sum_n 1/n^2 < \infty$$

$$\sum_n P(|y(t_o+n\theta)|>n\varepsilon) = \sum_n P(|y(t_o)|>n\varepsilon)\le E(|y(t_o)|)<\infty \text{(b)}$$

Now the assertions follow from the Borel-Cantelli lemma.

LEMMA 3.2. [11]. Let $\{y(t)\}_{t\ge t_o}$ be a θ-periodic real process with

$$\sup_t E(|y(t)|)<\infty .$$

Then $\frac{1}{n}\int_{t_o}^{t_o+n} y(s)ds$ converges a.s..

LEMMA 3.3. Let $\{y(t)\}_{t\ge t_o}$ be a H-valued progressively measurable process with

$$\sup_t E(|y(t)|^2)<\infty .$$

Then we have

$$\frac{1}{T}\int_{t_o}^{t_o+T} \langle y(s),dw(s)\rangle \xrightarrow{\text{a.s.}} o \quad \text{as} \quad T\longrightarrow\infty .$$

Proof. It follows from the strong law of large numbers for martingales (see [6; Theorem 9, p.118]).

PROPOSITION 3.4. Let $C\in C_b([t_o,\infty),L(H,U))$ and $d\in C_b([t_o,\infty),U)$. Assume (H1)-(H3) and

1) $E(|x_o|^4)<\infty$,

2) $L=A+BC$ generates a stable evolution operator U_L.

Define

$$\text{(3.16)}\quad V(t)=\int_t^\infty U_L^*(s,t)\left[M^*(s)M(s) + C^*(s)N(s)C(s)\right]U_L(s,t)ds$$

$$\text{(3.17)}\quad p(t)=\int_t^\infty U_L^*(s,t)V(s)\left[B(s)d(s) + f(s)\right]ds \quad ;\ t\ge t_o.$$

Then the feedback law $u=Cx+d$ is in U_1 and

$$\text{(3.18)}\quad J(u)=\overline{\lim_{n\to\infty}}\ \frac{1}{n}\int_{t_o}^{t_o+n} \left[2\langle p,Bd+f\rangle + \langle Nd,d\rangle + Tr(G^*VGW)\right]dt.$$

In particular $J(u)<\infty$.

Proof. We have

$$\text{(3.19)}\quad x(t) = U_L(t,t_o)x_o + \int_{t_o}^t U_L(t,s)(Bd+f)ds + \int_{t_o}^t U_L(t,s)G(s)dw(s)$$

Since $\int_{t_o}^t U_L(t,s)G(s)dw(s)$ has gaussian distribution with mean o and covariance $\int_{t_o}^t U(t,s)G(s)WG^*(s)U^*(t,s)ds$ and $E(|y|^4)\le 3\left[Tr(cov y)\right]^2$ for a gaussian random element y , then it is easily seen that

$$\sup_{t\ge t_o} E(|x(t)|^4)\le C'<\infty$$

for a constant C' independent on t_o, so that $u \in U_1$.
From (H3) it follows that V, p satisfy the following inner product equations on $[t_o, \infty)$

$$(3.2o)\quad \frac{d}{dt}\langle Vx,y\rangle + \langle Vx,Ly\rangle + \langle Lx,Vy\rangle + \langle (M^*M+C^*NC)x,y\rangle = o$$

$$(3.21)\quad \frac{d}{dt}\langle p,x\rangle + \langle p,Lx\rangle + \langle Nd,Cx\rangle + \langle V(Bd+f),x\rangle = o \;;\; x,y\in D.$$

Let x_λ be the strong solution of

$$(3.22)\begin{cases} dx_\lambda(t)=\{L(t)x_\lambda(t)+\lambda R_\lambda(t)[B(t)d(t)+f(t)]\}dt+\lambda R_\lambda(t)G(t)dw(t) \\ x_\lambda(t_o)=\lambda R_\lambda(t_o)x_o \end{cases}$$

It is known that

$$(3.23)\quad x_\lambda(t) \longrightarrow x(t) \text{ in probability as } \lambda\to\infty, \text{ for every } t.$$

By using Ito's formula for $\langle Vx_\lambda,x_\lambda\rangle$, $\langle p,x_\lambda\rangle$ and (3.2o),(3.21) we deduce

$$(3.24)\int_{t_o}^{t}\langle (M^*M+C^*NC)x_\lambda,x_\lambda\rangle ds = \langle V(t_o)x_\lambda(t_o),x_\lambda(t_o)\rangle - \langle V(t)x_\lambda(t), x_\lambda(t)\rangle + \int_{t_o}^{t}\left[2\langle Vx_\lambda,\lambda R_\lambda(Bd+f)\rangle + Tr(G^*\lambda R_\lambda^* V\lambda R_\lambda GW)\right]ds + 2\int_{t_o}^{t}\langle Vx_\lambda,\lambda R_\lambda G dw\rangle$$

$$(3.25)\quad 2\int_{t_o}^{t}\langle C^*Nd,x_\lambda\rangle ds = 2\langle p(t_o),x_\lambda(t_o)\rangle - 2\langle p(t),x_\lambda(t)\rangle - 2\int_{t_o}^{t}\langle V(Bd+f),x_\lambda\rangle ds + 2\int_{t_o}^{t}\langle p,\lambda R_\lambda(Bd+f)\rangle ds + 2\int_{t_o}^{t}\langle p,\lambda R_\lambda G dw\rangle$$

If we set $u_\lambda = Cx_\lambda + d$ and

$$(3.26)\quad J_\lambda = \int_{t_o}^{t_o+T}\left[|Mx_\lambda|^2 + \langle Nu_\lambda,u_\lambda\rangle\right]dt$$

by using (3.24),(3.25) we obtain

$$(3.27)\quad J_\lambda = \int_{t_o}^{t_o+T}\left[\langle (M^*M+C^*NC)x_\lambda,x_\lambda\rangle + 2\langle C^*Nd,x_\lambda\rangle + \langle Nd,d\rangle\right]ds =$$

$$\langle V(t_o)x_\lambda(t_o),x_\lambda(t_o)\rangle - \langle V(t)x_\lambda(t),x_\lambda(t)\rangle + 2\langle p(t_o),x_\lambda(t_o)\rangle - 2\langle p(t),x_\lambda(t)\rangle + \int_{t_o}^{t_o+T}\left[2\langle Vx_\lambda,\lambda R_\lambda(Bd+f)\rangle + Tr(G^*\lambda R_\lambda^* V\lambda R_\lambda GW) - 2\langle V(Bd+f),x_\lambda\rangle + 2\langle p,\lambda R_\lambda(Bd+f)\rangle\right]ds + 2\int_{t_o}^{t_o+T}\langle Vx_\lambda+p,\lambda R_\lambda G dw\rangle$$

By passing to the limit in (3.27) (as $\lambda \longrightarrow \infty$) we deduce

$$(3.28)\int_{t_o}^{t_o+T}\left[|Mx|^2 + \langle Nu,u\rangle\right]dt = \langle V(t_o)x_o,x_o\rangle + 2\langle p(t_o),x_o\rangle - \langle V(t)x(t),x(t)\rangle - 2\langle p(t),x(t)\rangle + \int_{t_o}^{t_o+T}\left[Tr(G^*VGW) + \langle Nd,d\rangle + 2\langle p,Bd+f\rangle\right]ds + 2\int_{t_o}^{t_o+T}\langle Vx+p,Gdw\rangle .$$

From (3.28) and Lemma 3.1 and Lemma 3.3 we get (3.18).

Remark 3.2. Under the hypotheses of Proposition 3.4 and if A,B,C,G, d,f are θ-periodic (extended by periodicity on R) then the process

$$x(t)=\int_{-\infty}^{t} U_L(t,s)\left[B(s)d(s)+f(s)\right]ds + \int_{-\infty}^{t} U_L(t,s)G(s)dw(s)$$

is θ-periodic and $u=Cx+d\in U_2$.

Similarly as in Proposition 3.4 (see also [11]) one obtains

LEMMA 3.5. Under the hypotheses of Theorems 3.1 and 3.2 we have for every $u\in U_i$, $t_o\le t_1\le t_2$,

$$(3.29)\int_{t_1}^{t_2}\left[|Mx|^2 + \langle Nu,u\rangle dt\right] = \int_{t_1}^{t_2}\left|N^{1/2}\left[u+N^{-1}B^*(Qx+r)\right]\right|^2dt + \int_{t_1}^{t_2}\left[2\langle r,f\rangle - \left|N^{-1/2}B^*r\right|^2 + Tr(GWG^*Q)\right]dt - \langle Q(t_2)x(t_2),x(t_2)\rangle + \langle Q(t_1)x(t_1),x(t_1)\rangle - 2\langle r(t_2),x(t_2)\rangle + 2\langle r(t_1),x(t_1)\rangle + 2\int_{t_1}^{t_2}\langle Qx+r,Gdw\rangle .$$

Proof of Theorems 3.1 and 3.2. Let $u\in U_1$. From Lemma 3.5 with $t_1=t_o$, $t_2=t_o+T$ and letting $T\longrightarrow\infty$ (by Lemma 3.1 and Lemma 3.3) we get

$$J(u) = \overline{\lim_{T\to\infty}}\ \frac{1}{T}\int_{t_o}^{t_o+T}\left[|Mx|^2 + \langle Nu,u\rangle\right]dt = \overline{\lim_{n\to\infty}}\ \frac{1}{n}\int_{t_o}^{t_o+n}\left[|Mx|^2 + \langle Nu,u\rangle\right]dt = \overline{\lim_{n\to\infty}}\ \frac{1}{n}\int_{t_o}^{t_o+n}\left\{\left|N^{1/2}\left[u+N^{-1}B^*(Qx+r)\right]\right|^2dt + 2\langle r,f\rangle - \left|N^{-1/2}B^*r\right|^2 + Tr(GWG^*Q)\right\}dt.$$

A similar computation holds for $u\in U_2$.
Now it is easy to conclude.

4. THE OPTIMAL CONTROL PROBLEM WITH TWO-SIDED AVERAGE PATHWISE COST

We assume (H1)-(H3) hold for J=R. Consider the controlled stochastic equation

$$(4.1)\quad dx(t)=[A(t)x(t)+B(t)u(t)]dt + G(t)dw(t)$$

Definition. A progressively measurable U-valued process $\{u(t)\}_{t\in R}$ is a mild solution of (4.1) if it satisfies

$$(4.2)\quad x(t)=U(t,s)x(s)+\int_s^t U(t,r)(Bu+f)dr + \int_s^t U(t,r)G(r)dw(r)\quad, t\geq s.$$

Define

$$(4.3)\quad \tilde{J}(u) = \overline{\lim_{T\to\infty}}\ \frac{1}{2T}\int_{-T}^{T}\left[|Mx|^2 + \langle Nu,u\rangle\right]dt$$

$\tilde{U}_1$ is the class of all progressively measurable U-valued processes $\{u(t)\}_{t\in R}$ such that

$$(4.4)\quad \sup_t\ E(|u(t)|^4) < \infty$$

and (4.1) has a mild solution with

$$(4.5)\quad \sup_t\ E(|x(t)|^4) < \infty .$$

$\tilde{U}_2$ is the class of all progressively measurable U-valued processes $\{u(t)\}_{t\in R}$ such that u,x are θ-periodic and

$$(4.6)\quad \sup_t\ E(|u(t)|^2) < \infty$$

$$(4.7)\quad \sup_t\ E(|x(t)|^2) < \infty .$$

We consider the optimization problems

$$(\tilde{P}_i)\quad \min_{u\in\tilde{U}_i}\ \tilde{J}(u) .$$

Remark 4.1. Let $C\in C_b(R,L(H,U)), d\in C_b(R,U)$. If U_L (L=A+BC) is stable and (H3) holds then

$$(4.8)\quad x(t) = \int_{-\infty}^{t} U_L(t,s)(Bd+f)ds + \int_{-\infty}^{t} U_L(t,s)G(s)dw(s)$$

is a mild solution of (4.1) and the control $u=Cx+d\in\tilde{U}_1$. Moreover, if A,B,C,G,d,f are θ-periodic then $u\in\tilde{U}_2$.

THEOREM 4.1. Assume (H1)-(H3) and (A,B) is stabilizable, (A,M) is detectable (for J=R). Then the optimal control for $(\tilde{P}_1)$ is given by

(4.9) $\tilde{u} = -N^{-1}B^*(Qx+r)$

where Q is the unique bounded solution of (2.1) and r is the unique bounded solution on R of (3.9). The optimal cost is

$$(4.10)\quad \tilde{J}(\tilde{u}) = \overline{\lim_{n\to\infty}} \frac{1}{2n}\int_{-n}^{n} \left[2\langle r,f\rangle - |N^{-1/2}B^*r|^2 + \mathrm{Tr}(GWG^*Q)\right]dt$$

and the optimal trajectory is

$$(4.11)\quad \tilde{x}(t)=\int_{-\infty}^{t} U_L(t,s)(f-BN^{-1}B^*r)ds + \int_{-\infty}^{t} U_L(t,s)G(s)dw(s),\ L=A-BN^{-1}B^*Q,$$

and is globally stable in mean square. Moreover, if A,B,G,f are θ-periodic , then $\tilde{u}$ is also optimal for $(\tilde{P}_2)$, $\tilde{u},\tilde{x}$ are θ-periodic and the optimal cost is

$$(4.12)\quad \tilde{J}(\tilde{u}) = \frac{1}{\theta}\int_{0}^{\theta} \left[2\langle r,f\rangle - |N^{-1/2}B^*r|^2 + \mathrm{Tr}(GWG^*Q)\right]dt .$$

Remark 4.2. Similar problems with expected cost and with state dependent noise has been considered in [1],[2],[3].

Remark 4.3. In the finite dimensional case results of the " law of large numbers" type has been considered in [4],[7].

REFERENCES

1. G. DA PRATO, A. ICHIKAWA :Optimal control of linear systems with almost periodic inputs.SIAM J. Control Optim.4(1987),1107-1119.
2. G. DA PRATO, A. ICHIKAWA :Quadratic control for linear periodic systems.Appl. Math. Optim. 18(1988),39-66.
3. G. DA PRATO, A. ICHIKAWA :Quadratic control for linear time varying systems.Report, Scuola Normale Superiore, Pisa(1987).
4. A. HALANAY, T. MOROZAN, C. TUDOR :Tracking almost periodic signals under white noise perturbations.Stochastics 21(1987),287-301.
5. P. KOTELENEZ :A submartingale type inequality with applications to stochastic evolution equations.Stochastics 8(1982),139-151.
6. R.S. LIPTSER, A.N. SHIRYAYEV :Theory of Martingales(Russian). Moscow.Nauka,1986.
7. P. MANDL :Limit theorems of probability theory and optimality in linear controlled systems with quadratic cost.In "Stochastic Systems".Lect. Notes in Control and Information Sciences 96. Springer-Verlag,1987.
8. A. MICHALIK :Stochastic Differential Equations in Hilbert Spaces.Banach Center Publications,Vol. 5,Warsaw,1979.
9. A. PAZY :Semigroups of Linear Operators and Applications to

Partial Differential Equations.Springer-Verlag,1983.

10. H. TANABE :Equations of Evolution.Pitman,1979.

11. C. TUDOR :Optimal control for an infinite dimensional periodic problem under white noise perturbations(to appear).

ADDRESSES OF CONTRIBUTORS

A.BENSOUSSAN	INRIA, Domaine de Voluceau, Rocquencourt B.P. 105, 78153 Le Chesnay Cedex FRANCE
T.BIELECKI	Institute of Econometries Main College of Planning and Statistics Al. Niepodległości 162, oo-554 Warsaw, Poland
T.BOJDECKI	Department of Mathematics Warsaw University PKiN, 9 p. 00-901 Warsaw, Poland
R.BUCKDAHN	Sektion Mathematik der Humboldt Universität, DDR 1086 Berlin, Postfach 1297 German Democratic Republic
A.CHOJNOWSKA-MICHALIK	Institute of Mathematics University of Łódź 90-238 Łódź, Poland
T.DUNCAN	Department of Mathematics University of Kansas, Lawrence, Kansas 66045, U S A
O.ENCHEV	Department of Mathematics "D.Blagoer" Institute of National Economy Varna - 9002, Bulgaria
H.ENGELBERT	Friedrich-Schiller-Univ.Jena, 6900 Jena, Universitätshochhaus German Democratic Republic
D.GĄTAREK	Institute of Systems Research PAN ul. Newelska 6 01-447 Warsaw, Poland
A.HALANAY	Fac.Mat.Univ. Bucuresti, Str. Akademiei 14, Bucharest, Roumania
R.MANTHEY	Friedrich-Schiller-Universität Sektion Mathematik Universitätshochhaus 17 6900 Jena GDR

B.MASLOWSKI
Matematicky ustav CSAV,
Zitna 25, 11567 Praha 1
Czechoslovakia

H.NAGAI
Department of Mathematics
and Computer Sci.
Tokushima University
Minemijosanjima 1-1,
Tokushima, Japan

B.PASIK-DUNCAN
Department of Mathematics
University of Kansas
Lawrence, Kansas 66045
U S A

E.PLATEN
Akad. Wissenschaften der DDR
Inst. für Math.,
1080 Berlin, Mohrenstr. 39
G D R

Z.POROSIŃSKI
Institute of Mathematics
Technical University of Warsaw
Wybrzeże Wyspiańskiego 27
50-370 Wrocław, Poland

A.ROZKOSZ
Institute of Mathematics
Nicholas Copernicus University
ul. Chopina 12/18
87-100 Toruń, Poland

A.SIEROCIŃSKI
Department of Mathematics
Technical University of Warsaw
Pl. Jedności Robotniczej 1
Warszawa, Poland

L.SŁOMIŃSKI
Institute of Mathematics
Technical University of Warsaw
Wybrzeże Wyspiańskiego 27
50-370 Wrocław, Poland

K.SOBCZYK
Institute of Fundamental
Technical Research PAN
ul. Świętokrzyska 21
00-049 Warszawa, Poland

Ł.STETTNER

Institute of Mathematics
Polish Academy of Sciences
Śniadeckich 8
00-950 Warsaw, Poland

K.SZAJOWSKI

Institute of Mathematics
Technical University od Wrocław
Wybrzeże Wyspiańskiego 27
50-370 Wrocław, Poland

D.SZYNAL

Institute of Mathematics UMCS
Pl. Marii Skłodowskiej-Curie 1
20-031 Lublin, Poland

C.TUDOR

Fac. Math. Univ. Bucuresti
str. Akademiei 14
70109 Bucharest, Roumania

K.URBANIK

Institute of Mathematics
University of Wrocław
Pl. Grunwaldzki 2/4
50-384 Wrocław, Poland

C.VARSAN

Department of Mathematics
INCREST, Bd. Pacii 220
79622 Bucharest, Roumania

J.ZABCZYK

Institute of Mathematics
Polish Academy of Sciences
Śniadeckich 8
00-950 Warsaw, Poland

Lecture Notes in Control and Information Sciences

Edited by M. Thoma and A. Wyner

Lecture Notes in Control and Information Sciences

Edited by M. Thoma and A. Wyner

Vol. 117: K.J. Hunt
Stochastic Optimal Control Theory
with Application in Self-Tuning Control
X, 308 pages, 1989.

Vol. 118: L. Dai
Singular Control Systems
IX, 332 pages, 1989

Vol. 119: T. Başar, P. Bernhard
Differential Games and Applications
VII, 201 pages, 1989

Vol. 120: L. Trave, A. Titli, A. M. Tarras
Large Scale Systems:
Decentralization, Structure Constraints
and Fixed Modes
XIV, 384 pages, 1989

Vol. 121: A. Blaquière (Editor)
Modeling and Control of Systems
in Engineering, Quantum Mechanics,
Economics and Biosciences
Proceedings of the Bellman Continuum
Workshop 1988, June 13–14, Sophia Antipolis, France
XXVI, 519 pages, 1989

Vol. 122: J. Descusse, M. Fliess, A. Isidori,
D. Leborgne (Eds.)
New Trends in Nonlinear Control Theory
Proceedings of an International
Conference on Nonlinear Systems,
Nantes, France, June 13–17, 1988
VIII, 528 pages, 1989

Vol. 123: C. W. de Silva, A. G. J. MacFarlane
Knowledge-Based Control with
Application to Robots
X, 196 pages, 1989

Vol. 124: A. A. Bahnasawi, M. S. Mahmoud
Control of Partially-Known
Dynamical Systems
XI, 228 pages, 1989

Vol. 125: J. Simon (Ed.)
Control of Boundaries and Stabilization
Proceedings of the IFIP WG 7.2 Conference
Clermont Ferrand, France, June 20–23, 1988
IX, 266 pages, 1989

Vol. 126: N. Christopeit, K. Helmes
M. Kohlmann (Eds.)
Stochastic Differential Systems
Proceedings of the 4th Bad Honnef Conference
June 20–24, 1988
IX, 342 pages, 1989

Vol. 127: C. Heij
Deterministic Identification
of Dynamical Systems.
VI, 292 pages, 1989

Vol. 128: G. Einarsson, T. Ericson,
I. Ingemarsson, R. Johannesson,
K. Zigangirov, C.-E. Sundberg
Topics in Coding Theory
VII, 176 pages, 1989

Vol. 129: W. A. Porter, S. C. Kak (Eds.)
Advances in Communications and
Signal Processing
VI, 376 pages, 1989.

Vol. 130: W. A. Porter, S. C. Kak,
J. L. Aravena (Eds.)
Advances in Computing and Control
VI, 367 pages, 1989

Vol. 131: S. M. Joshi
Control of Large Flexible Space Structures
IX, 196 pages, 1989.

Vol. 132: W.-Y. Ng
Interactive Multi-Objective Programming
as a Framework for Computer-Aided Control
System Design
XV, 182 pages, 1989.

Vol. 133: R. P. Leland
Stochastic Models for Laser Propagation
in Atmospheric Turbulence
VII, 145 pages, 1989.

Vol. 134: X. J. Zhang
Auxiliary Signal Design in Fault
Detection and Diagnosis
XII, 213 pages, 1989

Vol. 135: H. Nijmeijer, J. M. Schumacher (Eds.)
Three Decades of Mathematical System Theory
A Collection of Surveys at the Occasion of the
50th Birthday of Jan C. Willems
VI, 562 pages, 1989

Vol. 136: J. Zabczyk (Ed.)
Stochastic Systems and Optimization
Proceedings of the 6th IFIP WG 7.1
Working Conference,
Warsaw, Poland, September 12–16, 1988
VI, 374 pages. 1989